A–Z of Filtration and Related Separations

Ken Sutherland

Consultant Chemical Engineer,
Reigate, UK

ELSEVIER

UK Elsevier Ltd, The Boulevard, Langford Lane, Kidlington, Oxford OX5 1GB, UK
USA Elsevier Inc, 360 Park Avenue South, New York, NY 10010-1710, USA
JAPAN Elsevier Japan, Tsunashima Building Annex, 3-20-12 Yushima, Bunkyo-ku, Tokyo 113, Japan

First edition 2005

ISBN 1 8561 74077

British Library Cataloguing in Publication Data
Sutherland, Ken, 1931–
A-Z of filtration and related separations
1.Filters and filtration – Encyclopedias 2.Separation
(Technology) – Encyclopedias
I.Title
660.2'842'03

ISBN–10: 1856174077

Published by
Elsevier Advanced Technology,
The Boulevard, Langford Lane, Kidlington, Oxford OX5 1GB, UK
Tel: +44(0) 1865 843000
Fax: +44(0) 1865 843971

Typeset by Land & Unwin (Data Sciences) Ltd, Towcester, Northants, UK
Printed and bound in Great Britain by MPG Books Ltd, Bodmin, Cornwall, UK

Preface

This expanded glossary of filtration and other mechanical separation technologies arose from my editor's recognition of its need, as suggested in the second edition of the "Handbook of Filter Media", where I wrote of the: "now ageing, Filtration Dictionary and Glossary". That Handbook contained a fairly extensive Glossary of terms related to filter media, and this present volume has expanded from that to encompass the whole field of gas/liquid/solid separations by mechanical means.

Because the phrase "filtration and related separation technologies" is a bit of a mouthful, the word "separation" is used a lot in this work as a simple replacement – the context should make it clear that that is what is meant. Any reader desiring a more detailed explanation of the scope of this encyclopaedia is directed to the entry for "Separation" in the body of the text.

This A–Z of Filtration is not intended to be an exhaustive encyclopaedia of the end uses of such equipment, only of its technology. So there is no intentional coverage of equipment applications except where they impose some special features upon the equipment.

Neither is it put forward as a thorough text book on process theory or equipment design. It is hoped that the main features of such matters are indexed here, but for more understanding of underlying theory, the reader is directed to the books listed in immediately subsequent pages.

It is hoped that the A–Z covers the technology reasonably well, and that it will be a good source for the explanation of an unfamiliar term. It is also hoped that it will prove to have that serendipity of chance discovery that any good encyclopaedia should have.

I should like to think that it is complete, but fear of hubris suggests that I should not claim completeness. Indeed the publishers and I will welcome comment upon the text, especially suggestions as to what may have been missed.

My primary thanks must go to my editor, Geoff Smaldon, at Elsevier, whose idea this was, and who has guided and encouraged me throughout its preparation. Especial personal thanks are due to the late Derek Purchas, my mentor and guide in much of my later professional life – from whose work I have poached liberally for this one.

I trust that this "A to Z" will prove as useful in its reading as it has been to me in its compilation.

Ken Sutherland
(ken@sanspeur.demon.co.uk)
Reigate
2005

User's Guide

In principle, there is very little guidance that need be given to the reader of this book, other than to look up a subject of interest by its most obvious title. There is no editorial index because the book, like any dictionary or encyclopaedia, is written in alphabetical order, and the text acts as its own index.

However, some help may be needed in navigating these pages, in order to extract the most value from its use.

The text is fully cross-referenced so that the search may be made most efficiently: the first look-up may have used what is the most obvious form of subject description to the searcher, but the main entry may be found under another head word, to which reference is made wherever necessary. Cross-referencing takes three forms:

- direct referral, by means of emboldening the key word or phrase at its point of mention in the text: "A filter candle is a cylindrical filter **element**, long in relation to its diameter ... "
- supplementary information, by means of a "*See also:*" reference at the end of an entry, and
- redirection, by means of a "*See:*" reference immediately after a head word

[It has not seemed sensible to list every variation of a word, so the reader is expected to use a little lateral thinking in finding an entry: which will normally be in the singular rather than the plural, as the infinitive rather than an inflected verb, and so on.]

Not too surprisingly, in an encyclopaedia on the main topic of filtration, many of the potential indexed terms will start with the word "Filter ... ". This would mean that a large part of the book would be taken up by the "F" section. In order that the book should be better balanced, a significant number of such terms are described under the second word, unless the separation of the two words destroys the sense. Thus "Filter cake" is described under "Cake" (but with a cross-reference from "Filter cake"), while "Filter press" is indexed and described as such.

A few entries, relating to major parts of the separation spectrum, are given rather longer essay entries, but these will be found as close as possible to their place in the alphabetical order as possible.

The author is an unashamed believer that medium, as a noun, is a singular word, whose plural is "media", and that is how filter medium and filter media should be found in this work.

Reference Sources

Whilst most of the entries in this "A to Z" came from the head of the author, it must be acknowledged that a wide range of previously published texts have been used, both to check for less familiar entries, and to ensure commonality of use. This list gives the main books used as reference works during the compilation of the encyclopaedia.

1. *Texts that cover all or a large proportion of the scope of separation.*

Filters and Filtration Handbook
T Christopher Dickenson
4th Edition 1997 Elsevier Advanced Technology

Filtration Equipment Selection, Modelling and Process Simulation
R J Wakeman & E S Tarleton
1st Edition 1999 Elsevier Advanced Technology

Perry's Chemical Engineers' Handbook
Robert H Perry & Don W Green (Eds)
7th Edition 1997 McGraw Hill

6-50 to 6-54	Particle Dynamics
14-81 to 14-98	Gas-Liquid Phase Separations
17-19 to 17-59	Gas-Solid Separations
18-60 to 18-133	Liquid-Solid Separations
19-17 to 19-55	Solid-Solid Separations
19-56 to 19-65	Flotation
22-20 to 22-36	Alternative Solid-Liquid Separations
22-37 to 22-69	Membrane Separation Processes

Solid/Liquid Separation Technology
Derek B Purchas
1981 Uplands Press Limited

Solid-Liquid Filtration and Separation Technology
A Rushton, A S Ward & R G Holdich
2nd Edition 2000 Wiley-VCH Verlag

2. *Texts with a specialised coverage of some of the subject matter.*

Decanter Centrifuge Handbook
Alan Records and Ken Sutherland
2001 Elsevier Advanced Technology

Handbook of Filter Media
Derek B Purchas and Ken Sutherland
2nd Edition 2002 Elsevier Advanced Technology

Handbook of Industrial Membranes
Keith Scott
2nd Edition 1998 Elsevier Advanced Technology

Membranes for Industrial Wastewater Recovery and Re-use
Simon Judd and Bruce Jefferson (Eds)
2003 Elsevier Advanced Technology

Nanofiltration – Principles and Applications
A I Schäfer, A G Fane and T D Waite (Eds)
2005 Elsevier Advanced Technology

[Readers seeking a more theoretical background, and hence for elaboration of some of the theories just listed in this encyclopaedia should go to Perry, or Rushton *et al.*]

A

Abrasion. The wearing away of a material by moving contact with a hard surface, or by the impact of hard particles. This particularly affects the interior walls of pipes, vessels or machinery where the direction of fluid movement changes. It is not usually a serious problem in filter operation, but is serious within several different designs of centrifuge, where solids move over surfaces under high centrifugal forces, the worst problem being the outer edges of the **scroll** in **decanter centrifuges**. It is countered by the application of particularly hard wearing surfaces to those parts likely to suffer.

Abrasion index. A measure of the susceptibility of any material or piece of equipment to abrasion by contact with a suspension of solid particles. Used especially of **hydrocyclones**, where the index is heavily proportional to a characteristic velocity – raised to the power of at least 3 (in the case of gas **cyclones** this exponent is between 4 and 5). To minimise the effect of abrasion thus suggests the use of small diameter hydrocyclones.

Abrasion resistance. An important characteristic in any material to be used for the construction of process equipment, where fluids are being handled or granular solids are being moved. Especially important in the handling of suspensions of abrasive particles, and solutions to abrasion problems must be sought in both the choice of structural material, and design of those parts in contact with the wear-producing conditions.

Absolute. The word absolute as used in the **filtration** industry relates to the degree of **separation** of solid particles from their suspending fluid, and implies that such separation is 100% effective. In fact, no separation can be completely 100% effective, unless there is a significant difference between the smallest **particle size** in the **suspension** and the smallest **pore size** of the **filter medium** (or if there is a very sharp **cutoff** in **particle size distribution**). Thus, an absolute **filter** must be defined in relationship with the solid particles being separated, such that the filter can be said to remove particles down to a specified size with absolute certainty. The term is, however, commonly used for filters with a very high capture efficiency.

It should also be noted that the word implies only a complete removal of particles above a certain minimum in size, but says nothing about the filter's performance at other particle size levels – information better given by a **grade efficiency curve**.

Absolute rating. A term used by manufacturers of clarification filters to imply a size of particle above which no particles will be found in the filtrate. The term **absolute** filtering size is also used, but both terms must be regarded with caution.

Absorber. The process vessel in which an **absorption** process takes place. The absorbent is usually a liquid, and if the required separation is of a component of a gas mixture, then this would normally be done by bubbling the gas through the liquid as the continuous phase. If the gas is to be cleaned of a suspended solid, then this is as often done in a **wet scrubber**, where the liquid is sprayed as fine droplets through the gas phase, or run over an inert packing, with the gas flowing in the interstices.

Absorptimeter. An instrument used to measure the turbidity of a suspension by observing the transmission of light directly through the suspension, i.e. at 180° to the source.

Absorption. The entrapment of a particle, or a liquid droplet, or a gas within ("ab-") the body of a separating material. The term thus strictly only applies to absorption of any of these into a liquid, to form a solution or suspension, but is loosely used of the entrapment of particles or droplets within the depth of a thick filter medium.

The term should be distinguished from **adsorption**, although both are mass transfer processes, governed by the rates of diffusion of the separating species towards and away from a phase boundary. The process also reaches an equilibrium state, where the rates of absorption and desorption are the same, and the absorbent becomes saturated with the separated material. Before it reaches saturation, the absorbent must be removed from the absorber and replaced by fresh (or regenerated) material, although the absorber may be in a duplex system, of two absorbers in parallel, with one being regenerated while the other is used.

Absorptive capacity. The mass of a liquid that can be absorbed by any material, divided by the mass of the material.

Accelerated filtration. A term sometimes used for thin layer or **cross-flow filtration**.

Acceleration. Literally, a continual increase in velocity of any moving body, but mainly used in the filtration industry for the increase in velocity applied in an **accelerator** to the feed slurry entering a centrifuge.

Accelerator. A device installed in the feed zone of most **centrifuges**, especially disc and decanter types intended to accelerate the feed slurry up to the rotational speed of the bowl. It is usually a plate or target, with suitably shaped vanes, sitting in a direct line from, and close to the exit from the feed pipe.

ACFD. [*See Air Cleaner Fine Dust*]

Acicular. The word means needle shaped, and is used to describe **particles** that are very long with respect to their width, and so are almost **fibre**-like in shape,

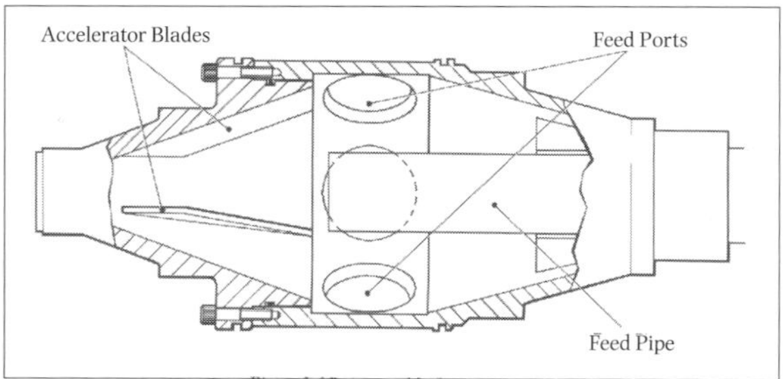

Decanter feed zone with accelerator blades

Acid, acidic, acidity. An acid is a chemical compound that generates hydrogen ions upon dissolution in water. The greater the degree of dissociation of the acid into ions, the greater the strength of the acid (and the lower the corresponding pH value of the solution). A material or a mixture may be referred to as acidic, and its acidity will be measured by its free hydrogen-ion content (pH).

An acid reacts with alkalis (bases) to form salts, and with salts of weaker (less acidic) acids to release those acids. Strong acids are often used for the chemical cleaning of **filters** and used **filter media**.

Acid cleaning. The cleaning of machinery by the use of **acidic** liquid formulations. This combines the flushing action of the bulk of the liquid, with the chemical action of the acid component in destroying potentially harmful residual matter, especially that affording the possibility of bacterial growth. Acidic liquids are the most common agents for **CIP** processes, with citric acid a common reagent in **membrane** cleaning.

Acid recovery, regeneration. The recovery of spent acids, from processes using acids, presents a very challenging application for **separation** processes, yet quite a demanding one in terms of degree of separation.

Acropore. A proprietary membrane material (Gelman, now Pall) made from a PVC/acrylonitrile copolymer, supported on a nylon base.

Acrylamide. An organic chemical, $CH_2=CH.CONH_2$, which polymerises though the double bond of the vinyl group, forming water-soluble **polyacrylamides**, for use as **flocculating agents**.

Acrylonitrile. A reactive chemical with several different industrial uses, which polymerises easily, and forms valuable copolymers, with significant usage as **filter media**.
[*See also Polyacrylonitrile*]

ACTD. [*See Arizona Control Test Dust*]

Activated carbon/charcoal. A surface is activated, i.e. made more effective in its adsorptive properties, by thermal (or occasionally chemical) treatment. The term is most used in relation to carbon, or charcoal, whose surface, usually in a very finely divided form, is energised to enable it to be used as an efficient **adsorption** agent, in the removal of odours and

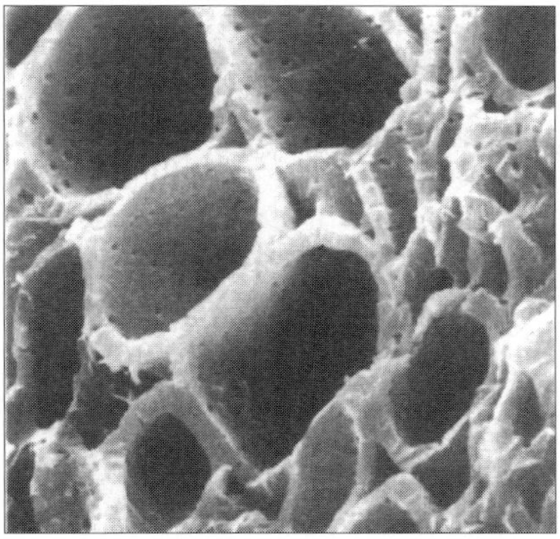

Micrograph of activated carbon

other gaseous impurities from air, or colouring agents or flavours from a liquid (especially in the production of potable water). The material is increasingly being used in a **combination filter** for impure gas streams, to separate particulate solids and gaseous impurities as well.

The raw material may be wood, peat, coal, or fruit or nut shells, which may be activated in carbonising ovens. This results in the creation of a highly porous material, with very high specific surface area (surface per unit volume).

Activated sludge. The activated sludge process is the most common **secondary treatment** process for waste waters from industrial or municipal sources. The process is an **aerobic** fermentation of the wastes, which results in the bacterial oxidation of carbon-containing matter (in suspension or solution). The rate of aeration controls the **digestion**, and some modern processes use oxygen rather than air for the aeration to improve performance. The sludge mass carrying the bacteria increases in amount during the fermentation, and after leaving the fermenter it is separated into two fractions, one to be returned to the process, and the other to be discharged for final disposal (usually after **dewatering**) on land or by incineration.

Active baffle. A device fitted in a **decanter centrifuge** to promote discharge of separated solids; a synonym for **Conveying baffle**.

Additive. Literally, some substance added to another, usually for some productive or characteristic-enhancing purpose. There are two main usages within **filtration**.

The first usage concerns those bulk materials that may be added to a wet **sludge** to decrease its average moisture content and/or change its properties, the latter including stabilisation of its bacterial content. The significant increase in the amount of material as a result of this kind of addition may count against the process where transport costs are high.

The other usage is for the addition of **coagulants** and **flocculants** to dilute slurries to improve the settling characteristics of the suspended solids.

The term is not normally used of **filter aids**.

Adhered liquid. That portion of the liquid in a suspension of a solid that is firmly adsorbed on the solid surfaces, and is therefore almost impossible to remove by mechanical **dewatering**. It exists as a very fine layer of liquid on the solid surface, of a thickness dictated by the nature of the liquid and the structure of the solid surface.

Adhesion. The process by which two solids adhere because of the attractive forces between them at the molecular level – the process of **adsorption** – but also applies to the conjoining of solids by means of the use of an adhesive liquid spread between them. The latter is of some consequence in the manufacture of filter **cartridges**, and in some other **filter media** formats.

There are two basic mechanisms of molecular attraction for adhesion:

- **Van der Waals forces** between molecules acting as dipoles; and
- electrostatic surface forces due to net charges on the particles and surface,

whilst a third mechanism is that of surface tension acting in a thin film of liquid between the particle and the surface.

Strictly speaking, the term adhesion should be used only of the attachment of a particle to a solid surface, whereas **cohesion** should be used of the conjoining of two or more particles.

One particular form of adhesion to be noted is the case where a particle is held on to a solid surface, or to another particle, by a thin layer of liquid at the point of contact. If this liquid is a solution, then, as it dries, the dissolved material will crystallise and form a very strong bridge between particle and surface. This can only be prevented by very thorough washing of a filter **cake**.

Adhesive forces. Those molecular-level forces that operate at phase boundaries to cause the adhesion of solid particles (or liquid droplets) to a solid surface, and so remove them from suspension.

Admix. A term sometimes used for **filter aid**.

Adpec filter. A proprietary design of **vacuum belt filter** that operates by means of a continuous belt, acting as both cake transport and **filter medium**. The belt travels over a long vacuum box, with an intermittent movement produced by a reciprocating movement acting on a roller at the discharge end of the filter. This roller creates a sharp movement of the belt for a short distance, while the vacuum is cut off, with the roller returning slowly during the rest of the cycle, with vacuum again applied. The cake is thus carried across the vacuum box in a series of short jerks, with cake falling off the belt during this motion.

Adsorbate. The material in a mixture or solution that is removed by adsorption.

Adsorbent. A material used in the separation of components of a mixture by the process of **adsorption**. **Activated carbon** is the most common such material in the **filtration** industry, for treating gases and liquids, although various clays, such as **fullers earth**, are also quite widely used in liquid processing. The use involves the removal of one or more components from a solution or mixture by the adsorbent, and the resultant removal of that component by filtration of the adsorbent.

Adsorber. The process vessel in which adsorption is undertaken. The adsorbent is usually a finely divided solid, and the adsorber holds this material as a **fixed (packed) bed**, a **moving bed**, or a fluidised bed.

Adsorbing colloid flotation. A process used for removing dissolved materials from solution by adsorbing them onto **colloidal** particles and removing these by **microflotation**.

Adsorption. The process whereby materials are caused to adhere to a solid surface so as to remove them from a mixture or solution. The mechanism involves bringing the material in question as close to the particular surface as possible, so that the attractive forces that exist between the material and the surface can take effect. The area of solid surface available to the mixture or solution needs to be very large, so as to provide as efficient a removal as possible, and this usually means that the solid is present in finely divided granular form, with each granule as full as possible of pores into which the material can diffuse.

Adsorption is one of the classic unit operations of chemical engineering, where it is usually employed for the separation of one component of a gas mixture from the rest. The process is controlled by the rate at which the required component can diffuse through the mixture and reach the adsorbent surface. This is not really a **filtration**-type of **separation** process, except where the gas removal is a process going on simultaneously with other physical separations, as in a **combination filter** used to separate both particles and odours from a gas stream.

However, adsorption is also one of the mechanisms whereby fine solid **particles** are trapped in a deep **filter medium** and held, not by the process of filling a **pore**, but by that of adhering to the nearby surface. The forces holding the separated material on the solid surface may be intermolecular, electrostatic, **van der Waal's**, and so on.

Chemisorption is adsorption at a surface, where chemical bonds are formed between **adsorbate** and **adsorbent**. This is an important component of catalytic processes.

It should be noted that the adsorption of solute molecules to the surface of the solids in a filter **cake** may make the solute more difficult to recover in cake washing.

(The entrapment of particles in a deep filter medium is also, confusingly, called **absorption**.)

Adsorption coagulation. A process of achieving aggregation of particles in which counter-ions adsorb and so reduce the charge on a particle surface.

Adsorption isotherm. An isotherm is a curve relating the pressure and volume of a gas at a constant temperature, there being a different isotherm for each temperature value. An adsorption isotherm is then the curve relating the amount of a vapour adsorbed onto the surface of an **adsorbent** to the partial pressure of that vapour, all at a constant temperature.
[*See also BET isotherm*]

Advancing knife. A **cake discharge** device used on a **drum precoat filter**. The filter collects dirt in the outer part of a thick layer of precoat, and this outer layer of precoat and collected dirt is removed by a knife blade that moves slowly towards the cloth and removes a thin layer of cake and precoat.
[*See also Blocking function*]

Aerated grit chamber. An air agitated vessel used for removing sand and grit from waste water. Air is introduced at the base to create a flow pattern that causes heavy particles to be separated by centrifugal action. The heavy particles settle to the bottom and are removed by a **screw conveyor**, and the cleaned water overflows from the top of the chamber. The air rate can be adjusted according to the specific nature of the particles to be removed.

Aeration. The controlled introduction of a gas (usually air) into a mass of liquid either for the achievement of a chemical or **aerobic** biological reaction (such as the **digestion** of wastes), or for physical reasons (such as the provision of bubbles to which particles can attach and so float to the surface). The introduction is usually as a finely divided stream of bubbles, though a **sparge** pipe or ring.

Aerobic processes. Any process taking place in the presence of air (or, more specifically, oxygen), for which the air or oxygen is usually a necessary reactant. Most often used of waste water treatment processes, which use air or oxygen finely dispersed throughout the bulk of the waste liquid, by means of **sparge** pipes or other dispersing means.

Aerodynamic cleaner. An **inertial separator** for the removal of dust from air. It takes the form of a wedge-shaped cell, with the dirty air entering at the widest part of the wedge. The side of the cell is made up of louvre vanes sharply angled away from the direction of flow of the incoming air. The clean air leaves the cell through these louvres, having made an abrupt, and almost complete change in direction to do so. The inertia of the suspended particles carries them straight on, as the air turns away, to the bottom of the cell and the collecting chamber there, from which the accumulated dirt is blown out by a bleed air stream.

Aerophilic, Aerophobic. Terms sometimes used in the mineral industry for **hydrophobic** and **hydrophilic** respectively.

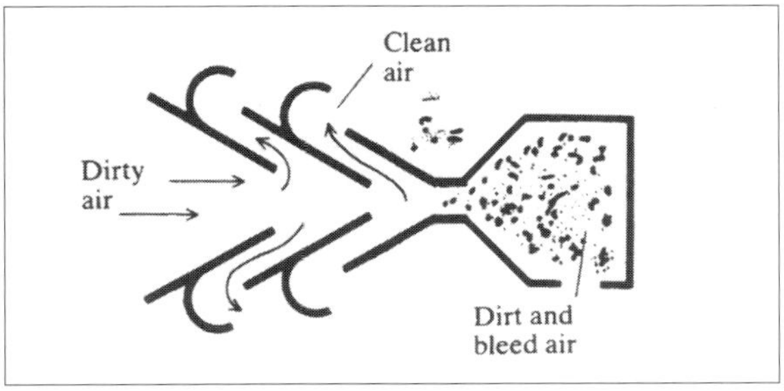

Aerodynamic cleaner

Aerosol. A **dispersion** of very small solid particles or liquid droplets in a gas, which may be created intentionally (as in sprays for de-odourising, for example) or arise as a result of other activities. The **particle size** of an aerosol is usually so small that the aerosol is very stable, and settling out by gravity effectively does not happen. The cleaning of a gas carrying such a suspension therefore requires very specialised techniques, involving **coalescers** and **depth filters**.

Aero-Vibe screen. A **vibrating screen**, made by Allis Chalmers, used in the mineral industry for wet or dry screening. It consists of a coarse wire screen, suspended from a set of cables, in a nearly horizontal position. Solids to be screened are moved down the slight slope of the screen by the vibration induced by a belt-driven shaft carrying eccentric weights.

Affinity membrane. A membrane that has been chemically treated, as a result of which special molecular groups are attached to its surface, which are capable of forming complexes with specific compounds in the feed.

Affluent. A term infrequently used for the suspension fed to a **separation** device, equivalent to **influent**, and the opposite of **effluent**.

Ageing. The holding of a **suspension** in an unseparated state, for a period of time, during which the **separation** characteristics of the suspension will improve. This is usually because of particle size growth, such as by crystal growth or **agglomeration**. Ageing is undertaken in holding tanks, which are stirred just enough to overcome the natural settling tendency of the solids, but not enough to interfere with whatever process is causing the improvement in separability.

Agglomeration. Agglomeration is the bringing together of a number of finer particles to make a coarser particle, or of creating a larger drop by bringing together several droplets of liquid. The purpose is to make **separation** of the particle or drop easier by virtue of its larger size, because settling velocities increase with particle or drop diameter, or because the particle or drop is now large in relation to the pores of a **filter medium** being used to filter it from the fluid in which it is suspended. The larger particle sizes resulting from any agglomeration process may also be helpful in increasing the **porosity** of the resultant filter **cake**.

Agglomeration is achieved by a number of distinct processes: coalescence of liquid droplets by passage through a coalescing "filter"; **flocculation** or **coagulation** of suspended solids by the addition of coagulating chemicals, such as inorganic salts or **poly-electrolytes**; and aggregation of fine dry solids by gentle agitation so as to enable physical forces to create the larger agglomerates.

These resultant larger groups of particles, or larger drops, must then be processed in ways that ensure that they are not broken up again and dispersed back into the suspending fluid.

Agglomeration skin flotation. A **flotation** process used to separate **hydrophobic** and **hydrophilic** particles. The hydrophobic particles attach to air bubbles to form **aggregates**, which are denser than water but less dense than the hydrophilic particles. The two species are then separated by **gravity concentration**.

Agglutination. The clustering together of particles because they are sticky, or because of the presence of a glue.

Aggregate. In **separation** terms, an aggregate is any assemblage of finer particles forming a larger particle, as a result of **agglomeration**. The word is also used, especially in the construction industry, to refer to the coarse bulking material, such as gravel, broken rock or a comminuted waste material, which is added to a cement mix to form concrete.

Air blowing. In **filtration** terms, blowing with air is a filter **cake** treatment process, in which the residual moisture in the cake is reduced by a stream of air. This can be done in many types of filter, but not all, because efficient air blowing requires good sealing between parts of the filter to avoid bypassing of the air.

Air classifier. [*See Elutriation*]

Air Cleaner Fine Dust. One form of the name in which test powders are made available for challenge tests for **filters** or **filter media**. ACFD (also called Air Cleaner Fine Test Dust, ACFTD) is usually paired with a coarse equivalent (ACCD or ACCTD). Increasingly being replaced by ISO standard dusts.
[*See also Arizona Control Test Dust, Bacterial challenge test*]

Air cleaning. Any process by which an air (or gas) stream is cleaned free of suspended dust. This can be done by **filters** and **cyclones**, but can also be effected by **electrostatic separators** and **scrubbers**, wet or dry. Some air cleaning duties, such, for example, as ensuring freedom from bacteria and viruses, require exceedingly efficient **filtration**.
[*See also Dust, Dust collector*]

Air conditioning. The processes used to create a working or living atmosphere that is free from dust and harmful micro-organisms and from obnoxious odours, and also has an acceptable temperature and humidity.

Air core. The central zone, at the axis of a **hydrocyclone**, inside the **secondary vortex**.

Air dry. A term used to describe the condition of a mass of wet porous material that has been exposed to the ambient atmosphere, and is in equilibrium with it, such that no further loss of moisture will occur on further exposure. The condition obviously depends on the relative humidity of the ambient air, but is generally accepted to be about 10% moisture content.

Air filter. Any of a wide range of filters used to remove particulate material from air, either as part of a process to produce clean air for living or working spaces, or for breathing, or as an exhaust air cleaner. A prime characteristic is an acceptable **dirt holding** capability, so that the time intervals between **element** changes are as long as possible, but an air filter should also have an adequate efficiency, coupled with as large a filtration area as possible, to accept high air flow rates at economic pressure differentials.
[*See also Air cleaning*]

Air-filtrate separation. Many types of **filter**, and also **centrifuges**, produce a filtrate (centrate) that is a highly aerated liquid, if not actually a two-phase flow. The air will quite often interfere with subsequent processing (even if this is just being pumped away), so most filtrate **receivers** are designed to encourage separation of air and filtrate, and allow separate removal of each phase from the receiver.

Air flotation. [*See Flotation*]

Air flow rate. Where filter **cakes** are to be dewatered by passage of air (or any other gas) through the cake, to blow the **mother liquor** or **wash liquid** out of the pores, the air flow rate will increase as more and more pores are cleared, and will reach a maximum value, for a given overpressure, once the cake is as clear of liquid as it can be by this method.

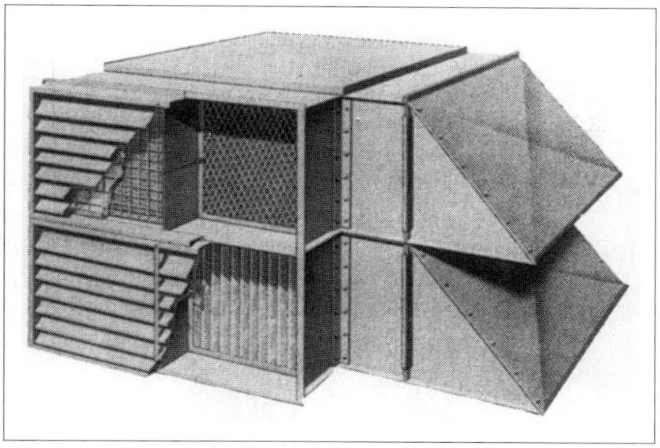

Engine air intake filter

Air intake filter. An important sub-set of the **air cleaning** category, in which air destined to be used to burn fuel in an engine is freed of dust, prior to combustion, to prevent the damage to the interior of the engine (especially the blades of a gas turbine) that would be caused by the abrasion of ash particles.

Air-line filter. A **cartridge** (or similar) **filter** installed in the delivery line from an air compressor, to filter the pressurised air and protect any equipment in which the compressed air is to be used.

Air permeability. A method for characterising clean **filter media**, by determining the volumetric flow of air through them. This can be useful for comparison within a range of media, but says very little about **filtration** behaviour.

Air purification. Can refer to **air cleaning** by **filtration**, etc, but usually refers to the separation from an air stream of gaseous impurities, such as **VOCs** or odours, a process usually requiring **adsorption** or **wet scrubbing**.

Air sparge. [*See Aeration*]

Air sparged hydrocyclone. A design of **hydrocyclone** in which the wall is made of a material porous to an air flow, and a jacket holding compressed air surrounds the body of the hydrocyclone. The air flows at a controlled rate into the **separation** zone and creates a **froth** that is carried out as the **secondary vortex**, leaving as a two phase mixture from the **vortex finder**. This design is used to achieve oil/water separations, and in mineral flotation.

Ajko filter. An early form of **diaphragm filter press**.

Akins classifier. A separating device used to remove light solids, which will float, from dense solids, which will not, by gentle agitation of a **suspension** of the mixture in a separating liquid. A sloping trough has a long screw rotating so as to drag the solids up the slope, for discharge of the dense fraction at the top, while the liquid, carrying the light fraction, overflows at the bottom end.

Algae. Algae are aquatic micro-organisms, neither plant nor animal, which are the source of much of the earth's oxygen, and which form the base for the food chain of almost all aquatic life. They range in size from around 1 μm up to giant seaweeds (and **filtration** has a significant part to play in the processing of many seaweeds), but it is in their finer forms that they most concern the filtration technologist. Algae can multiply rapidly ("blooms") and cause problems with cooling water intakes, for example, and they can also grow on **membrane** surfaces and thus form one cause of membrane fouling.

Alginates. Natural polymers (polysaccharides) used occasionally as a flocculating agent. The most common is sodium alginate, extracted from seaweed, and used particularly for the treatment of drinking water, and some beverages.

Alkali cleaning. The use of alkali, most often sodium hydroxide solutions, as a cleaning agent for **filter media**, and **separation** equipment. Organic fouling of **membranes** is most often treated under alkaline conditions.

Allen cone. A version of the conical settling tank, or **sand cone**, much used in mineral processing, for the **separation** by **sedimentation** of rapidly settling materials such as sand. The main problem with such devices is the removal from them of the settled solids, and the Allen cone achieves this by means of a valve at the base (i.e. apex of the inverted cone), activated by a mechanical linkage from a float chamber in the body of the settling zone.

Aloxite. A proprietary form of abrasive powder, made from fused alumina, also used as a **standard test dust**, in challenge tests for **liquid filters**.

Alpha value. A term used by some **filtration** practitioners to characterise the **permeability** of a filter **cake**. It derives from a particular interpretation of the **Darcy** equation, and gives a mean value over the whole filter cake depth. Alternatively, spot values can be estimated from the **Kozeny-Carman** equation. Theoretical derivation is not usually possible, experimental evaluation from a test rig being the normal method of determination.

Alum. [*See Aluminium sulphate*]

Alumina. The common name for aluminium oxide, one of the most important ceramic materials, and the basis for many **filters** used for high temperature gas **filtration**. It is also the basis for most **ceramic membranes**, and is used in some **filter test dusts**.

Aluminium chloride. Before the advent of polymeric flocculants, aluminium chloride was a major flocculant chemical, and still has use where the cost of synthetic materials is not justified.

Aluminium hydroxide. A **sludge** frequently produced in water treatment processes, which is difficult to **dewater**. Its **separation** is made much easier if it is subjected to a **freeze-thaw process** prior to separation.

Aluminium silicate. A ceramic material, frequently used in ceramic media for high temperature **filters**.

Aluminium sulphate. A traditional, inorganic reagent for achieving **flocculation**. To a major extent, replaced by synthetic polymeric chemicals. Commonly (although, strictly, incorrectly) known as alum.

Ama KDF filter. [*See KDF filter*]

Ambient. A word describing the prevailing conditions (mainly temperature and pressure) of the environment around a **separator**, but can also be used to refer to the environment in general.

Amicron. A rarely used word referring to a particle with dimensions of about 1 nm.

Amides. Compounds (mainly organic) formed from ammonia, by the substitution of an organic acid for one or more of the hydrogen atoms. They take the form $RCONH_2$, $(RCO)_2NH$ or $(RCO)_3N$, and, chemically, are weakly basic. They are primarily of interest here as the monomers used in the formation of **polyamides**.
[*See also Acrylamides, Nylon*]

Amines. Organic chemicals formed from ammonia, by the substitution of a hydrocarbon group for one or more of the hydrogen atoms. They are strongly basic in chemical character, but are of interest as the monomers from which polyamine **flocculants** are made.

Amorphous. A particulate material that has no special shape or crystal structure is said to be amorphous.

Ampholyte. A chemical, usually an **electrolyte**, that exhibits **amphoteric** properties.

Amphoteric flocculant. A flocculant which is either **anionic** or **cationic** depending on the pH of the suspending liquid.

Amphoteric material. Chemical materials, usually oxides or hydroxides, that behave as alkalis in the presence of strong acids, and as acids in the presence of strong alkalis. Thus zinc oxide will dissolve in acids to form zinc salts, and in alkalis to form zincates. Aluminium hydroxide shows similar behaviour, as do amino acids.

AMU. [*See Atomic mass unit*]

Anaerobic processes. Any process, either chemical, biological or physical, which takes place in the absence of air (or, more specifically, of oxygen). It is most often applied to biological processes for which the absence of oxygen forces the consumption of oxygen-containing materials, and hence their destruction.
[*See also Digestion*]

Analytical cut size. That particle size on the **cumulative oversize** particle size distribution of the feed material that corresponds to the total efficiency of the **separator**.

Anaphoresis. [*See Electrophoresis*]

Andreasen pipette. A test device that enables the observation of a **sedimentation** process. A wide glass jar is stood vertically, and a **suspension** allowed to settle in it under the influence of gravity. The settling behaviour can thus be observed visually, but the pipette is closed at the top by a stopper, carrying a valve connected to a vacuum source, via a small collection chamber, which can then be emptied by turning the valve the other way, to release a small sample from a particular layer within the settling slurry. The samples are taken from a fixed position within the body of the pipette. The settling behaviour in the pipette is assumed to obey **Stokes' law**.

Anemometer. A device for measuring gas flow rates, for instance through a **clean room**, or through a **filter**. It can work mechanically, such as a rotating vane, or thermally, such as by determining the cooling effect on a hot wire.

Angle of contact, Angle of wetting. [*See Contact angle*]

Angularity. Descriptive of **particle shape**, an angular particle is well removed from spherical in shape, and its angularity is a measure of that divergence, being the reciprocal of **sphericity**. Also called rugosity.

Angular momentum. The momentum of a moving body is the product of its mass and its velocity. In the case of a rotating body, the angular momentum is the product of its moment of inertia and its **angular velocity.**

Angular Press. A proprietary version of the **double belt press**, featuring a vertical wedge-shaped compression zone (the "angle"), following a horizontal draining zone, and followed by further compression around a series of rollers.

Angular velocity. The rate at which a body rotates about its axis, normally expressed in radians/second. If a particle is moving with an angular velocity ω, at a distance r from the centre of rotation, then the **centrifugal acceleration** acting on it will be $r\omega^2$. A feature of the liquid held in a **centrifuge** rotating at constant speed is that the angular velocities are constant for all of the liquid (unlike a **hydrocyclone**, where the behaviour approximates to constant **angular momentum**).

Anion. In solution a salt will dissociate into two parts, called ions, one positively charged (the **cation**), and one with an equal negative charge, the anion. Because of their negative charge, anions will be attracted to the positive electrode in an electrolytic cell, and they will be removed from solution by the appropriate **ion-exchange** medium. Anionic materials, generally, carry a negative charge, and anionic **flocculation agents** would be employed to promote the **agglomeration** of positively charged suspensions.

Anionic flocculant. A negatively charged flocculant, used in the **agglomeration** of positively charged particles.

Anionic membrane. A membrane for use in an electrolytic cell or **fuel cell**, or in **electrodialysis**, which is permeable to **anions**, but not to **cations**.

Anode. The positively charged electrode in an electrolytic cell, to which the anions are attracted.

Anthracite. A commonly used material for **deep bed filters**, because it is hard and relatively non-porous (and so does not provide places where bacteria can collect and grow). It has a density that is significantly different from that of sand, and can thus easily be segregated during cleaning and resettling of the medium in a filter.

Anti-vibration mount. A mechanical device, fitted below many vibrating or rotating machines, especially **centrifuges**, to isolate the machine vibrations from the surrounding environment. Also called a vibration isolator.

Aperture. Literally any opening or hole in a material, the term has a particular connotation in **filtration** technology in referring to an opening, described as a two-dimensional shape (square, rectangle, circle, diamond, etc) formed either between the threads or filaments of a woven material, or as a hole in a thin

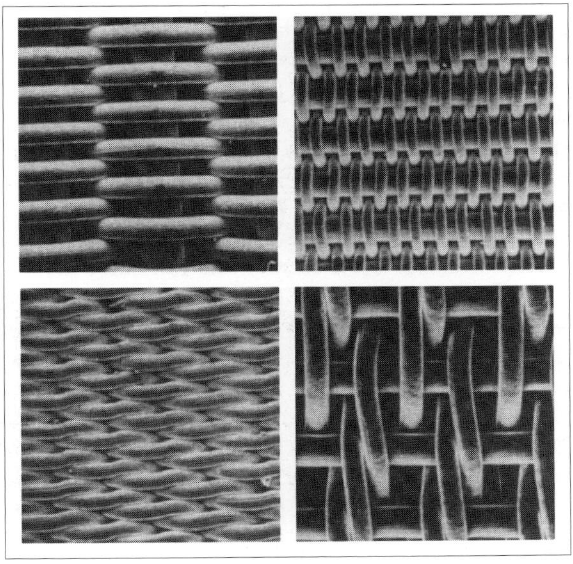

Apertures of different shapes

sheet produced by a variety of methods (pressing, drilling, electrolytic erosion, etc). There is an element of dimensional control implicit in the use of the word, namely that the manufacturing process for the material is able to produce apertures of quite closely defined sizes and uniformity of size. This all contrasts with the randomly sized pores in a **packed bed** of granules or fibres.

Apex. Although normally thought of as the highest point of some feature, the word is used in its alternative meaning of "tip or point" in the **separation** world, when referring the lowest point of a **cyclone** or **hydrocyclone**. It is through this conical zone that the separated solid leaves a cyclone, at its axis.

Apex nozzle. A nozzle attached to the apex of a **cyclone** so as to control the rate of discharge of the separated solids.

Aphron flotation. A flotation process in which small bubbles are produced mechanically, outside the cell, under high shear conditions.

Aqueous slurries, suspensions. Suspensions of solid **particles** in water, or a **solution** in water as the solvent. The processing of such suspensions is thus susceptible to the **hydrophilicity** (or **hydrophobicity**) of, for example, a **filter medium**.

Aramid. (Sometimes aramide). An aromatic nylon, which is made by the condensation polymerisation of a diamine with terephthalic acid. Best known are Du Pont's Nomex and Kevlar, both very tough, if expensive, thermoplastic polymers.

Archimedes number. A dimensionless number, specifically containing the particle diameter, mainly used in the Archimedes-Omega diagram, as a means for estimating the settling velocity of particles undergoing gravity **sedimentation**, under conditions where the **Reynolds number** is not known, and therefore the flow regime is uncertain, and **Stokes' law** cannot easily be applied. The diagram relates the Archimedes number to the **Omega number**, which specifically contains the settling velocity, so that, given the particle diameter, the settling velocity can be calculated. The Archimedes number is given by:

$$Ar = gd^3(\rho_s - \rho_l)/\nu^2\rho_l = gd^3\rho_l(\rho_s - \rho_l)/\eta^2$$

where g = gravitational acceleration
d = particle diameter
ρ_s = particle solid density
ρ_l = density of suspending liquid
ν = kinematic viscosity of suspending liquid
η = (dynamic) viscosity of liquid

The Archimedes number (also called the Galileo number) is also used in the design of **deep bed filters**.

Area bonded. In the production of **spun bonded** materials, a random layer of extruded **fibres** is laid down on a moving belt. This layer is then thermally bonded to give cohesion to the material. If this bonding occurs all over the material, it is said to be area bonded.
[See also Point bonded]

Arizona Control Test Dust. A standard test material for the evaluation of a **filter medium** by a capture efficiency test. ACTD is used in one of two size fractions: Fine (0 to 80 μm), and Coarse (0 to 200 μm).

The nominal particle size distribution of the fine dust is:

Size range (μm)	Mass fraction (%)
0 – 5	37 – 41
5 – 10	15 – 21
10 – 20	13 – 19
20 – 40	15 – 21
40 – 80	6 – 12

and of the coarse dust:

Size range (μm)	Mass fraction (%)
0 – 5	10 – 14
5 – 10	9 – 15
10 – 20	11 – 17
20 – 40	20 – 26
40 – 80	27 – 33
80 – 200	6 – 12

Arrestance. A measure of **filtration efficiency**, employed, for example, in the **ASHRAE** filter test procedures. If a weight w_0 of solid particles is injected into the air flow upstream of the **filter** on test, and a weight w_1 of particles pass through the test filter (and are captured on a sample filter), then the arrestance:

$$A = 100(1 - w_1/w_0)$$

expressed as a percentage figure.

Artisan filter. An early, and successful form of a **filter thickener**, operating effectively by **cross-flow**,

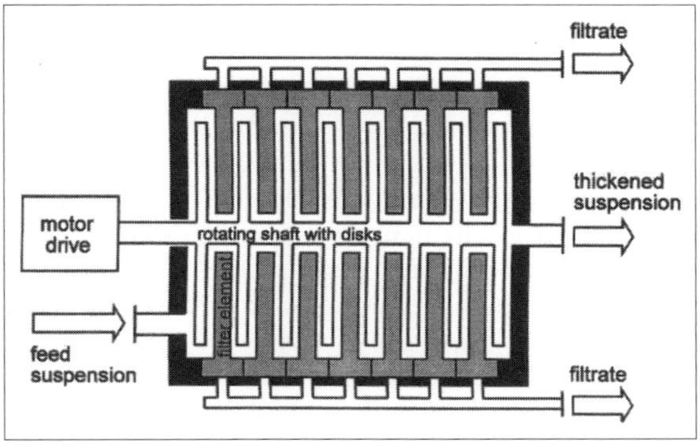

Artisan filter

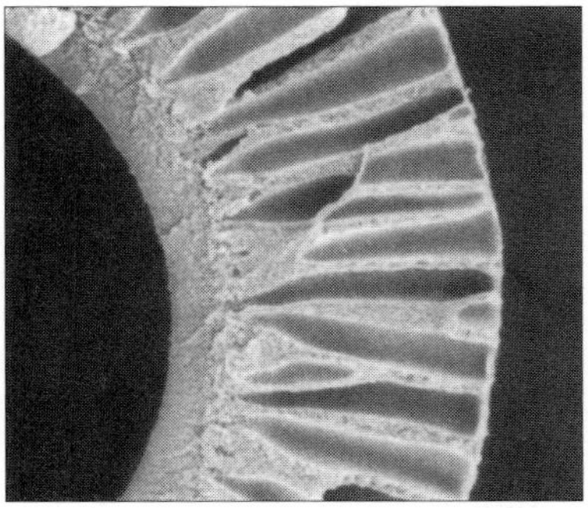

Asymmetric membrane

in which the **cake** thickness is kept low by means of the shear produced by rotating discs between stationary plates carrying the **filter medium**. In a cylindrical pressurised shell are a number of annular hollow plates, fixed to the shell, to form a number of spaces extending radially outwards from the axis. Filter medium is supported on each side of these plates. At the axis is a rotating shaft on which are mounted the shearing discs, one in each space. When **filtration** begins, a thin cake is formed on either side of each plate, which acts as the effective filter medium for the rest of the operation. The shear created by the rotation of the discs restricts the thickness of the cake, and the remainder of the separation produces a thick slurry, and a clear **filtrate** drains from the centre of each filter plate, out of the **casing**.

Asbestos. A natural fibrous material, which, for many years was used as an excellent **filter medium**, in such applications as final **clarification** of beer; the filter **sheet** was, for most of its existence, made from a mixture of **cellulose** and asbestos. The slow realisation of the toxic nature of asbestos has led to its complete elimination as a filter medium material.

ASHRAE. The American Society of Heating, Refrigeration & Air conditioning Engineers is a major source of air **filtration** standards, its 52.2P classification being a well-used standard, while its 52–76 specification for a filter test rig is also widely used.

ASTM standards. The American Society for the Testing of Materials has set a number of standards for test procedures for **filters**, especially **membrane** filters.

Asymmetric media. The term is applied especially to **membrane** media, which then operate effectively as a **surface filter medium** supported by a **substrate** of some kind. The medium has different pore sizes across its thickness, with the finest at the side facing upstream, and the largest downstream. The variation may be gradual, or, more commonly, it may be a sharp change, with a very thin, very finely porous layer on top of a thicker, stronger and much more coarsely porous layer that acts as a support for the thin layer. The degree of asymmetry is the ratio of the size of the surface pores on the two surfaces of the membrane.

It should be noted, however, that some asymmetric membranes are made in the opposite fashion, with the coarsest pores facing the slurry, and the finest at the permeate side. This enables the membrane to act as its own prefilter, and brings an element of **depth filtration** to the use of a membrane.

Atmosphere. A historical unit of pressure, expressed in absolute (including the pressure of the atmosphere) or gauge (excluding the actual atmospheric pressure) terms. Equivalent to 101.3 kPa.

Atomic mass unit. The AMU is a measure of atomic and molecular masses. The basic unit is one twelfth of the mass of an atom of the C^{12} carbon isotope, and this is most often in **filtration** terms referred to as one **Dalton**. The main use is in defining the **separation** ranges of **membrane** processes – such as quoting **ultrafiltration** as being effective from molecules of the weight 10,000 Daltons (sometimes 10 kD), down to 200 Daltons.
[*See also Molecular weight cut-off*]

Atomisation. The formation of a liquid into a fine spray, or mist, of droplets, as might be done, say, for oil into a combustion chamber. Such a mist could then be used to challenge a **coalescing filter**.

Attachment efficiency. A term used in the description of **flotation** separations, in which rising gas bubbles become attached to particles of suspended solid and so carry them upwards out of the body of the liquid. The particle and bubble must first collide, and having collided, the particle must attach itself to the bubble, not all collisions being successful capture events. The attachment efficiency describes the number of collisions that do result in successful capture as a fraction of all collisions.

Atterberg limits. Parameters of a porous mass of solids containing some liquid, originally derived in order to classify soils containing a substantial amount of clay, and which therefore exhibit some plasticity. The same terminology is now used of filter **cakes**.

There are two Atterberg limits: the liquid limit and the plastic limit. If the moisture content of the cake is higher than that at the liquid limit it is said to be flowable (like a pumpable paste). If the moisture content is lower than that at the plastic limit it will be hard and brittle. Between the two limits the cake is described as plastic (a "sticky cake"). A **plasticity index** is used to quantify the behaviour of cakes between the liquid and plastic limits.

Augmentation factor. Where **candle**, **cartridge** or **bag filters** are operating with fluid flow from the outside towards the interior of the **element**, than the separated solids accumulate as a **cake** on the outside. As it is the cake which is the actual **filter medium**, this means that the **filtration** area will increase as filtration proceeds. The augmentation factor describes this increase in area – which helps offset the increase in **pressure drop** across the thickening cake.

Autocoagulation. The **coagulation** of particles of the same species by the attraction of positive and negative sites existing on different parts of the particle.

Automatic filter. A filter, usually with an **element** in the form of a cylindrical **cartridge**, which is cleaned automatically, either by **backflushing** or by some form of mechanical cleaning device, such as a brush or scraper. The cleaning cycle is then instituted either by a timed programme, or by a signal that the **pressure drop** across the filter element has reached a prescribed value. The mechanical devices usually operate while the filter is working, while backflushing requires a short cessation of the filtering process. Usually operate with one element, but may have two in parallel (**duplex**) so that the cleaning can be done off-line.
[*See also Self cleaning filter*]

Automatic filter press. A filter press in which the opening of the **plates**, discharge of solids, and possibly cloth washing are carried out automatically, although there is a range of degrees of automation.

Autopress. A proprietary version of the plate **filter press** (by BHS) with a plate package that is compressible, and which has an impermeable membrane surrounding the plate package, to create a hermetic seal.

Average particle size. Any real mass of particles will have its constituent particles arrayed in a wide range of particle sizes (and, possibly, shapes). If this distribution is determined (say, by a sieve analysis), and the resultant graph plotted of frequency of occurrence against corresponding particle size, then the resultant size distribution curve will most often adopt a bell-shape, falling away sharply on both sides of some main value, with, perhaps small tails on each side.

The main value is normally expressed either as a mode, or a median, or an arithmetic mean. The mode is the size that occurs most frequently in the mixture (although a real mixture can have more than one mode).

The median is the particular size that divides the mixture exactly in half, with as many particles in the one half as in the other, while the arithmetic mean is the sum of the products of all of the particle sizes and their frequencies, divided by the total number of particles. In a symmetrical distribution these three measures will be equal.

The distribution is usually defined by quoting the modal value (assuming just one peak) and some measure of the width of the "bell".
[*See also Normal distribution*]

Axial filter. A term used to describe a **dynamic filter** in the form of a porous cylinder. The cylinder rotates about its axis, and the feed slurry flows across the outer face of the cylinder. **Permeate** flows through the cylinder wall and out of the filter from its centre, while concentrate leaves the housing at the opposite end from the feed zone.

Axial flow. The flow of a fluid along the axis of rotation of a machine, or in a direction parallel to that axis, or the axis of a body of fluid within the machine. Refers especially to the flow of clarified liquid within a **centrifuge**, which, in the case of a **decanter** needs to be enabled by **windows** or holes in the **conveyor flights** (as distinct from the normal **helical flow** pattern).

Axial dispersion coefficient. A term used in filter **cake** theory, to describe washing effectiveness. It measures the changes in concentration in a direction parallel to that of the main flow of fluid.

Axial velocity. A feature of **axial flow** and thus relating to any movement parallel to the axis of rotation of an item of **separation** equipment, but of particular interest in the operation of **cyclones** and **hydrocyclones**, where the axial velocity reverses direction, and must therefore have a zero value at one point inside the cyclone. There is still some doubt over exactly where the reversal occurs.

Azeotrope, Azeotropic mixture. An azeotrope, or azeotropic mixture, is a homogeneous mixture of two liquids whose composition is exactly the same as that of the vapour formed by distillation of the mixture, and so cannot be separated by fractional distillation. **Membrane separation processes** may well provide a method of separating the two components.

B

Bacillus. A rod-shaped form of **bacteria**.

Back diffusion. A term generally referring to the movement of molecular species in a fluid, by **diffusion**, in a direction opposite to that of the fluid flow. Refers specifically to the movement of solute molecules from the **mother liquor** held in a **cake** backwards across the boundary with the following wash liquor. This back diffusion reduces the efficiency of **displacement washing**.

Back-drive. Any system, either electrical, hydraulic or mechanical, used to effect control of the **conveyor** differential speed in a **decanter centrifuge**. Most often this is achieved by means of a variable speed gear box driven by the main drive motor for the centrifuge, but it can also feature a quite separate drive.

Backflushing. A method for cleaning **filters** of all kinds that involves the reverse flow of the filtration fluid through the medium to carry away any collected **cake** of solids. The removed cake may then fall to the bottom of the **filter housing** for automatic or manual removal, or it may be carried out of the housing in the reverse flow. The process is also used to purge collected solids from the filtration layer of a granular **depth filter**. Where the **filtrate** is a liquid, this is also called **backwashing**.

Back flush filter. The patented Bokela RSF back flush filter is a plate-and-frame **filter press** with travelling slot-shaped suction devices that are moved at regular intervals through each filter chamber to backflush the **media** free of gel-like or other soft deposits on the media.

Backing cloth. A coarsely porous and robust material, usually a **woven** fabric, which is mounted downstream of a **filter medium**, to support the medium, and sometimes to protect the equipment from the consequences of a rupture in the medium. It has a specific function also in protecting the **cells** of **vacuum filters**. Other purposes include the easier sealing between filter components, or improving **filtrate** flow away from the filter medium, or the protection of the medium from damage.

Back mixing. This refers to the recycling of material from one part of a process back to an earlier stage, either intentionally or unintentionally. Its primary feature in the **filtration** industry is in the recycling of a relatively dry material to be mixed with a wet material, either to make the latter more easily fed to a **separator**, or more easily dried, or more easily shipped to another process.

Back pressure. At any point in a process, the back pressure is the **differential pressure** between that point and the end of the process. It can also refer to an applied pressure, necessary to maintain a required condition (such as to keep carbon dioxide in solution during the final **filtration** of carbonated beverages).

Backpulse. A term used in **cross-flow membrane filtration** to refer to the removal of a **fouling** layer from the membrane surface, by means of a pulse of fluid (liquid or gas) from behind the membrane.

Backwashing. Specifically used to refer to the regeneration of the granular media in a **deep bed filter** by passing liquid up through the bed (i.e. in the opposite direction from that of the **filtration** flow). This is done quickly, at a higher flow rate than that used in filtration, so that the bed of granules expands to release trapped material. (Used generally as a synonym for **backflushing**.)

The term is also used occasionally to mean the washing of a **cake** off a **filter medium** by means of a reverse flow of liquid through the medium.

Bacteria. Unicellular microscopic organisms, that are prokaryotic (having no separate internal structures), unlike _all_ other living forms, which are eukaryotic, containing within their cells separate membrane-enclosed organelles, including a nucleus. Bacteria occur in spherical (coccus), rod-shaped (bacillus) or curved (spirillum, the helical spirochete or the comma-shaped vibrio) forms. Some are pathogenic, but most are harmless and/or beneficial to life in general. Bacteria exist as discrete particles in the diameter range 40 μm down almost to 0.1 μm, so are removable by **microfiltration**. For example, the common rod-shaped *E. coli* bacterium typically has a diameter of 0.5 μm and a length of 2 μm.

Bacterial challenge test. An efficiency test for **filter media** (also known as a bacteria retention test) that involves counting the bacteria content (frequently *P. diminuta*) in a feed stream and the resultant **filtrate**, by the usual bacteria growth methods. The result is usually expressed as a log reduction for the filter medium. The test becomes very important when the filter medium is being considered for use as a sterilizing filter – for the removal of **pathogenic** materials from drinking water, for example.

BAF. [*See Biological Aerated Filtration*]

Baffle. Any obstruction placed in the path of a flow of fluid, intended to make that fluid change its direction of flow. It can also refer to obstacles to the flow of solids in some designs of **centrifuge**.

In the case of a **decanter centrifuge** a baffle is used to control the movement of solids by the **scroll**. It takes the form of a restriction, usually a disc or cone, fixed to the **hub** of the conveyor, which either impedes the movement of the solids, or redirects the movement of the liquid flow. (Also known as a cake baffle.)

Baffles are also used to restrict flow within a centrifuge **casing**, to prevent or minimise cross contamination of discharged products.

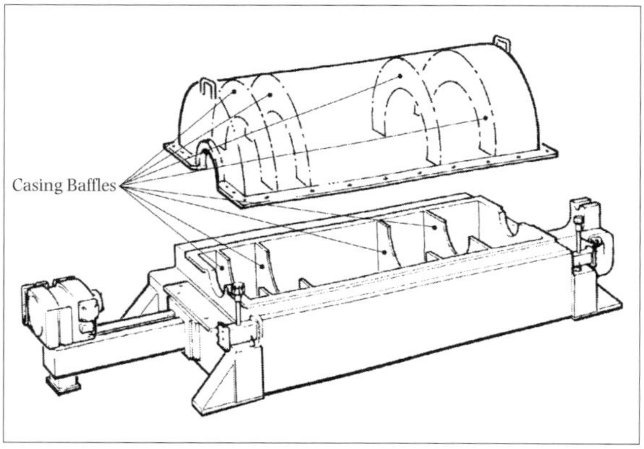

Decanter casing showing baffles

Baffle collector. A gas cleaning and dust removal system, using one or more plates or **louvres** mounted so as to cause a sharp change in the direction of flow of the gas. The inertia of the dust particles causes them to continue in their original direction of motion, whereby they become separated from the gas flow. Such devices can operate in laminar flow, removing particles to below 100 μm, and more complicated designs are effective to 20 μm or less.
[*See also Aerodynamic cleaner*]

Baffle cone. A conical baffle, built onto the **conveyor hub** of a **decanter centrifuge**, shrouding the feed zone, and impeding the flow of solids travelling from the **bowl's** cylindrical section onto the **beach**.

Baffled flocculator. Once a **flocculating agent** has been added to a slurry, then a finite period of time must be allowed for the flocculation to occur. This can be achieved in non-agitated equipment, such as a trough, which has been fitted with a set of baffles, such as the over-and-under or slotted types, so as to increase the residence time of the flocculating suspension.

Baffle disc. A disc built onto the **conveyor hub** of a **decanter centrifuge** to restrict the flow of scrolled solids.

Baffle-ring centrifuge. A type of **cone screen centrifuge**, developed by Krauss Maffei, in which the progress of the dewatering solids across the conical screen is hindered by a succession of circumferential baffles and/or steps in the screen surface. At each baffle or step, clumps of the cake disintegrate and release more mother liquor. The centrifuge has a vertical axis of rotation, with its feed zone at the bottom, dry solids overflowing from the top of the basket. There may be a cylindrical section of the screen at the feed point to deal with thin slurries. Some models are fitted with fans that induce an air flow, which helps to dewater the cake. Very low moisture contents are claimed for coarse particle separations.

Baffle plate. A plate, usually circular in shape, that is suspended in a pipe to promote turbulence in, and therefore mixing of a fluid.

Bag. [*See Filter bag*]

Bagasse. The fibrous material of sugar cane, which is the residue of the extraction of cane sugar. Most bagasse is burned as fuel, but a fraction of fine fibres can be extracted in a **cyclone**, and used as a **filter aid** in some of the sugar factory's filtration processes.

Bag filter. The bag filter is a separating device for removing solid particles from liquids or gases. It exists in several forms, with the common feature that all employ **filter bags** as the active element (although systems indistinguishable in any other way from the formal bag filter do use **candles** or **pockets** as the filtering **element** rather than bags). In the case of solid/liquid separations, the bag filter may have just one bag mounted in a co-axial cylindrical housing, or it may have several bags suspended from a single plate, usually mounted in a hexagonal array.

The most common format in solid/liquid separation is as a simple clarification filter consisting of a gravity or pressure fed bag, with filtration from the inside outwards.

Because of the much larger fluid volumes being handled in gas cleaning, the bag filter used for solid/gas separation will normally be a very large structure (or baghouse), containing a multitude of bags operating in parallel. The bags may be fitted outside a supporting cage, with filtration flow from outside the bag to its interior, or inside a similar cage, with the flow now from inside out. In the former case, **backflushing** will blow the filter cake off the outer surface of the bag, to fall into the bottom of the housing. In the latter case, separated cake will collect in the bottom of the bag, which may then need to be removed manually from the housing for emptying.

Baghouse. [*See Bag filter, Fabric filter*]

Balance line. A small pipe, used in a **vacuum filter** system, that runs from the suction eye of the **filtrate** transfer pump back to the evacuated part of the filter. This creates a U-tube into which the filtrate flows by gravity, thus making the filtrate pumping a much easier task. It also allows vapour to return to the filtrate vessel.

Ballotini. Small spherical glass beads, mainly made to form reflective surfaces on road signs and cinema screens. They can also be used as a good material for **challenge tests** of **filters**.

Band filter. [*See Belt filter*]

Band strainer. [*See Travelling screen*]

Bar. A measure of gas pressure, defined originally as 1 kg/cm^2, and much used because of its similarity to a standard atmosphere (= 1.034 bar). Now defined as 10^5 Pa, but the millibar is still much used in meteorological reporting.

Barometric leg. A tall liquid column, whose height is equivalent to the local atmospheric pressure as measured in terms of the height of the liquid being filtered. It is formed in a vertical pipe, which is attached to the outlet of a continuous **vacuum filter** and which dips into an open topped seal pot at its base. Through this pipe the **filtrate** can then leave the filtration zone, overflowing from the seal pot. This is a simple means of filtrate transfer, but it needs adequate height in the filter installation to enable it to be fitted.

Barrer. A unit used in measuring the gas **permeability** of **membranes**.

Bar screen. A **filter medium** created from a set of metal bars (each usually of wedge shaped cross section) arranged parallel one to another, with the set either flat, or curved, or in the form of a cylinder. The wider faces of the bars form the **filtration** surface, and are mounted in such a way as to give a precise aperture between each pair of bars, which, because of the wedge shape, widens as it moves away from the filter surface.

As used in mineral processing applications or water treatment, a bar screen will usually be a flat structure, inclined slightly to the horizontal, and made of stout metal bars, for use either in **size classification**, or the removal of **tramp material**. [*See also Sieve bend*]

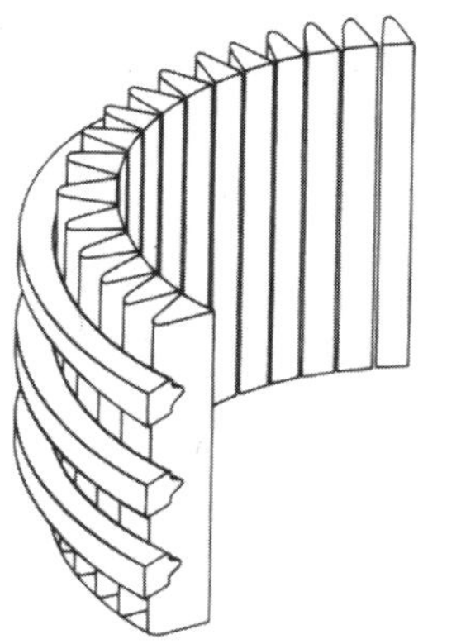

Section of cylindrical bar screen

Basket strainer

Basket. A simple filter element, consisting of a coarsely porous or perforated material arranged in a cylinder, usually closed at one end (possibly by the same material as the wall) and open at the other. As used in an **in-line strainer**, the basket will be made of a coarse wire mesh or perforated metal or plastic sheet, mounted in a cylindrical housing. This housing may be perpendicular to the main flow line, or inclined to it (Y-type), so as to reduce the pressure loss caused by the sharp changes of flow direction in the right-angled mounting.

The basket will eventually become blocked, and need to be cleaned. This can be done manually, by opening the housing and washing the basket, or by backflushing, or by some kind of automatic mechanical system within the housing.
[*See also Basket centrifuge*]

Basket centrifuge. A term used generally to refer to any of the **centrifugal filters** in which the separating surface takes the shape of the cylindrical wall of a drum or basket. This wall is usually made of a perforated material, which can either act as the **filter medium**, or as a support for a finer material.

Basket strainer. [*See Basket*]

Basket weave. A version of **plain weave** in a fabric where two or more **weft** threads are interwoven evenly with two or more **warp** threads.

Batch centrifuge, Batch filter. A **centrifuge** or **filter** designed for **separation** as a **batch process**, and thus not equipped with a system for the continuous removal of separated solids.

Batch filtration. [*See Batch process*]

Batch process. Any solid/liquid **separation** process can be carried out continuously, or as a batch process. In the latter case, a quantity of slurry is prepared, and then passed through the **filter** or **centrifuge**, with the **filtrate** or **centrate** collected in a **receiver**. The separated solids accumulate in the separating device until all of the batch is processed, whereupon the device is opened, all of the accumulated solids are removed, the device is cleaned, re-assembled, and the whole process is repeated. Solids removal may be done manually, or mechanically. **Washing** and **dewatering** of the solid **cake** are possible during the process at times best suited to each step, and can thus be independent of the separation step. Batch processes are used when the production rate is small, or where

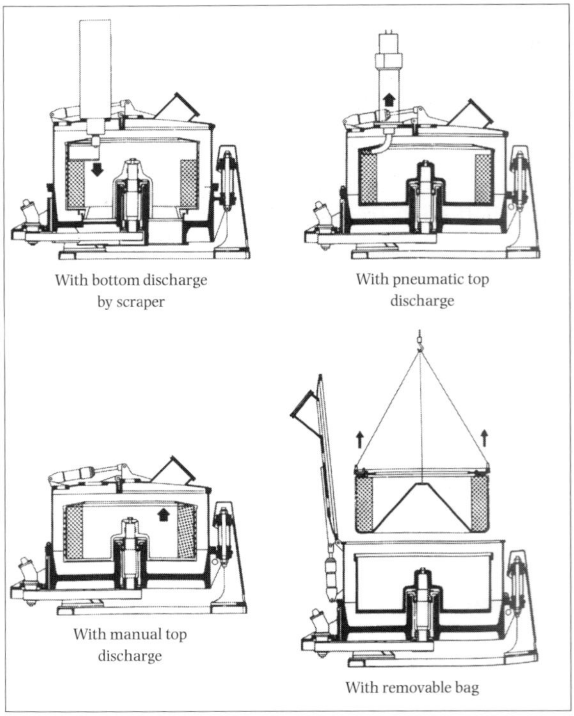

Basket centrifuge

a range of products is to be produced in succession. Gas/solid separation rarely involves batch processing.

Batch separation. [*See Batch process*]

Batch settling curve. A graphical representation of **sedimentation** behaviour, in the form of a plot of sediment height against time obtained in a **jar test**.

Batch test. Any laboratory test of a **separation** system, which uses a relatively small sample of the **suspension** to be separated, and is undertaken to study the relevant separation behaviour. More specifically, the term refers to the tests done on samples of suspensions of solid particles in order to be able to forecast their performance in a **sedimentation** process. These would include a dilute settling test, and a test under **hindered settling** conditions.
[*See also Jar test*]

Bathtub strainer. A temporary strainer fitted into a flow circuit, after it has been assembled for the first time, or after a period of maintenance, to ensure that it is free of **tramp material**. Such strainers have to be fitted into any convenient point in the circuit, and are often in the form of a trough, when they are called bathtub strainers.

Batt. Refers either to a type of **felt**, or to the pile of carded **webs** set up prior to consolidation into a felt.

Beach. The conical section at the solids discharge (rear) end of the **bowl** of a **decanter centrifuge**, up which the solids are scrolled before discharge. In some older and smaller decanters, the bowl consists only of beach. (Sometimes referred to as the extension.)

The term is used similarly in the case of an area of exposed solids adjacent to the pond of a **mechanical classifier**. In either usage, it is a zone where the solids are draining free of liquid.

Beach angle. The semi-included angle of the cone that forms the **beach** of a **decanter centrifuge**. It is the angle subtended by the inner surface of the beach to the decanter axis.

Beach length. The axial length of the beach in a **decanter centrifuge**.

Beach liner. A replaceable covering fitted to the inner surface of the beach of a **decanter centrifuge**. Its purpose is either to prevent the weakening of the structure of the centrifuge by erosion, or as an aid in the scrolling of solids up the beach, in which case it will be roughened, ribbed or grooved.

Beaker centrifuge. A high speed centrifuge used mainly for testing and other laboratory work. It may have a vertical rotor, in which case the beakers are held in arms that are vertical at rest, but swing out to the horizontal in rotation. If the axis is horizontal, then the beakers are in fixed vertical holders, but then must be sealed. The beakers are usually designed for **sedimentation**, when they are often in the form of test-tubes with conical bottoms, but they can also include some **filter medium**, when the test is for **filtration** characteristics. Beaker centrifuges are also used for very small scale production processes.

Beam thickener. [*See Bridge thickener*]

Beater. A device sometimes fitted in the **dewatering** zone of **rotary drum vacuum filters**, whereby the accumulated filter **cake** is beaten to help to liberate the moisture trapped in it. The device usually consists of a rotating shaft, parallel to the filter's axis, carrying two or more flexible strips that stretch across the whole width of the filter surface. The beater is pressed against the cake such that its rotation flexes the strips against the cake, and so cause the cake to move, and release trapped moisture.

Bed porosity. [*See Porosity*]

Belt. A continuous strip of material running between two rollers (and frequently over others). If the material is porous, then the belt is usable as a **filter**; if it is non-porous, then it acts as a conveyor, or it is used above a deposit of wet solids to squeeze them through a porous belt, to improve **dewatering**.

In the present context, a belt is mainly thought of as a **filter medium**, primarily used for harvesting solids from liquid suspension, usually moving from a feed zone, through a filtering zone, possibly through a **cake washing** zone, a **dewatering** zone and on to a cake removal zone, after which the belt returns through a cleaning zone to the start. The filter belt may then run together with an impervious belt that squeezes the cake to improve dewatering.

Belt control. A device fitted to any continuously operating belt machine to ensure that the belt does not move out of its correct running position. It will normally comprise a detector that registers sideways movement, and a deflection roller or other means of restoring the belt's proper position.

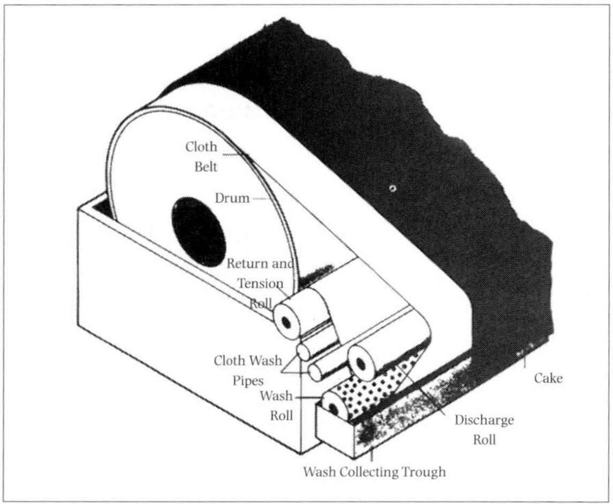

Drum filter with belt discharge

Belt discharge. A **rotary vacuum filter** can be fitted with a belt as the **filter medium**, which goes round most of the drum but then diverts over a roller with a much smaller diameter than the drum, so that the tighter turn over the roller causes the separated filter **cake** to leave the belt and be discharged from the filter. The medium can then be washed on both sides to clean it ready for the next filtration cycle.

Belt extractor. Similar in many ways to a **belt filter**, the De Smet belt extractor is a device for carrying out **percolation leaching** of a porous solid mass. It consists of an enclosed, discontinuously moving perforated belt. The belt is made up of flat surfaces of **wedge wire** or mesh cloth. The material to be treated is fed into the belt at one end, and held on it to a depth of 1.5 m by the side walls. The leaching liquid is sprayed on countercurrently. After each wash, the bed is allowed to drain. The top of the bed is scraped with a rake which breaks up the surface and maintains the **permeability** of the bed.

Belt filter. Any one of a number of types of continuous **filter** employing an endless belt as the **filter medium**, onto which the feed slurry is fed. The driving force for the **filtration** may be by simple drainage under gravity, or by suction from underneath the belt, or by applying pressure above the pre-drained cake by squeezing with another belt. The most common form of the vacuum belt filter operates horizontally, occupying a considerable floor space, but it enables very thorough washing of the cake before final **dewatering** and discharge.

Belt press. The **belt filter** operating by squeezing the wet **cake** by means of a second belt is also called a belt press. The belt press can be arranged horizontally, like a vacuum belt filter, but can also take the form of a **tower press**, with the belts running between sets of rollers, one above the other, to create a very tortuous path, but occupying much less floor area. Whatever the design, the feed to such a device will normally need **thickening** to make it stiff enough not to wash off the belt in the feed zone. (Sometimes called a double-belt press.)

Bending length. A measure of the stiffness of a material, and especially of a **filter medium**. The

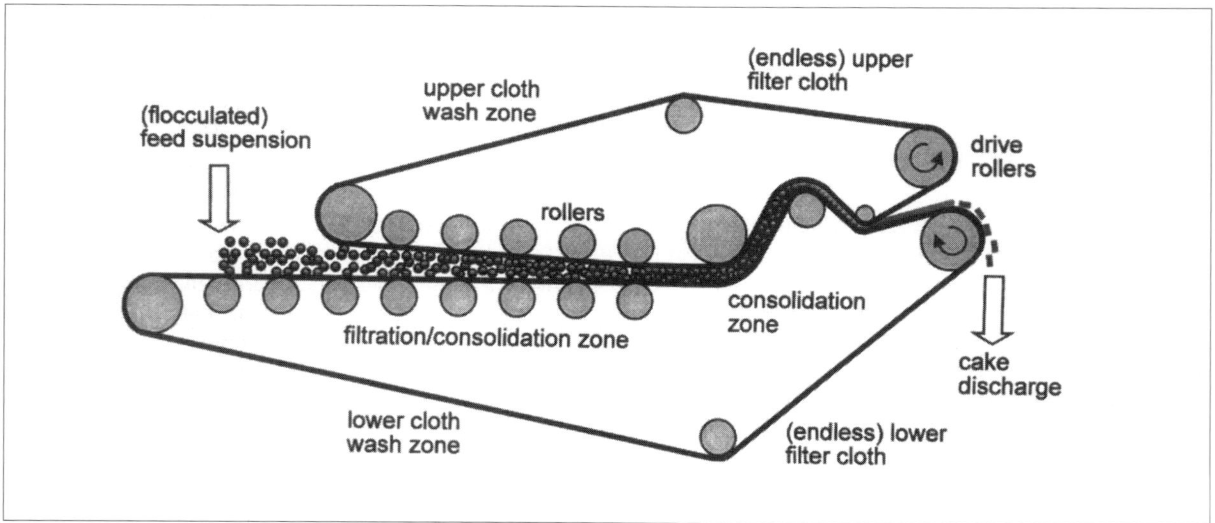

Belt filter

textile industry test for stiffness is an empirical one, recording the length of overhang of a strip of material necessary for it to bend through an angle of 41.5° from the horizontal, under its own weight. The bending length is then half of the overhang length.

Bentonite. A group of clay minerals that are hydrous aluminium silicates, also containing iron and magnesium, as well as either sodium or calcium. The chief component is **montmorillonite**. Sodium bentonites can absorb large quantities of water, thereby swelling to many times their original size, and are widely employed because of their adsorptive characteristics. Among many other uses are the clarification of wine and the softening of water. Calcium bentonites do not swell, but can be made into a fine adsorbent powder, also known as **fuller's earth**.

Bernouilli equation. Bernouilli's theorem is effectively the principle of conservation of energy in the movement of an ideal fluid. It states that the sum of :

- the energy associated with the fluid's pressure,
- the potential energy associated with its elevation above some datum level, and
- the kinetic energy of its movement

is constant. For such a fluid, flowing in a horizontal line (so that there is no change in potential energy), the fluid pressure must be inversely proportional to its velocity.

Thus, where the velocity changes because the cross-sectional area of the carrying pipe changes, the smallest area corresponds to the highest fluid velocity, and therefore the lowest pressure (this is the Venturi effect, employed in water jet pumps).

A practical use for Bernouilli's theorem occurs in the calculation of **trans-membrane pressures**, which can be obtained from a consideration of the energy balance in the liquid passing through the **membrane**, as described by Bernouilli's equation relating pressure, velocity and density.

BET isotherm. Brunauer, Emmett and Teller developed a theory of **adsorption** of gases on solids, suggesting that when a gas is brought into contact with a solid surface, adsorption occurs in discrete layers, each one molecule thick. The first layer thus becomes the location for the adsorption of the second layer, and so on. The theory assumes that the heat of adsorption is the same for each layer, which implies that the heat of liquefaction or evaporation for the second and successive layers is the same as that for the liquid state. The BET isotherm describes this multilayer behaviour, and is a reasonably good basis for the determination of the surface area of finely divided powders.
[*See also Surface area*]

Beta factor, ratio, value. The word "beta" has several uses in **separation** technology. Its most common usage is as a measure of the efficiency of a **clarification** operation. It is then defined as the ratio of the number of particles per unit volume larger than a given size, in the suspension upstream of a **filter**, to the number per unit volume in the **filtrate** larger than the same size. Removal efficiency may then be represented by:

$$100(\beta - 1)/\beta$$

The beta value is also a frequently used term for the filter **cloth resistance**. It can be established on the basis of **permeation** experiments with a particle free liquid or directly from a filter experiment via the **Darcy** equation.

In the development of the theory of solids separation and movement within a **decanter centrifuge**, beta is also used as a scrolling capacity function

Finally, the beta index is a measure of **cake resistance** used in the water industry.
[*See also Cloth index*]

BH. The product of magnetic flux density, B, and the magnetic field strength, H, which defines the performance of a magnet to be used in a **magnetic separation**. Measured in tesla amperes per metre.

BHS-Fest filter. A design of continuous pressure, **rotary drum filter**, from the BHS company. It consists of a drum rotating in a horizontal pressure shell, which has a slot along its length from about 90° to 150° (of clockwise rotation). The drum is compartmented into **cells** by partitions which make a close fit with the inside of the shell. The vessel is under pressure from about 150° until 90°, and can be divided into sections by internal seals to allow **cake** formation, **washing** and **dewatering** by gas blowing. The cake is discharged in the open section at 120° by gas blowback and a spring loaded scraper which digs the cake out of the cell. The cloth can be washed in the open section. The filter can be jacketed for operation at high or low temperature. Cake thickness is not variable but is fixed by the depth of the cells. The maximum size of filter is limited by the ability to hold a seal at the long junctions between the pressurised and ambient sections at the open slot.

Bib. An outlet pipe from the bottom corner of a **filter press plate**, which directs the flow of **filtrate** or **wash liquor**, flowing from the drainage surface of the plate, into the filtrate collection channel.

Binding. Used occasionally as an alternative to **weave**.

Bingham plastic. The Bingham plastic model is a two-parameter rheological model widely used in the oil drilling muds industry to describe the behaviour of many muds. Fluids obeying this model are called Bingham plastic fluids, and they exhibit a linear relationship between **shear** stress and **shear rate**, after an initial shear stress threshold (sometimes called the plastic viscosity) has been reached.

Biological Aerated Filtration. The treatment of waste water flows by contact with active biological material held on a **packed bed** of inert material, which also acts as a **filter** for fine solids. BAF has a much smaller footprint than **activated sludge** processes, and is also able to convert nitrogenous contaminants to gaseous nitrogen.

Biological filter. A considerable volume of fresh and waste water processing is undertaken on biological or trickle filters. These are **packed bed**s of inert material, in quite coarse sizes, on which a film of **bacteria** is established. The flow of liquid over the packing allows the bacteria to react with organic matter dissolved or suspended in the water, and so purify the water. There is very little actual **filtration** involved.

Biological Oxygen Demand (BOD). A measure of pollution of a waste liquid stream, in the form of its ability to deplete the dissolved oxygen in a river or other body of fresh water. BOD_5 is the amount of oxygen a given volume of effluent will consume under standard conditions over 5 days. The effluent is diluted with water containing suitable **bacteria** and saturated with oxygen and the oxygen consumption determined after 5 days have elapsed. The Ultimate Oxygen Demand (UOD or BODU) test is similar but occurs over a longer period – up to 90 days. If the composition of the effluent is known, UOD = 2.67C + 4.57N where C and N are the concentrations in ppm of carbon and nitrogen respectively. (Also known as biochemical oxygen demand.)
[*See also Chemical Oxygen Demand*]

Binder-free media. Filter media made without any binding material being used to hold the fibres or filaments or particles of the medium together. The **adhesion** is achieved by natural interlocking, or by thermal processes (such as **sintering**).

Blackness test. One of a number of test procedures that involve the comparison of differing standard qualities of blackness, including **Ringelmann charts** (used for comparison of densities of black smokes), atmospheric dust spot tests (in which **filter** efficiencies are determined, based on the degree of blackness on a special filter paper downstream of the filter under test), and a qualitative test for the fit of dust masks (where black streaks are made on the wearer's face after carbon black is blown around the face, while the mask is being worn).

Blade. In **filtration** usage, this has the same meaning as **knife**, although **doctor blades** are sometimes referred to – these being fixed blades mounted close to a moving process surface, to control a layer's thickness or to remove material from the surface – most often used in papermaking or printing.

Blaine number. A measure of the volume-based or mass-based surface of a granular material. The Blaine procedure measures the gas permeation through a tablet made from the solid in question, and evaluates the pressurised gas volume flow rate according to the **Kozeny-Carman equation**. The required parameters are the solid density, the porosity of the tablet, the applied pressure, the volumetric gas flow rate and the gas viscosity. The Blaine number is a good comparative measure for specific surface areas, and is a useful tool for determining the filterability of a suspension. It is also used as a major indicator in the cement industry.

Blake number. A **dimensionless number**, akin to the **Reynolds number**, used in the study of **deep bed filters**, where the various flow channels may be of very different sizes. The Blake number:

$$Bl = xU_0\rho/\mu(1-\varepsilon)$$

where x = average diameter of bed particles
U_0 = superficial liquid velocity through space occupied by packed bed
ρ = liquid density
μ = liquid viscosity
ε = bed porosity

Blanket clarifier. [*See Sludge blanket clarifier*]

Bleed. The intentional extraction from any flow circuit of a proportion of the flow for any purpose, such as sampling, process control, or recirculation.

Bleeding. A term used for the loss of particles of **filter media** into the clean **filtrate**, as a synonym for **shedding**. It can also refer to the initial passage of fines in a suspension through the filter medium, before the **cake** has become established.

Blinding. The progressive blockage of a **filter medium**, and corresponding gradual reduction in its **porosity**, as its **pores** fill with trapped particles that cannot be removed by **back flushing**. Blinding may result in the medium's having to be discarded, or it may be possible to clean the medium physically or chemically to make it suitable for re-use, so at some pre-determined point in the blinding process, set by the **pressure drop** across the medium, flow is stopped and the medium is cleaned or replaced.

Blinding index. A term sometimes used as a measure of the ability of particles to migrate through a **cake** during compression and so blind the lower levels. It is determined from the difference between the **cake resistance** of two cakes formed from slurries at two different concentrations.

Blockage, Blocking. The specific act of filling a **pore**, at its upstream entry point, or somewhere within its length, by a particle from the **suspension**, which progressively leads to the **blinding** of a **filter medium** (and occasionally used as a synonym for blinding).

Blocking filtration. An alternative term for surface **straining filtration**, where a particle passes through the **filter medium** unless it is too large for the perforation, when it either goes across the surface or blocks the hole.

Blocking function. A parameter involved in the use of a **precoat filter** for the **clarification** of a liquid stream. In a continuously operating filter, such as a **rotary drum**, a thick layer of precoat material is laid down on the **filter medium**. This precoat cake removes suspended solids by **depth filtration**, the dirt being trapped in the outer zone of the cake. This outer layer is then cut away as a very thin layer, by means of an **advancing knife**. The blocking function, which is a combination of clean **cake resistance**, resistance of the trapped solids, and the knife advance rate, is used to determine the optimum rate of advance of the knife (an example being 0.2% of the precoat cake thickness per revolution of the drum).

Blowdown. The emptying of slurry or **cake** from a pressure vessel by the use of gas pressure.

Blow off timer. A device used in **rotary drum** or **disc vacuum filters** to introduce a sharp burst of air to each cell or compartment as it enters the cake discharge zone. It controls a valve leading from a high pressure air reservoir, and it makes the knife action in removing the cake more effective.

Blowback. The removal of a filter **cake**, after **filtration** of a liquid slurry, by the passage of air through the **filter medium**, in the opposite direction to the flow of liquid.

BOD. [*See Biological Oxygen Demand*]

Body feed, Body aids. [*See Filter aids*]

Body-feed filtration. The addition of filter aids (body aids) to a suspension to improve the filterability of the suspension by increasing the **permeability** of the resultant filter **cake**. This is not the same as **precoat filtration**, where similar materials are laid down on the surface of the medium to prevent the **bleeding** of fines at the beginning of the filter cycle. [*See also Filter aids*]

Bohna filter. The Bohna filter is a novel form of **deep bed filter** in which the liquid flows through a **packed bed** of sand from one side to the other. It is con-ceptually similar to a **sheet filter**, with a **filter press** configuration, but the flow is across the width of the sheet rather than through its thickness. It consists of a series of 10 mm thick rigid sand sheets, which operate in parallel. The sand is compacted and does not move during **backwashing**, neither is it much affected by surges in flow.

Boil point. In a **bubble point** test for a **filter medium**, as the differential pressure across the test piece is further increased, the point is reached at which air bubbles are profusely emitted across the entire surface of the medium.

Bolting cloth. A very fine mesh fabric, made from **wire mesh** or, increasingly, from polymeric **filaments**, in **plain weave**, with apertures down to 5 μm. Used for the dry **sieving** of fine powders – originally for the sieving of flour – and also as a **filter medium** in vacuum filtration, because of its low operating pressure drops, and easy **cake release**.

Boltzmann constant. A fundamental physical unit, which is the gas constant per molecule, equal to the universal gas constant divided by Avogadro's number. [*See also Stokes-Einstein equation*]

Bond diagram. A two-dimensional curve relating the **residual moisture content** of a **cake** undergoing **dewatering** to the appropriate Bond number, on a log/log scale. This is a special form of the **capillary curve**, and is especially applicable to the centrifugal dewatering of wet cakes, to define the residual moisture content that can be expected when the various forces involved are at equilibrium. The Bond curve usually has four zones. The first, for the lowest Bond numbers, covers the state of full saturation, up to the point where capillary entry pressure is reached, the curve being flat in this zone. In the second zone, the curve drops, sharply at first, then more slowly, until reaching the **Bond plateau**. In this zone, the capillaries are emptying. The third zone is the Bond plateau, which is flat for smooth particles (i.e. there is no dewatering as the forces increase), but for rough particles slopes gently downwards as the Bond numbers increase. The fourth zone marks the point where the forces are high enough to begin the removal of **interstitial liquid** and the moisture content therefore falls again.

Bond filter. An intermittent **vacuum belt filter**, supplied by Delkor. In the Bond filter the cloth indexing is done by electrical control of the end pulley. This allows good control and minimises **cake** cracking and extends cloth life. Not so suitable as the **Adpec filter** for solvent applications, which also allows more positive discharge over a small diameter roller. The Bond filter has a longitudinal vacuum box with moveable partitions. The cake may be compressed using rollers.

Bond number. A dimensionless number (sometimes called the **capillary number**) employed in **centrifugation**, to relate the forces expelling moisture retained in a solid **cake** to those forces retaining it. There are two slightly different versions of the Bond number, according to the appropriate zone of the Bond curve:

for the capillary dewatering zone,

$$Bo_c = d_h h_c \rho_l g C / \gamma_l \cos \delta$$

and for the interstitial water removal zone,

$$Bo_i = d_h^2 \rho_l g C / \gamma_l \cos \delta$$

where: d_h = hydraulic pore diameter
h_c = cake thckness
ρ_l = liquid density
g = gravitational acceleration
C = centrifugal C-value (ratio of centrifugal force to gravity)
γ_l = surface tension of the liquid
δ = wetting angle of liquid at the solid surface

Bond plateau. The zone on the **Bond diagram** in which the **capillary** moisture has been completely drained away, leaving only the **interstitial liquid** and other adsorbed liquid in the **cake**. A significant increase in the liquid expelling force is needed before the **interstitial liquid** starts to be expelled.

Bonded, Bonding. When fine **fibres** or **filaments** are held together in a mass, they are said to be bonded – by their own mechanical structure (curls in the fibres or hairs on their surfaces), or by thermal fusion at the points of contact, or by an adhesive. Bonded fibres are commonly used as **filter media** for **clarification** filters, the use of resin adding strength and avoiding fibre shedding during **filtration**.

Bonding does not usually include **sintering**, but it should.

Boot, Bosch. These are names sometimes given to the trough holding the feed slurry for a **drum filter**.

Boozer filter. A proprietary design of **rotary vacuum disc filter**, supplied by the Bokela company.

Boss. A protrusion on the drainage surface of a **filter press plate**, which buts up against a similar boss on

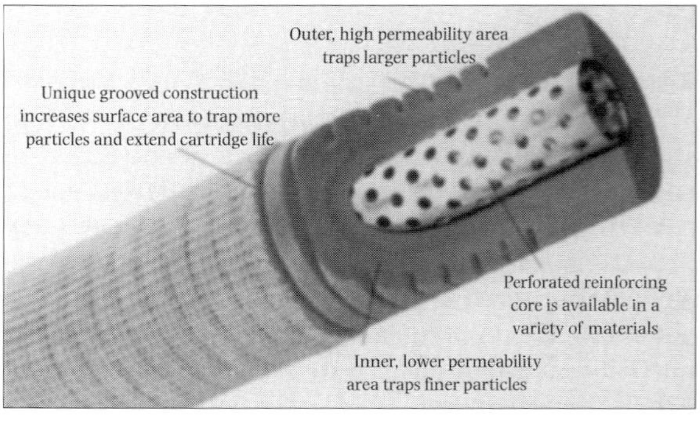

Bonded cartridge

the adjacent plate, and so maintains a separation between the plates, and resists distortion of the plates should the filter **cake** build up unevenly.

Bottleneck. A restriction in a fluid flow line that holds back the fluid and reduces the process throughput. Specifically a restriction in the **filtrate** take-off system, which can reduce the performance of a **filter** or **centrifuge**.

Bottle centrifuge. A laboratory centrifuge used for the spinning of test tubes containing suspensions.
[*See also Beaker centrifuge*]

Bottom fed filter. Any **filter** in which the feed slurry is fed below the filtering surface thus including most types of **vacuum drum filter**. The term is also used for **vacuum disc filters** even though the feed actually enters from the side of the filter element.

Boucher filterability index. A parameter used to characterise granular media **deep bed filters** used for water treatment. The index is given by the slope of a graph plotting the natural logarithm of the ratio of the head loss at any time to the initial head loss, against volume filtered at that time.

Bounce. A small particle being carried along the streamline of a fluid may carry on and hit a solid surface if its inertia is high enough, while the fluid diverts around the obstacle. The collision may then result in **adhesion** of the particle to the surface, so that the particle is removed from suspension. The system inertia is characterised by the **Stokes' number.** The higher the Stokes' number, the greater the chance of inertial impaction – but also the greater the chance that the particle will bounce off the collector.
[*See also Impingement*]

Bowl, Bowl shell. The hollow shell of a **sedimenting centrifuge**, which holds the slurry being processed. This can be the drum of an **imperforate basket centrifuge**, the tube of a **tubular centrifuge**, the double-conical shape of the **disc-stack separator**, or the combined cylinder and cone of the **decanter centrifuge**.

Bowl assembly. In the case of a **decanter centrifuge**, where the **bowl** operates inside a stationary **casing**, the bowl assembly refers to the assembled bowl, **end hubs**, and **conveyor**.

Bowl classifier. A circular **mechanical classifier** with a rotating rake for moving the solids to a central **underflow**. It takes the form of a shallow conical bowl, but as it does not dewater the coarse solids themselves, it is usually mounted above a rake or spiral classifier. The bowl section acts as a fine **classifier** and wash liquid can be added to achieve a sharper classification. The feed enters at the centre of the bowl and fines overflow at the periphery. The coarse solids settle and are raked to the centre where they discharge to the rake or spiral classifier for dewatering.

Bowl desilter. Similar to a **bowl classifier**, but with peripheral discharge of coarse solids (again to a separate rake classifier for dewatering).

Bowl filter. A continuous **vacuum filter**, characterised by a horizontal circular filtration surface, divided radially into segments, like a **table filter**, but with the outer flange attached to the filter surface and turning with it. The discharging filter **cake** has thus to be lifted over this flange by a special discharge screw, which means that a **heel** of dewatered material must be left on the **filter medium**, to protect it. The bowl filter, in common with many other vacuum filters, is well suited to the filtration of crystals and other rapid settling materials.

Bowl liner. A formed metal sheet, inserted into the bowl of a **decanter centrifuge**, to act as a scrolling aid, or for wear protection. It can cover the whole of the bowl wall, or only part of it. It could be plain, ribbed, grooved or given a special surface finish.

Bowl wall. The inside surface of the bowl of a **centrifuge**, on to which any solids in the material being separated will settle. In the case of a **decanter centrifuge**, these separated solids are then conveyed along the wall by the **scroll**.

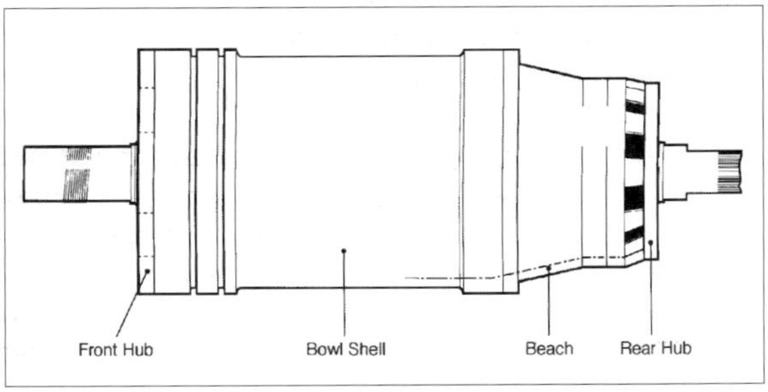

Decanter bowl assembly

Box press. A batch press used for the **expression** of a liquid from a mass of porous solid. It consists of a stack of open-topped boxes, inside each of which is a bag, usually made of canvas, containing the material to be pressed. In the box, the bag rests on a perforated base, supported on a drainage grid. The stack of boxes is placed below a mechanical or hydraulic ram, which presses down on the stack and compresses it as a unit, at pressures up to 100 bar.

Boyland cone. A variant of the **deep cone** settling tank, in which the cone is suspended from a knife edge in such a way that the increase in weight of the cone as the settled sand accumulates ultimately causes the discharge valve in the base of the cone to open and release the solids.

Breaking extension. The fractional strain of a textile at its **breaking load**.

Breaking load. In a test of the strength of a textile, the load required to cause a textile to fail. Expressed in kg/mm width.

Breakpoint chlorination. [*See Chlorine*]

Breakthrough. A term sometimes used to describe the unwanted passage of solid particles through a **filter medium**, or of air through a filter **cake**.

Breakthrough pressure. The gas pressure upstream of a filter **cake** that is needed to initiate **dewatering** by gas blowing.

Breathing apparatus. Equipment carried that enables the wearer to breathe independently of the **ambient** air, or to breathe ambient air after it has been rendered safe to breathe. In the latter case it will be fitted with **filters** that will adequately remove dusts and toxic or noxious gases.

Breathing air filters. Filters designed to remove impurities from **ambient** air that may be making it unfit for breathing. This means separation of suspended dust to very fine **cut points**, and also the removal of gaseous impurities, for which a **combination filter** will be required, able to adsorb gases.

Breathing air systems. [*See Breathing apparatus*]

Breathing cartridge filter. The name given to a special **wedge wire** filter **cartridge**, developed by the Brieden company, whose wire-to-wire spacings can be increased during **backflushing** to release stuck particles.

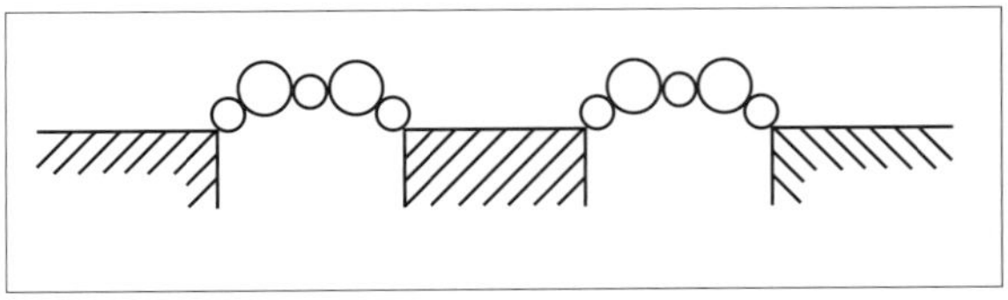

Bridged pores

Brewing. One of the major industrial applications of separations equipment, especially **filters** and **decanter centrifuges**, is in the brewing of beer. The use of such equipment is growing, as brewing becomes ever more fully automated.

Bridge thickener. A **gravity thickener** in which the **rake** mechanism, for moving the settled solids towards the discharge point, is supported by a superstructure spanning the tank. (Also called a beam thickener.)

Bridging. The formation of a bridge across the mouth of a **pore**, by the simultaneous arrival at the pore of several small particles, any one of which would individually have passed on through the pore. This is the first stage in the formation of a **filter cake**, with the bridging particles then effectively acting as the **filter medium**. Before the bridges are fully formed, fine particles will pass (or bleed) through the medium. Once the bridges are in place, filtration proceeds efficiently. The bridging effect may be achieved by the use of a **precoat** layer, which avoids **bleeding** by the suspended solids.

Brittle cake. A **filter** or **centrifuge cake** that has been so well dewatered as to be rigid and breakable into smaller pieces.

Brownian diffusion. Movement of species through a mixture or suspension driven by the random **Brownian movement**.

Brownian movement. The movement observed under the microscope of the small particles of a **colloidal** suspension, caused by collisions with the particle of molecules of the suspending fluid. If the particle is small enough, then the collisions all around it do not even out and there is a residual force on the particle. The effect is readily observable with particles of 1 μm or less in diameter, but is not observable at 5 μm or more. The effect of Brownian motion is to set a lower limit to the size of particles that will settle under the force of gravity.

Brush. One of the methods of cleaning **strainers** is to brush the surface free of accumulated solids – either manually or automatically.

Brutsaert equation. A relationship used in determining the **permeability** of a filter **cake**, involving the degree of saturation, during gas pressure **dewatering** of the cake. It gives a value for the relative liquid permeability of the **Wyckoff and Botset equation**:

$$p_{rl} = [(S - S_r)/(1 - S_r)]^n$$

where: p_{rl} = relative liquid permeability of the filter cake
S = relative degree of saturation of the complete cake
S_r = residual saturation after complete mechanical dewatering
n = an experimentally determined index, dependent upon the nature of the solid in the cake.

Bubble flocculation. Mechanical **flocculation** induced by the agitation caused by bubbles passing through the slurry.

Bubble fractionation. A **flotation** process in which the flotation column operates without a layer of froth or foam. Separation is achieved by the concentration gradient formed.

Bubble point. If a piece of **filter medium** is mounted horizontally, completely saturated with a standard liquid, a layer of which sits on top of the medium, and a gas pressure is applied to the lower side of the medium and slowly increased, then a point is reached when bubbles of gas start to appear from the medium in the supernatant layer of liquid. The first bubble appears from the top of the largest pore, whose **capillary pressure** is the lowest. The gas pressure corresponding to the appearance of this first bubble is the bubble point of the medium.

Bubble point test. The **bubble point** of a **filter medium** is easily observed, and it makes the basis for a quick and accurate assessment of pore diameter. The **pore diameter** can be calculated from the liquid viscosity and the differential pressure across the medium at the bubble point:

$$d_{bp} = 4\gamma_l/\Delta p$$

where: d_{bp} = the equivalent pore diameter determined by the bubble point test
γ_l = the viscosity of the liquid
Δp = the differential pressure at the bubble point

The test is applicable in the pore diameter range 0.1 µm to 100 µm.
[*See also Laplace equation*]

Bubble size. A key parameter in the process of **flotation**. Bubbles form and grow on nucleation sites until **buoyancy** forces are sufficient to release them into the liquid flow. New bubbles then become the focus for the release of **dissolved air**, and so the bubbles grow.

Bubbling zone. A feature of **rotary vacuum drum filters** that achieves the cleaning of the **filter medium**. A zone within the **control head** at the leading edge of the **filter cell** is equipped for air blowing at low pressure, from inside the filter cell out through the filter medium. This loosens deposits in the pores and on the medium surface.

Buchner funnel. The simplest form of **vacuum filter**, in wide use in laboratories, which uses **filter paper** as the separating medium. It takes the form of a cylindrical vessel with a coarsely perforated base, on which a disc of filter paper is placed. At the bottom is a discharge nozzle, which is normally fitted through a rubber bung into a suction flask – and the vacuum is usually created by a water jet pump. For very easily filtered materials, the Buchner funnel can operate as a gravity driven filter.

Bucket centrifuge. A laboratory centrifuge that holds small sample tubes, and is used for predicting performance of full scale **sedimenting centrifuges**.
[*See also Beaker centrifuge*]

Buffer chamber, Buffer zone. A compartment within the **conveyor hub** of a **decanter centrifuge**, lying between feed and flocculant zones, to prevent splashing from the **feed chamber** into the **flocculant chamber**.

Bulk density. The mean density of the solid material and the interstitial fluid in any mass of material – a quantity of granules, for example, or a piece of thick **filter medium**. It is determined from the weight and volume of the mass, and the ratio between the bulk density and the true density of the solid alone is a measure of the empty space within the solid mass. If the **void fraction** (or **porosity**) of the mass is known, together with true densities of solid and fluid, then the bulk density can be calculated from:

$$\rho_b = \rho_s(1 - \varepsilon) + \rho_f . \varepsilon$$

where ρ_b = bulk density
ρ_s = true density of solid
ρ_f = true density of fluid
ε = porosity of mass of solid

If the fluid is a gas, then the second term can usually be ignored.

Bulk oil flotation. A process used for the separation of minerals by allowing some to attach onto oil droplets and thereby float.

Bulk settling test. A laboratory test used for sizing **gravity clarifiers**.

Buoyancy. The buoyancy of a body immersed in a fluid expresses the tendency of that fluid to exert a lifting effect on the body. Archimedes famous principle states that such a body experiences an upward force (i.e. a force opposed to that of gravity) equal to the weight of the fluid that would fill the space occupied by the immersed part of the body. **Stokes' law** is an expression of the balance between buoyancy and friction forces.

Burden. A term sometimes used for the solids loading on a gas filter.

Bursting distension. The height to which a standard disc of material is raised before its failure in a dry or wet bursting test.

Bursting ratio. The bursting strength of a sample of **paper** divided by its **grammage**.

Bursting strength. A measure of the ability of any material to resist a potential bursting force when pressure is applied to one side of the material. Specifically, it is resistance of a **filter medium** to breaking under the effect of a force perpendicular to the medium, applied to the centre of a test piece, under carefully prescribed conditions. The sample is normally circular, clamped into the test rig around its periphery, and the force is applied by an air pressure on one side of the sample. The medium may be tested in the wet or dry state.

Bypass. The passage of a **suspension** being filtered through the **filter** unaffected, because of a crack in the filter **cake** or **filter medium** is an example of an unwanted bypass. The word is also used of an intentional bypassing of the medium in a **vacuum filter**, to control the suction pressure.

Bypass filter system. Most **filtration** processes take place on-line, with the full process flow being treated. However, there are occasions when a small, highly efficient **filter** installed on a by-pass line to the main flow is able to remove sufficient suspended material, such that the recombined flows have an adequate purity.

C

Cabin air. A particular segment of the **HVAC** industry, which has only come into prominence of late, with the realisation that air being breathed by several people needs to be treated to avoid the acquisition of germs. The problem first came into prominence in the case of airplane cabins, where considerable numbers of people are enclosed together in quite a small volume, for quite considerable periods of time, with the same air being recirculated all the time. The problem has been extended to the interior of automobiles, where air-borne dust, pollen and diesel fumes have become of increasing concern. The term now refers to any transport means where any number of people are confined for any length of time (including agricultural tractors), breathing air that needs to be conditioned for safety and health reasons.

Cabin (air) filter. Any **filter** intended for the treatment of **cabin air**, to render it safe for breathing. The **filtration** requirements are quite severe, as **bacteria** and **viruses** have to be removed, as well as plant pollen and diesel residues, at least to $\mathbf{PM_{2.5}}$ limits. The accepted method of cabin air filtration employs **depth filters**, with very high separation efficiency figures.
[*See also HVAC filters*]

Cabinet. In **filtration** terms, this refers to a working space, such as a bench, which is fully enclosed to protect the work being done in it from the impact of a dusty or dirty environment. Sealed doors are provided for the movement of tools and equipment in or out, and glove ports for the operators' arms. The cabinet is fed with filtered air, and kept under a slight positive pressure to prevent in-leakage of dust. A cabinet can be used for delicate work that does not justify the setting up of a whole **clean room**.

Cable thickener. A form of **gravity thickener** with a hinged **rake** attached to the bottom of a central shaft. The rake is pulled by cables.

Cage. An array of **wire**, **rod** or coarse **woven wire mesh**, usually cylindrical in shape, used to support a flexible **filter medium**, usually in the form of a **bag** or a **sleeve**. The more common arrangement is for the medium to be placed over the cage, with fluid flowing from the outside of the bag or sleeve to the inside of the cage, and then out of the **filter**. The cage provides support against the pressure differential resulting from fluid flow.
[*See also Bag filter, Core, Gas cleaning*]

Cage press. Type of hydraulic press used for **expression**. Similar to a **curb press**, but has fine internal grooves for **filtrate** flow, and so is more suitable for more oily and less fibrous materials.

Caisson thickener. A **gravity thickener** that has a central column containing a pump for removing the thickened **underflow**.

Cake. A filter cake is the more or less porous layer of collected solid particles that forms on the upstream surface of a **filter medium** during a **cake filtration** process. Once formed, the cake takes over the role of filter medium from the original material. Cakes may be formed in batch or continuous filtration, and thus will need to be removed from the filter either batch-wise or continuously. Before discharge from the filter the cake may need to be washed free of the original **mother liquor**, and/or **dewatered** as thoroughly as possible. The cake may be the desired product of the filtration process, so that complete solids recovery is sought, usually as free of mother liquor as possible, or it may be a waste product, removed to make subsequent treatment easier, in which case washing will not be required, but effective dewatering may be.

The word 'cake' also refers to any reasonably rigid material discharged from, for example, a **centrifuge**.
[*See also Cake filtration*]

Cake baffle disc. A recent development for the **decanter centrifuge**, in which a disc is mounted on the **conveyor hub**, generally at the junction between the cylindrical part of the **bowl** shell and the **beach**, to restrict the flow of cake, and provide a back pressure for the scrolled solids. It can also be fitted anywhere up the beach or into the cylindrical section.

Cake compression. Most filter cakes are compressible, and this has two important effects on the **cake filtration** process, which work against one another in seeking efficient filtration. The first effect relates to the filtration process itself, wherein the cake may be compressed as it forms, thereby reducing its **permeability**; the pressure differential across cake and restraining medium then has to be increased to maintain a required fluid flow rate – and the increased pressure causes greater compression of the cake. The less compressible a cake, therefore, the longer the cycle time will be in a batch filter, or the greater the amount of solid that will be collected per unit area of a continuous filter.

The other effect comes once filtration is complete and the cake is to be **dewatered**. An important method for achieving dewatering is the squeezing of the cake (either centrifugally or by the application of mechanical force) – and the more compressible the cake, the easier will be dewatering by squeezing.

Cake demoisturing, Cake dewatering. The removal of **mother** or **wash liquor** from the interstices of a filter cake that has been produced by **separation** from a liquid. This can be done by draining (either under **gravity** or by vacuum or under the effects of **centrifugal force**), by **gas displacement**, or by the use of **mechanical compression**. The removal of mother liquor is more usually achieved by the use of thorough **cake washing**, as a result of which it can be effectively completely replaced by wash water. However, the removal of all moisture is unlikely, with there being a **residual moisture content** for any method of demoisturing, even in the interstices between particles, let alone in the pores within the particles.

Cake discharge. The removal process for a finished filter **cake** from the **filter**. After batch **filtration**, discharge can be effected manually or automatically, but continuous filtration will almost always need automatic discharge, and it is this need to mechanise the discharge that has created the wide range of process filters, some of considerable complexity, especially when they operate under full vacuum, or moderate pressures.

Cake filter. Any kind of filter in which a cake of separated solids is formed as part of the **separation** process. The cake can form directly on the **filter medium**, or after a **precoat** layer has been laid down.

Cake filtration. A filtration process, in which the bulk of the work of separating the suspended solid from the suspending fluid (either liquid or gas) is done by the growing cake. Cake filtration is one of the three main practical filtration processes (the others being **depth filtration** and **cross-flow filtration**, with surface **straining** being a less-used fourth process), but it can only occur after an initial surface straining has taken place. This latter process involves the **bridging** over of the pores of the basic **filter medium**, in order to prevent the passage of fine particles through the medium. Once the bridging is complete, bypassing of fines ceases, and the cake starts to build. A certain minimum concentration of solids will be necessary in the suspension for bridging to occur.

Cake filtration is the normal mechanism involved when the concentration of solid particles in the feed suspension is high, and when the separated solid is valuable, so that as little loss of the solid as possible is sought. The main exception to this generalisation occurs with the filtration of waste **sludges** (e.g. those from municipal waste water treatment). The **filtrate** from cake filtration may also be valuable, or it may be a waste product.

Where no loss of suspended solids can be accepted, or where clarity of the filtrate is important, then the initial bypassing of fines as the particle bridges are established cannot be allowed, and then a **precoat** stage will be necessary.

The operation of cake filtration is a result of the combination of the required **filtration purpose** and the concentration of the solids in the suspension, but it is widely practised, and can be achieved under gravity, pressure, vacuum or centrifugal force. It can easily lead to effective **cake washing** and **cake demoisturing**.
[*See also Filtration*]

Cake formation. Refers to the stage in a batch or semi-continuous **filtration** process during which the separated solids start to lay on the bridges across each of the pores of the **filter medium**.

Cake formation angle. The angular sector available for cake formation in a **drum, disc** or **table filter**. The cake formation angle, α (measured in degrees), is related to the cake formation time, t, and the number of revolutions, n, of the filter by:

$$t = \alpha/360n$$

A cake formation angle starts at the position where the **filter cell** is completely immersed in the **suspension**, and ends where the cell emerges from the suspension.

Cake formation equation. The cake formation equation describes the formation of the filter cake in a way that derives from a combination of **Darcy's** equation with the mass balance around the **filter**. Under conditions of constant filtration pressure, the equation relates **cake thickness** to its formation time as:

$$h = \{(R_m/r_c)^2 + 2\kappa.\Delta p.t/r_c.\eta\}^{1/2} - R_m/r_c$$

where h = cake thickness
R_m = filter medium resistance
r_c = specific cake resistance
κ = **kappa-factor**, a dimensionless constant
Δp = differential pressure across cake plus medium
t = cake formation time
η = dynamic viscosity of the fluid

As the cake forms, a plot of its thickness against the square root of the elapsed time will be linear.

Where filtration is taking place under conditions of constant feed rate, the equation relates to changes in differential pressure with time:

$$\Delta p(t) = V.\eta.A^{-2}\,(r_c.\kappa.V.t + A.R_m)$$

where A = filter area
V = cumulative filtrate flow

Cake heel. [*See Heel*]

Cake height. [*See Cake thickness*]

Cake moisture content. An important parameter in the **cake filtration** process, where the objective is usually to achieve as low a moisture content in the final cake as possible.
[*See also Cake washing, Dewatering*]

Cake permeability. The freedom of fluid flow through a filter cake is measured by the cake permeability, which is the reciprocal of the **cake resistance**:

$$p_c = 1/r_c$$

where p_c = cake permeability, measured in **Darcies** [cm^2]
r_c = cake resistance.

Cake permeability is determined experimentally, for example by means of the **t/V = f(V) method**.

Cake release. Whilst it could also refer to the **cake discharge** characteristics of a **filter**, cake release more specifically refers to the ability of a **filter medium** to part company from the accumulated cake, at that part of a filter, or in that part of the filtration cycle, where such separation is required. This ability is a surface phenomenon: the more hairy the surface of the medium, the more difficult the release, whereas a very smooth surface will release the cake more easily. Accordingly, the materials from which filter media are made will be treated ("finished") to give them the required surface finish. The ability of a filter medium to release a collected cake is an important part of its filtration performance characteristics.
[*See also Heel*]

Cake resistance. The cake resistance measures the resistance to the flow through the cake of any particular fluid. The specific cake resistance, r_c, is independent of **cake thickness** and can be directly determined by the application of **Darcy's law**: if a clear fluid is passed through two similar cakes of different thicknesses, at a known pressure differential, the rates of **filtrate** flow for the two cases will generate two equations from which the resistance can be calculated. It can also be determined by the **t/V = f(V) method**.

The cake resistance is an inverse measure of filterability, and, if expressed in terms of the volume of cake, it is the reciprocal of **permeability**.

Cake reslurrying. The remixing of a separated **cake** with a liquid, in order to carry out a subsequent process. This will most often be a wash liquor as the first stage of **re-slurry washing**.

Cake thickness. The thickness of the filter cake produced by the filtration of a given suspension can be calculated from the constant pressure **cake formation equation**. A cake may vary in thickness from a few mm to over 1 m, depending upon the nature of the suspension and of the filtering system.

Cake thickness indicator, limiter. In a **pressure leaf filter**, for example, it is necessary to stop the **filtration** process once a pre-determined thickness of cake has been reached, in order to prevent the damage that could ensue in the event of the over-filling of the cake space. This can be done by means of observation of the pressure differential across the cake, but is also achievable by means of a mechanical device such as a diaphragm or a slowly-moving propeller, whose movement is affected by the proximity of the cake.

Cake volume. The external volume of a filter cake, including both the volume of the solid matter of the cake, and that of the voids within it.

Cake washing. A filter cake is produced when a **suspension** in a liquid flows through a **filter medium**, and however easily drained the cake, some of the original **mother liquor** will remain within the cake when the **filtration** is complete. If the cake is required for further processing, then the presence of even a small amount of residual liquor may be unacceptable, and so it must be removed. The removal can be effected by gas displacement, by evaporation, or by mechanical compression, but the most usual method is by washing, using a **wash liquor** that is completely miscible with the mother liquor. The displacement of mother liquor is a diffusion-controlled process, and may not be perfect.

Cake washing can be done *in situ* by means of **displacement washing**, when the wash liquor is passed through the cake in the filter, usually in the same direction as the original filtration flow. This is the easier method, but is usually less effective than **re-slurry washing**, in which the cake is discharged from the filter into a mixing tank, wash liquor added, to re-slurry the cake, and the suspension in the wash liquor is agitated for a period of time consistent with the degree of washing required. The resultant slurry is then filtered again through the same filter. This results, of course, in the residual moisture in the cake not being pure wash liquor, but a mixture of wash and mother liquor, even if very dilute.

More effective washing is effected by means of repetition of the washing process. Re-slurry washing can be carried out in a series of tanks, with gravity settlement of the solids, until adequately washed – when they can be filtered again.

Some designs of filter, in particular those known as **filter-dryers**, can be adapted to allow re-slurry washing to be undertaken in the filter chamber, without the need to discharge the cake from the filter.

The efficiency of a washing process is measured by the freedom from mother liquor after washing, and the amount of wash liquor used, often called the **wash ratio**.

Cake yield stress. The minimum pressure that can express moisture from the **cake**, especially as produced in a **decanter centrifuge**. The cake yield stress depends upon the solids concentration in the cake, and increases rapidly as that concentration increases.

Caldecott cone. A settling tank in the shape of an inverted cone. The simplest type of continuous settler, with the sloping wall of the cone leading the settled solids to the central output at the base. The Caldecott cone has a sliding gate valve to control the rate of discharge. It has one or two central disc baffles above the exit, to promote annular flow in this region and so avoid bridging over the discharge. Such devices rely upon the ease of flow of the settled solids, for which sand is the best example – an alternative name for them being **sand cones**.

Calder-Fox scrubber. A device used in the **demisting** of gases, especially where acid mists are involved. It features demisting by impaction on a vertically slotted orifice plate, down which the separated liquid runs, to a collection sump beneath the plate. Re-entrainment of larger droplets is possible, and these are caught by a second series of plates.

Calender, Calendering. A calender is a device for modifying the surface of a fabric (woven or non-woven), usually (in the **filtration** business) to make it more smooth than that before such finishing, and so improve its ability to release a filter cake. Calendering may be effected by passing the material over a heated roller, but is more usually achieved by passing the material between two (or more) heated rollers, which

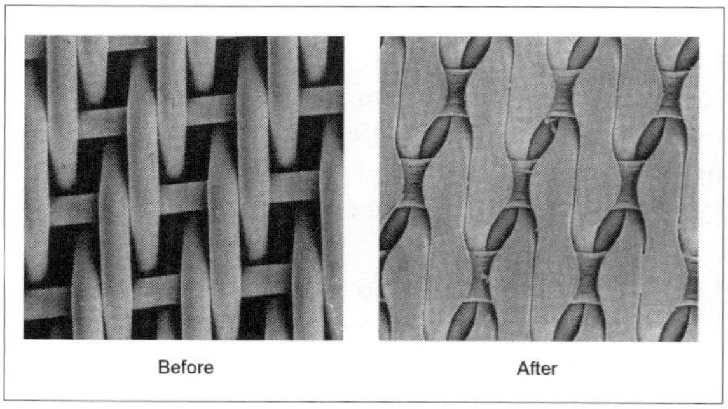

Calendered material

apply pressure to the material as well as heat. The rollers can be embossed so as to produce a pattern on one surface (or both) of the material.

Whilst calendering can be beneficial for some filtration characteristics, it must be noted that it will almost certainly reduce the average **pore** size on the calendered surface.

[*See also Cake release*]

Calendered wire cloth. Woven wire cloth is calendered by passage between two heavy rollers in order to flatten the points of overlap of the wires, in order to reduce the thickness of the cloth, and to provide a smoother surface. The product is sometimes called rolled wire cloth.

Calibrating tank. A vessel installed in a process plant or test rig to determine liquid volumetric flow – as delivered by a pump, or as passed by a **filter**.

Callow cone. A version of the **sand cone** in which the underflow is taken off continuously by means of a flexible hose, the exact position of which is decided by the balance between the head of dense underflow in the hose and that in the whole cone.

Camp-Hazen model. A model of the ideal continuous **sedimentation** tank. It equates the residence times derived from the settling motion of a particle in the vertical direction, and the horizontal plug flow movement caused by the drag of the liquid.

Camp number. A number used in **coagulation** process calculations, equal to the product of **flocculation** time and **shear rate**.

Camp-Stein equation. A relation that gives the **shear rate** in a stirred vessel, and hence the **Camp number**, in terms of power input to the stirrer, vessel filled volume, and liquid viscosity:

$$(\text{shear rate})^2 = \text{power/volume.viscosity}$$

Candle. In **filtration** terms, a candle is one form of replaceable **filter element**. A very common element, it takes the form of a rigid tube, open at one end and closed by a hemisphere of the tube material at the other. It is usually long in relation to its diameter, with a flange moulded on the open end from which the candle can be suspended in its housing. Like **filter bags**, candles can be used singly, although they are much more often used in arrays, suspended from a tube-plate. Filtration flow is from the outside inwards, and accumulated solids are blown off the outer face, to drop into the base of the housing.

Originally made from a white ceramic, their appearance giving them their name, candles are now made from any fairly rigid material, but ceramic candles are a major means of dealing with dust-laden hot exhaust gases.

The material of the candle can be self-supporting (such as a ceramic), or can be mounted on a supporting **cage** (such as **bonded** granules or fibres). The medium used in a candle is relatively thick, and may work either by **depth filtration** or by **cake filtration**. In the former case, the material of the candle will fill with retained solids, and will usually then be removed for cleaning. In the cake filtration case, the collected solids will usually be blown off the surface of the candle (by **back-flushing** or by a **reverse pulse** of gas).

Candle filter, Candle filtration. A candle filter is a batchwise filter, designed to use one or more **candle**s to achieve a relatively high efficiency filtration process. In its most usual form, a large cylindrical tank is mounted vertically, a tube-plate is fitted by flanges on the top of the tank, and a cover is then fastened above the tube-plate. An array (usually hexagonally close-packed) of holes in the tube-plate holds one candle per hole, to create a large **process** or

Candle filter

utility filter, normally operated under an imposed pressure, and providing quite a high filter area per unit volume of the vessel. Fluid flow into the vessel starts below the tube-plate, around the outsides of the candles, through the medium of the candles, to collect in the region above the tube-plate, and leave the filter from that region. If operating as a cake filter, the accumulated cake may either be blown off the candles dry, by gas pressure after **demoisturing**, to collect in the lower part of the **filter vessel**, from which they will normally be scraped out through a solid retaining valve; or the cake can be washed off the candles, in which case the cake falls into the bottom of the tank as a thick slurry, which can be drained more easily from the vessel.

If the candle filter operates in the **depth filtration** mode, then it will normally have to be opened up, by removing the cover, and all the candles removed for cleaning, once the dirt load reaches a critical value.
[*See also Fundabac filter*]

Candle press filter. [*See Tube press*]

Canister. A somewhat old-fashioned term for a container of **adsorbent**, such as **activated charcoal**, used mostly for purifying breathing air. Used in association with a **respirator**.

Cant. A term used of the **conveyor** in a **decanter centrifuge** to describe the angle between the tips of the **flights** on the conveyor and the **bowl** wall. This is more usually referred to as the **rake** of the scroll, but cant indicates a positive rake, i.e. that the rotation of the scroll tends to lift the solids that are being conveyed.

Cantilevered decanter. The normal arrangement of a horizontally mounted decanter centrifuge is for it to be supported on bearings at both ends of the **bowl assembly**. However, in some cases where the bowl is quite short, the bowl assembly is supported by bearings at one end only, and is cantilevered from those bearings. This makes the maintenance of the centrifuge easier, but increases the potential for out-of-balance vibrations.

Cap. [*See Endcap*]

CAP. [*See Continuous Area Press*]

Capacity. In general, the volume of fluid that a vessel or other item of equipment will hold when full. Can refer to the amount of slurry that a **batch filter** can process before it is necessary to remove the collected **cake**, or replace the **filter medium**. Can also relate to a throughput in volume or weight terms, per unit time.

Capillarity. A relatively obsolete term, relating to the behaviour of liquids in small channels, dictated by the particular property of **interfacial tension**, operating usually at the three-phase contact surface of a liquid, a solid and a gas.

Capillary. A very fine channel through any material, comparable in size with a hair (from the Latin word for hair), and capable of allowing fluid flow through the material. The capillary may be tubular, as in a blood vessel, or irregular in shape, formed by a group of particles pressed together. Some capillary pores may be closed at one end, allowing them to fill with liquid, but then being difficult to empty.

Capillary belt filter. A gravity **filter** in which the medium is a continuous belt, made of an absorbent material, into which the mother liquor is absorbed, driven by the capillary forces. The liquid is than separately squeezed out of the belt.

This can also be a **belt filter** in which the liquid drains from the cake through the openings of a vibrating screen to be absorbed by a porous belt running below the screen, in a direction opposite to that of the cake transport. The belt of absorbent is then squeezed between two rollers and returned to the filter.

Capillary condensation. The condensation of a vapour in the capillary pores of a material. There is a strong relationship between the **relative humidity** of the vapour in the gas phase and the **capillary pressure** in the pore.
[*See also Adsorption*]

Capillary diameter. The **equivalent diameter** of a **capillary** as determined by the **Laplace equation** relating the capillary diameter, d_c, with the **capillary pressure**, p_c:

$$d_c = 4\gamma.\cos\delta/p_c$$

where γ = **interfacial tension**
δ = **wetting angle**

This is determined experimentally by means of a **bubble point test**.

Capillary drainage. The discharge of liquid from the interstices of a porous mass, such as a **filter cake**, under the influence of gravity or **centrifugal force**. The amount draining is then determined by the **interfacial tension** between the draining liquid and the gas above it.

Capillary drain height. As a saturated porous mass drains free of liquid, under the influence of gravity or centrifugal force, a layer remains saturated at equilibrium, because of the effect of **capillary pressure**. The capillary drain height is then related to the interfacial forces between liquid and solid.

Capillary element. This term is sometimes used to describe a cylindrical **membrane element** with a diameter in between that of **hollow fibres** and tubular membranes.

Capillary entry pressure. Since **capillary pressure** is inversely proportional to the effective diameter of a capillary, then, as pressure is increasingly applied on one side of a porous mass, it will be the largest pore which blows clear first, and the corresponding applied pressure is called the capillary entry pressure. It defines the lowest value of applied pressure needed to dewater a porous mass by blowing with a gas.
[*See also Bubble point*]

Capillary filter. A term sometimes used to describe a vacuum filter whose **filter medium** has pores so fine that gas will not flow through them under vacuum differential – but can also mean a **capillary belt filter**.

Capillary liquid. The liquid held in the interstices of a porous mass that are interconnected such that this liquid can, in principle, be expelled mechanically. It is sometimes divided into coarse capillary liquid – that which is easily expelled by gas pressure or centrifugal force, and fine capillary liquid – which is not.

Capillary model. A theoretical model of **filtration**, which assumes that the filter **cake** is made up of a series of parallel capillaries. The Zeitsch model for **centrifugal filters** is also a capillary model.

Capillary module. [*See Hollow fibre*]

Capillary moisture. A term relating to the moisture content of a **filter cake**, and corresponding to greater than 90% **saturation**.

Capillary number. A **dimensionless group** sometimes used to characterise the **dewatering** of a **filter cake** by drainage or gas blowing.
[*See also Bond number*)

Capillary pressure. Refers specifically to the pressure differential across the curved boundary between a liquid and a gas in a capillary pore. The capillary pressure holds the liquid in the pore, and is balanced by the **interfacial tension**. The interface between the two fluids in the pore will adopt a shape that is dependent on the pore geometry.

The capillary pressure may be positive or negative in value. It is positive if the pressure in the more dense phase (usually the liquid) is lower in value than that in the less dense phase. Thus, the capillary pressure of a liquid droplet suspended in a gas is negative, whilst it is positive for a gas bubble in a liquid.

Capillary pressure curve. Describes the relationship between the **capillary pressure** and the **degree of saturation** within a porous mass, and so describes the capillary pressure distribution through the mass, deriving from the pore size distribution within it. The curve shows the minimum degree of saturation that can be reached at equilibrium for each level of gas pressure applied externally to the mass. A set of curves can be produced for **dewatering** times between zero and infinity. All these curves pass through the point representing full saturation and the **capillary entry pressure**, and each is asymptotic to a particular value of saturation degree for high values of capillary pressure. The infinite time curve defines the saturation degree corresponding to the **residual moisture content**, at the limit of mechanical dewatering.

Capillary pressure curve

The curve can be developed using applied gas pressure or **centrifugal force**, and the two situations should produce the same curve. The curve for a homogeneous cake is independent of the cake thickness.

Capillary rise. If a tube containing a fine **capillary** is immersed at one end in a body of liquid, then the liquid will rise in the capillary to a level governed by the **interfacial tension** between the liquid and the gas above it. The height to which the liquid rises, h_c, is proportional to the **capillary pressure**, p_c, and inversely proportional to the liquid density, ρ:

$$h_c = p_c/\rho g$$

where g = gravitational acceleration

The relationship between capillary pressure and effective pore diameter enables capillary rise also to be related to diameter:

$$h_c = 4\gamma.\cos\delta/\rho g.d_c$$

Capillary suction. Because of the forces resulting from **interfacial tension**, a dry porous body put in contact with a mass of liquid will suck the liquid into the pores within the body, i.e. the body will absorb a quantity of liquid into itself.

Capillary suction (time) test. A laboratory test of the **filterability** of a **suspension**, which involves placing a sample of the suspension into an open-bottomed vessel that is sitting on a sheet of filter paper or similar material. The liquid of the suspension is absorbed by the paper and moves outwards in a circle from the base of the vessel. The speed with which the wetted circle expands is a direct measure of the filterability. The speed can be measured by observing the time taken to pass two marked limits, or it may be indicated automatically by electrical signals. The test really only works for **sludges** or thick **slurries**.

Capsule. A small, usually self supporting, filter **element**, shaped like a lens, with **filter media** as both upper and lower surfaces. May be mounted as a set, one above the other, on a central supporting core, and then used as a replaceable **cartridge**.

The word is also used to describe a small self-contained and disposable filter unit, employed in laboratory, medical and pharmaceutical applications. It is formed usually of a piece of filter medium sealed into a casing fitted with inlet and outlet fluid connections; usually discarded when "full".

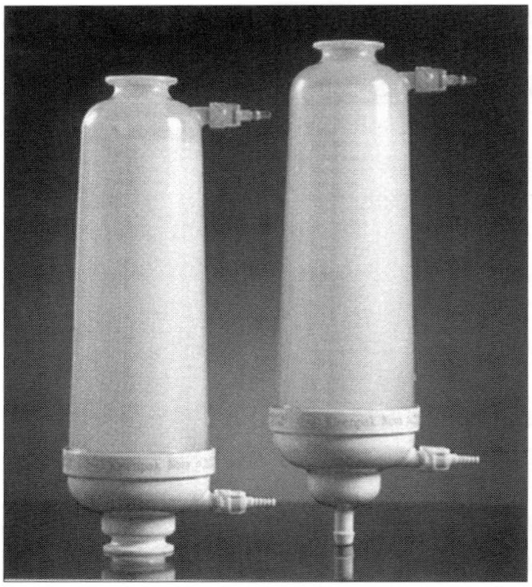

Capsule filters

Capture. The entrapment of a solid particle or liquid droplet out of suspension in a fluid, by a **filter medium**. There are several physical mechanisms, any one or more of which may be involved in the capture process.

Carbon, Carbon fibre. Elemental carbon exists in three basic forms: amorphous (carbon black powder and activated charcoal), graphite and diamond. It is an excellent material of construction, with high levels of thermal and electrical conductivity, and high resistance to thermal shock. It is now available as **filter media** in the form of **membranes**, and **non-woven** materials, especially those using very fine fibres.

Carbon (activated). [*See Activated charcoal*]

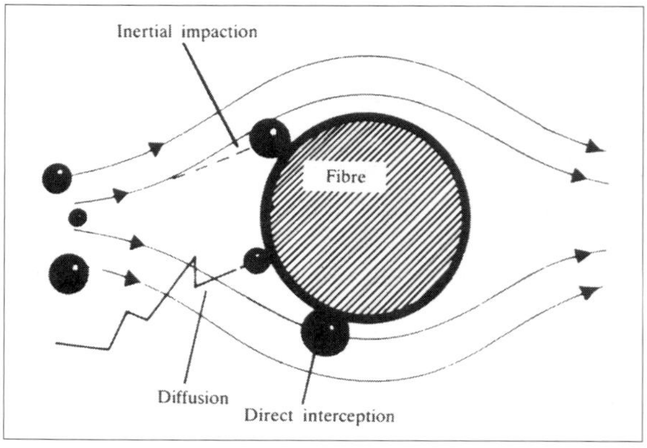

Particle capture mechanisms

Carding. One of the major process steps in the preparation of **staple fibres** for **spinning** into a **yarn**, or for the production of a **felt**. Cotton, wool, waste silk, and man-made staple are carded to separate individual fibres and cause many of them to lie parallel, and also to remove most of the residual impurities. Carding produces a thin sheet (or **web**) of uniform thickness that can be laid down as part of a felt (made from a set of such layers), or that can be condensed to form a thick, continuous, untwisted strand called a **sliver**.

Carman-Kozeny equation, relation.
[*See Kozeny-Carman equation*]

Carousel. A circular arrangement of **magnetic separators**, developed by Sala, in which a series of compartments containing a matrix passes through one or more water cooled solenoid magnets. The matrix is expanded metal or metal wool. The magnetic field is parallel to the flow which limits the depth of the matrix. Feed slurry flows downwards through the matrix, and wash water can be added as necessary.

Cartridge. A replaceable filter **element**, generally cylindrical in shape and long with respect to its diameter, which operates by filtering a fluid from the outside of the cartridge to its inside. It normally comprises a central open-structured **core**, on which is placed the **filter medium**. The latter can be a thin flat sheet, or, much more likely, a **pleated** sheet, to maximise the filtration area, or a thicker layer of **bonded** granules or fibres. The element can also be made up from a stack of flat rings, or from a continuously wound yarn, wire or ribbon. Coarse filtration (straining) cartridges use wire mesh or perforated sheets as the filter medium. Cartridges are made to a set of generally accepted standard dimensions, so as to be interchangeable as to source, and they represent the largest class of **utility filter**.
[*See also Candle*]

Cartridge filter, Cartridge filtration. Any **pressure filter** using a **cartridge** as a replaceable **element**. This will be fitted in a cylindrical housing, with appropriate fluid connections, which is easily opened for element replacement. Fluid flow is normally from the outside of the element to its inside, with separated solids accumulating on the outside or within the depth of the **filter medium**.

Cascade. A flow system consisting of a series of vessels such that flow between them is by overflow from one down to the next in line.

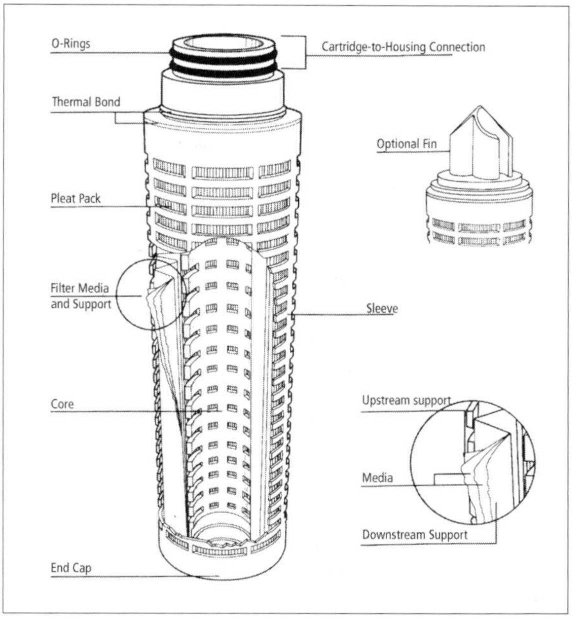

Cartridge filter element

Cascade cell. A particular kind of **flotation cell** into which no air is directly introduced. The flotation air is, instead, entrained by turbulence.

Cascade impactor. A multi-stage sampling device used for the sampling of **aerosols**. The May design features four plates in series: the largest droplets (above, say 10 μm) are caught on the first plate, successive plates capturing smaller droplets – with the sampling velocity being increased at each stage, so that the smaller droplets will impact on them.

Cascade washing. A particular feature of **horizontal plate filters**, in which the cake is washed by allowing the wash liquor to overflow from one plate to the one directly below it.
[*See also Washing*]

Casing. In a **centrifuge**, the separation takes place in the **bowl assembly**, from which the separated phases are emitted at high speed. The assembly must be surrounded by a casing structure that captures the two phases (**centrate**, and thickened slurry or **cake**), keeps them apart, and directs them appropriately for subsequent treatment. The casing must fit closely to the exterior of the bowl, without creating the chance of contact, and consequent friction problems.

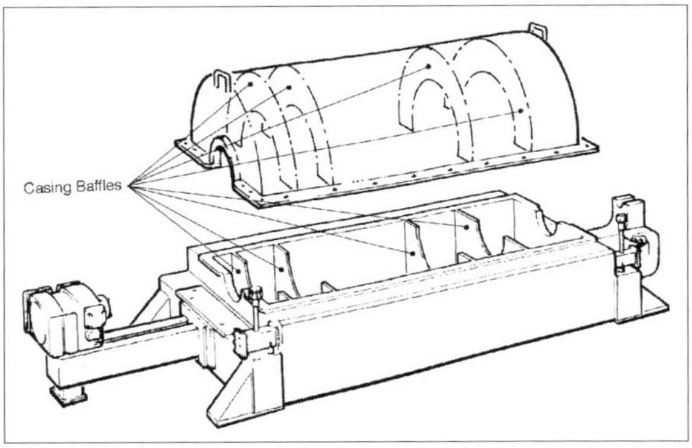

Decanter centrifuge casing

Casing baffle. A partition plate within the **casing** of a **centrifuge** that enables the two discharges from the centrifuge to be kept apart.

Casing gutter. A ledge fixed to the inner edge of a **casing baffle** on a **centrifuge**, which prevents any splashings of liquid inside the **casing** from running onto the outer **bowl** surface.

Cassette. [*See Panel*]

Casting. The production process in which a molten solid is poured into a mould, to achieve the shape of the mould when set. In the particular case of **membranes**, some processes involve the pouring of the molten mixture onto a flat surface, followed by wiping to produce a constant thickness.

Cataphoresis. [*See Electrophoresis*]

Catchpot. A vessel used in a **vacuum filter** system to prevent the carry-over of **filtrate** into the vacuum pump. It is placed, usually high up, between the filtrate receiver and the pump, with a **barometric leg** discharge system.

Cathode. The negatively charged **electrode** in an electric cell, which attracts **cations** towards it.

Cation, Cationic. A cation is a positively charged **ion** in a **solution**, usually deriving from an ionisable **solute**, and being matched by equivalently charged **anions**. A cation moves towards the **cathode** once electrical flow in the cell has been established.

The materials used as **flocculants** to promote **agglomeration** in suspensions normally work by carrying an electric charge. Accordingly a cationic flocculant carries a positive charge.

Caulking (thread). Caulking is a method of holding a **fabric**-like **filter medium** onto a component of a **filter**, or around the **filtrate drainage** system. The material is placed over a rectangular or dovetail groove, and forced into the groove and held there by the thread (which may be a rope, a rubber band or even a strip of lead).

Cell. Three different connotations to the word "cell" exist in the filtration business. The first concerns those cells that are the basic unit of living beings, which may need separation from a fluid, especially the unicellular bacteria and viruses that act as pathogens. The second meaning refers to the **filter cell** that is the separating unit in a number of automatic filters. Thirdly, there is the **electric cell**, where **filter media** are used as separators between the electrodes.

Cell belt filter. [*See Tray belt filter*]

Cell drum filter. [*See Drum filter*]

Cell foot. A part of the **filter cell** in a **disc filter** that connects the cell to the **filtrate** offtake pipe.

Cell insert. A structure fitted inside a **filter cell** to aid its resistance to the applied differential pressure, and also to improve the freedom of **filtrate** drainage from the cell. Inserts vary in shape according to mechanical need, cost and **permeability**.

Cell-less drum filter. A **drum filter** without partitions inside the drum, so as to be fully exposed the imposed vacuum. **Cake** removal is effected by means of a control shoe with sliding seals, held against the inside of the drum in the recovery zone, so that the cake may be blown off the medium by means of a series of compressed air pulses. The **filtrate**, having flowed through the medium to the inside of the drum, is removed from the lowest part of the interior by means of a filtrate suction pipe.

Cellulose. A polymer of glucose, differing from starch in that cellulose is a β-glucoside, whereas starch is an α-glucoside. Cellulose is the most abundant organic compound in the world, being the main constituent of the cell walls of all plants. As generated from wood pulp or cotton linters by mechanical or chemical disintegration, it forms the fibres that are the basis of most of the world's **paper**. It can be dissolved, usually by the viscose process, and regenerated as an extrudable polymer (**Rayon**). It can also be dissolved in an acetic acid/anhydride mixture to form the tri-acetate, another extrudable thermoplastic.

Cellulose is thus important in the **filter medium** business as **filter paper**, woven **filament**, and **membrane**, but also makes a good **filter aid**.

Cellulose acetate. An artificial, silk-like material, with a good lustre and tenacity, and which is insensitive to moisture (unlike **cellulose** itself). It is a good basic material for **filter media**, as a fibre in woven or non-woven fabrics, or as a **membrane**.

Cellulose paper. [*See Paper*]

Cellulose (regenerated). [*See Rayon*]

Celtic weave. [*See Plain weave*]

Cemented cake. A filter cake that has solid bridges between the particles, normally created by drying, which evaporates liquid in the interstices and leaves solute as the bridge.

CEN classification. A classification of air and gas filters produced under the auspices of the Comité Européen de Normalisation. Standard EN779 defines four ranges, coarse dust, fine dust, HEPA and ULPA (G, F, H and U respectively), graded by separation efficiency.

"Censor" decanter. A special **decanter centrifuge** (KHD) intended for the separation of a mixture of suspended solids into two **fractions**, by virtue of their different densities (NB not different particle sizes). The **bowl** has a conical section at each end, and a **conveyor** that moves solids from the centre of the bowl (where the **feed zone** is situated) to each end of the bowl, for separate discharge. The suspending liquid is prepared so that one part of the solids sinks and the other part floats. The Censor is mainly used in the sorting of plastics.

Central system. A term usually employed to refer to a main **filtration** system serving many different work sites, from which slurries are gathered together for processing, and to which the clean **filtrate** is returned. Particularly used of engineering workshops where arrays of machine tools use cutting and lubrication fluids, which need to be cleaned of machining residue before they can be re-used. Such a system provides economy of scale in the **filter**, but loses the flexibility that comes with having a filter at each machine tool.
[*See also Cutting fluids, Swarf*]

Centrate. The clarified liquid leaving a centrifuge.
[*See also Filtrate, Permeate*]

Centre feed. A specific feature of recessed plate **filter press**es, wherein the feed slurry enters through a feed channel near the geometrical centre of each plate.

Centre pipe. [*See Core*]

"Centridry". A complete process system (KHD) for the dewatering and drying of **sludge**s. It includes a **Centripress decanter centrifuge**, followed by a pneumatic dryer. The solids discharged from the centrifuge go straight into the hot air flow of the dryer. The system is particularly intended for the processing of sludges from waste water treatment.

Centrifugal acceleration. If a particle is constrained to move at constant **angular velocity** in a circle around a fixed point, then a force must be acting on it in a direction towards the centre of the circle, and there must be an acceleration in this direction. This is the centripetal ("centre seeking") acceleration. If the particle is free to move through a rotating mass of liquid, then the acceleration is in the opposite direction. This centrifugal acceleration is given by the product of the radial distance from the centre and the square of the angular velocity ($a = r\omega^2$).

Centrifugal discharge (filter). A means of discharging accumulated **cake** from the plates of a **horizontal plate filter**, achieved by spinning the plates about a central vertical axis. The shaft can be driven from above, which leaves a clear zone at the base of the housing vessel for discharged cake removal. An underdriven shaft, however, allows easier access to the drive for maintenance, and gives a more stable rotating system.

Centrifugal drainage. The removal of mother or wash liquor from a cake of solids in a **filtration centrifuge**, under the effect of the **centrifugal force**.

Centrifugal extractor. A special design of **disc-stack separator**, used to separate two immiscible liquids of different densities. The term is also used to describe a centrifugal device that is divided into concentric zones, and is used for liquid/liquid extraction (**Podbielniak extractor**).
[*See also Centrifuge*]

Centrifugal filter, filtration. A **centrifuge** whose rotating **drum** is perforated and carries **filter**

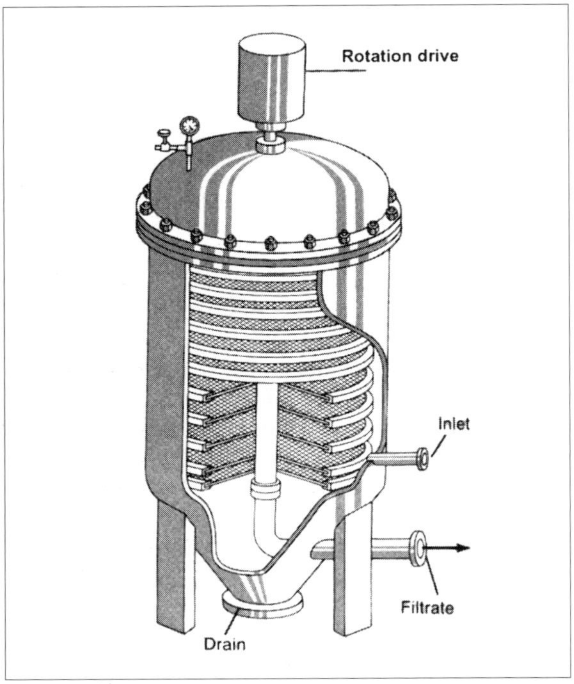

Rotating plate filter with centrifugal discharge

medium on its inner surface, supported on an open-structured frame, through which the **centrate** flows. It is thus able to effect a separation by **filtration** under the much greater driving force caused by the centrifugal action. However, there are some designs (e.g. from Heinkel and Krauss-Maffai) which have the centrifuge in a pressurised housing so that a gas overpressure can be added as driving force to the centrifugal force, especially in the cake dewatering stage (at the expense of a much more complicated mechanical structure for the centrifuge system – the overpressure can be as much as 6 bar)
[*See also Centrifuge*]

Centrifugal filtration number. Describes the force acting on the **mother liquor** in a filtration centrifuge.
[*See also G force, C-value*]

Centrifugal force. If a particle is constrained to move at constant **angular velocity** in a circle around a fixed point, then a force must be acting on it in a direction towards the centre of the circle. This is the **centripetal** ("centre seeking") **force**. The balancing force in the opposite direction is the centrifugal ("centre fleeing") force, and it is this force that causes **centrifugal separation**s to occur.
[*See also Centrifuge*]

Centrifugal sedimentation. A centrifuge with a solid, imperforate **bowl** achieves separation by **sedimentation**, much more rapidly than in the case of gravity sedimentation, because of the extra impetus from the **centrifugal force**.
[*See also Centrifuge*]

Centrifugal separation. Any separation achieved by the effect of **centrifugal force**, whether **filtration** or **sedimentation**, and whether in a **centrifuge** or in a **cyclone** or other kind of **inertial separator**.
[*See also Centrifuge*]

Centrifugal settling velocity. The speed at which a particle settles through a **suspension** in a **sedimenting centrifuge**. Expressed as **Stokes velocity** multiplied by **G force**.

Centrifugal time. A term used occasionally to refer to the residence time of solids being separated from a **suspension** in a **centrifuge**.

Centrifugal value. [*See C-value*]

Centrifugation. [*See Centrifugal separation*]

Centrifuge. *See* pages 41–43

Centrifuge rotor. [*See Centrifuge*]

Centripetal force. [*See Centrifugal force*]

Centripetal pump. A device for converting the rotational energy of a liquid, such as that separated in a **centrifuge**, into the pressure energy necessary to deliver the liquid a required distance through a pipe.
[*See also Peeling disc*]

"Centripress". The brand name of a special design of **decanter centrifuge**, made by KHD, intended to achieve a high degree of efficiency in the **dewatering** of wet **sludge**s, and used in KHD's **"Centridry"** process. The decanter operates with an extra-deep layer of separated sludge, and a special design of conveyor screw able to cope with the resultant high **torque**.

"Centrisizer". A special design of **decanter centrifuge**, developed by KHD, for the production, from a single **suspension** of a mixture of solid particles, of two **fraction**s of different average particle sizes. The **cut point** between the two fractions can be

Centrifuge. Simply defined, the centrifuge is a device for achieving mechanical **separations** of liquid/liquid or liquid/solid mixtures under the accelerated gravitational force achieved by rapid rotation. The separation can be effected by **sedimentation** or **filtration**.

In practice, the term covers a wide range of equipment, with uses throughout industry, which represents a major part of the mechanical separations business. The first significant event in the history of the centrifuge was Gustaf de Laval's invention of the centrifugal cream separator in 1878, but simple spinning drum washers and filters were in use before that date (and live on to this day in the domestic or commercial laundry equipment).

The separation achieved in a centrifugal device is similar in principle to that achieved in a gravity sedimentation or filtration process, but with the much higher driving force resulting from the rotation of the liquid. In the case of sedimentation, where the driving force results from the difference in density between the solid particles and the suspending liquid (or between two immiscible liquids), the separation is achieved with a force from 1000 to 20,000 times that of gravity. In the case of filtration, the **filtrate** is driven through the **filter medium**, and any cake built up on it, with a force of 100 to 2500 times gravity.

Basket centrifuge

The simplest device using centrifugal force to achieve separation is not normally thought of as a centrifuge: it is the **hydrocyclone**, in which the centrifugal separation is produced by the motion of the slurry, a motion induced by the tangential introduction of the feed material. The normal use of the word centrifuge is to refer to a more or less complex rotating assembly, which is driven by a motor at constant speed, or at speeds which vary from one part of the operating cycle to another.

The types of centrifuge used for sedimentation include:

- **tubular bowl**,
- **chamber bowl**,
- **imperforate basket**,
- **disc stack separator**, and
- **decanter**.

This list is in order of increasing complexity of design – as with any centrifuge (or indeed any filter), design complexity increases according to the increasing concentration of solids in the feed, and to the wish to discharge these solids on a continuous basis.

Thus the tubular bowl centrifuge and the simplest version of the disc stack separator are used primarily for the separation of immiscible liquids. In this duty, they can accept a small solids content in either or both liquids, which will accumulate inside the centrifuge, until it is necessary to stop operation, open the centrifuge and clean it free of accumulated solids.

These two types, together with the chamber bowl design, are also used for liquid clarification, where the dirt content of the liquid is low (so that stoppages for cleaning are infrequent). If the solid content of the suspension is higher, then the imperforate basket centrifuge may be used to separate the solids, and produce a reasonably clean liquid (separation efficiencies are not all that high with this type). The accumulated solids in the basket may be removed manually (by scraping or shovelling, or by lifting out a bag), or semi-automatically (by knife or skimmer), but in either case these solids will not have been dewatered at all.

Higher solids contents can be accepted by the disc stack centrifuge, when it is fitted with a means for automatic discharge. Separation of the solids from the liquid is much more efficient, because of the higher centrifugal forces, but the solids are still fairly saturated with liquid at the point of discharge. The methods of solid removal include the use of

Disc-stack separator

nozzles, which run continuously, allowing a thick slurry to discharge; **valved nozzles** that open automatically when the solid depth in the bowl reaches a pre-ordained figure; and the **opening bowl** design, in which the shell splits at the widest part of its periphery for a short period, also controlled by solid depth.

The decanter centrifuge is intended to handle significant solid concentrations in the feed suspension, and, at the same time achieve quite good degrees of clarification of the liquid **centrate** (or, in the **three-phase decanter**, of both liquids). The decanter has a drainage zone, so that the discharged solid is partially dewatered, with lower moisture levels possible in the **screen bowl decanter**.

Centrifugal filters are intended to handle high concentrations of feed solids, but are best suited to solids that filter and drain easily. As already noted, their design complexity is very much the result of providing ease of solids removal from the **filter medium**. The types of centrifugal filter include:

- fixed bed (batch and semibatch)
  - manual discharge
    - scraped through open base
    - removable bag
  - automatic discharge
    - knifed from vertical basket
    - **peeler** (horizontal basket)
    - **inverting basket**
- moving bed (continuous)
  - cone screen
  - pusher

The simplest type of centrifugal filter has a drum rotating about a vertical axis, with motor and shaft either below the drum (underdriven) or above it (overdriven), the latter making for easier solids removal from the manual discharge ranges. The drum may be in the form of a perforated basket (coarse wire mesh or perforated plate), or it may have no base, just perforated walls. When the accumulated solids reach a certain level, the feed is stopped and the cake of accumulated solids can be washed and then allowed to dewater by drainage. The rotation is then stopped, and the solids either dug out, or lifted out in a bag of filter medium. The drum (or filter bag) is washed and the process repeated.

The same type of machine can be fitted with a knife that will cut the cake out, without the rotation having to be completely stopped (although this will need to leave a heel of cake, to protect the medium's surface). This automatic knife function is developed in the peeler centrifuge, where the drum is now mounted horizontally, and the filtration, washing, dewatering and cutting out all occur at the same rotational speed, on a constant cycle.

A quite different form of automatic discharge is achieved with a horizontal basket by having it lined with a fabric to act as the filter medium, which, when the cake is sufficiently dewatered, is pulled out of the basket by its centre and is thus turned inside out, releasing the solids into the casing of the centrifuge.

The most complex of the centrifugal filters are those in which the separated solids move along the filtering surface, from the feed end of the basket to the opposite end, where they are discharged into the surrounding casing. To make this movement easier, the screens that do the filtering are usually made of **wedge wire** bars, mounted lengthways in the screen. During the passage along the screen, the solids are filtered, washed and dewatered, and the system must ensure that sufficient time is available for these all to occur to the necessary degree.

The cone screen centrifuges are all, as their name implies, fitted with a screen in the shape of the frustum of a cone, and all rotate about the axis of this cone. They can be mounted with the axis of rotation horizontal or vertical. The slurry is fed to the narrow end of the screen, and the solids tumble "downhill" to the widest end. The component of the centrifugal force that is perpendicular to the screen at any point provides the filtering force, and the component parallel to the surface controls the movement of the solids across the screen. This component is obviously the greater, the wider the angle of the cone, and if this angle is greater than the angle of repose of the wet solids, then they will move. One specific cone angle may only be suitable for a limited range of solids, so cone screen centrifuges exist with an ingenious range of devices to slow down the movement of the solids (such as the **baffle ring centrifuge**), or to speed it up (a screw conveyor or various vibrational motions).

The other main form of continuous centrifugal filter is the pusher centrifuge. Here, the basket rotates about a horizontal axis, and the screen is in the form of an open drum, but the drum is in two or more annular segments, of increasing diameter away from the feed end. These segments move backwards and forwards across one another, pushing accumulated heaps of solid ahead of them. In this way, the solids are moved along the totality of the screen, and the pusher can provide very effective washing and dewatering of the solids.

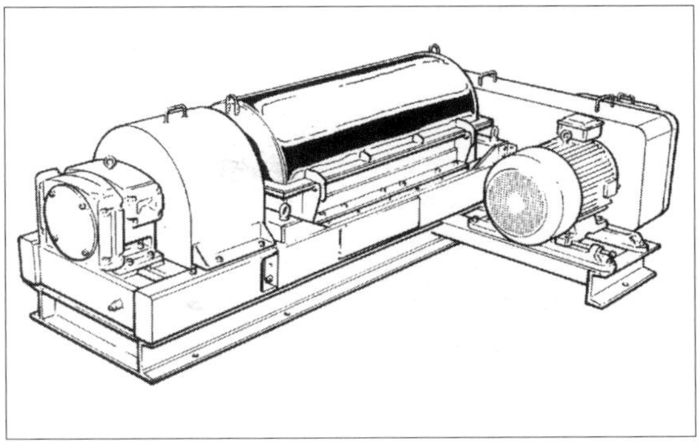

Decanter centrifuge

Centrifugal filters generally cannot handle solids as fine as those covered easily by many pressure filters, and another drawback, especially of the moving bed machines, is that they need a fairly stiff feed, to stop it running straight out of the basket unfiltered – i.e. it needs to be thickened prior to filtering.

adjusted according to the nature of the feed slurry, and the required size of each fraction.

"Centritest". A laboratory test centrifuge, developed by Bokela, as a **quick start centrifuge**. It is a **beaker centrifuge**, designed to enable scale-up measurements for processes with short residence times in the centrifuge.

Ceramic, Ceramic media. A ceramic is an inorganic material that has been fired to a high temperature. It is thus able to resist moderately high temperatures in use. (The term does not usually encompass carbon or glass.) Because of their suitability for use at high temperatures, ceramic materials have been very successful at solving some of the major problems of gas cleaning.

Ceramics are mainly oxides of metals or metalloids such as silicon, and are very resistant to corrosion and abrasion. As such, they play an important part in the solid discharge process of a number of filters, and especially of centrifuges. The tips of the flights of the scroll of a **decanter centrifuge** are subject to great force whilst moving the separated solids along the bowl. Accordingly these tips are covered with small **tiles** made of ceramic to resist the resultant wear.

In the form of stoneware they are one of the oldest forms of **filter media**.

Ceramic filter. One of the oldest forms of filter would have been plates or tiles of porous pottery, and the material has continued to play an important part in the **filtration** business. Modern filtration systems will normally use ceramics as the medium, and not as a material of construction.

Ceramic disc filter. A continuously acting **disc filter**, developed by Outokumpu. Instead of the normal filter cells, the filter elements are double ceramic discs, sealed at their circumference, filtering through both faces. The discs have a pore size of about 1 μm, but are several millimetres thick, and so have a low **permeability**, but produce a very clear filtrate. The **capillary pressure** of the **hydrophilic** material is larger than the applied pressure differential, so the pores are always wetted.

Ceramic foam. A foam can be formed from molten ceramic by aerating it thoroughly and then freezing the mass while it is full of bubbles. This then makes a suitable **medium** for the **filtration** of liquid metals free of unwanted solid contaminants, prior to their being cast or extruded.

Ceramic foam filters

Ceramic magnets. These rely on a ferrite component to give the magnetic properties that are used in **magnetic filters** and separators. They mainly are ferrites of strontium or barium, with high coercive force, high electrical resistance and light weight.

Ceramic membrane. A finely porous **filter medium** made from sintered **ceramic**, whose pore sizes are such as to allow it to work in the **microfiltration** and upper **ultrafiltration** ranges – and so it is called a **membrane**, even though being the very opposite of the materials first used as membranes. Ceramic membranes are available as hexagonal or cylindrical "bricks", with numbers of holes parallel to the long axis. The filtering surface is then coated with a thin, very fine layer of ceramic powder, which is sintered on the tube surfaces. These structures are essentially rigid, although some flexible ceramic media materials are being developed.

CETOP RP70 system. A simple method of expressing the composition of a mixture of solid particles by means of two numbers, used to define contaminant levels in hydraulic oils. Each number represents a range of particle counts per 100 ml of **suspension** sample. The ranges start at 1 to 2 particles, and increase in 24 steps up to 8 millions to 16 millions. Each range covers a factor of two in number, and is twice the number of the preceding range:

Range	1–2	Range number	1
	2–4		2
	4–8		3
	...		...
	4 m – 8 m		23
	8 m – 16 m		24

The system then applies these range numbers to two different particle sizes: total count of all particles over 5 μm in size, and total over 15 μm, to give a specification such as 16/10 (which means a contaminant level of between 32,000 and 64,000 particles over 5 μm, and of between 500 and 1000 over 15 μm).
[*See also Contamination classes, profile*]

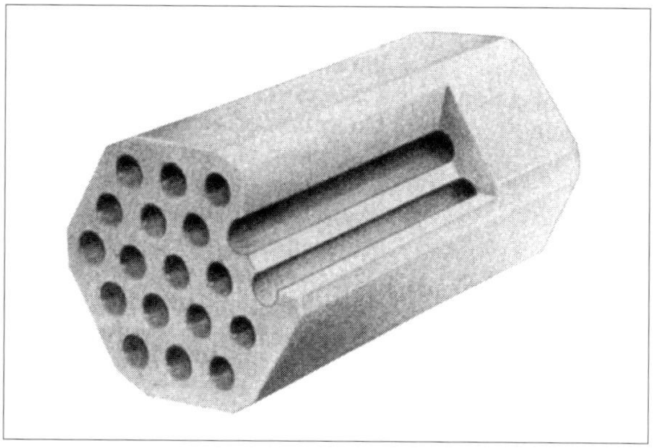

Ceramic membrane monolith

Chain assisted cake removal, Chain discharge. A version of a **drum filter**, in which a set of chains (or strings) lie on the upstream surface of the **filter medium** and move parallel with it. At the end of the filtration process, the medium is taken away over a roller, but the chains continue further in the same direction, so lifting the formed cake from the surface of the medium. The cake is then broken free of the chains by the passage of the chains over a second roller. This system is best suited to a fibrous solid, which forms a felted **cake**, rather than a solid made up of crystals or granules.

Chain weave. A type of heavy **twill weave**, employing plied **yarn** for both **warp** and **weft** threads. Features the twill pattern of two-under, two-over. Not particularly strong, but with good **filtration** performance at high fluid fluxes.

Challenge test. A test procedure intended to measure the efficiency of a **clarifying filter** (or, more particularly, the appropriate **filter medium**) in capturing particles of a given size. The test involves feeding a **suspension** of particles of a known size (and/or size distribution) to the filter under known conditions.

Chamber bowl (centrifuge). A simple type of **centrifuge** consisting of a cylindrical bowl rotating about a vertical axis, and used for the clarification of liquids. Inside, the bowl is divided into a number of

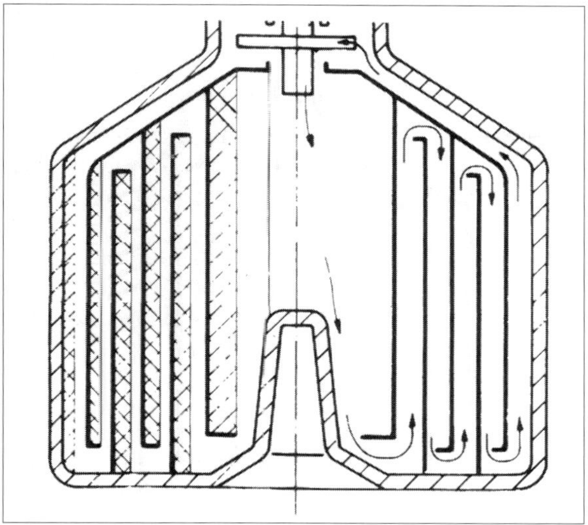

Chamber bowl centrifuge

concentric compartments or chambers, such that the liquid being clarified flows through each chamber, in succession from the centre outwards, i.e. from a zone of relatively low **centrifugal force** to zones of progressively increasing force. The solids separated in each chamber settle to the bottom of the chamber, and the unit has quite a high solids storage volume. Because of the increasing centrifugal force, the coarsest solids settle in the central chamber, and finer and finer solids settle in the sequential passage through all chambers. In operation, the chamber bowl operates for a long period before it is necessary to open it to remove the accumulated solids. (Also known as a multichamber centrifuge.)

Chamber filter. A **batch pressure filter** comprising a set of chambers in parallel, each chamber having **filter medium** on each side, such as a **filter press** or **diaphragm press**. The chambers can be horizontally or vertically orientated.

Chamber filter press. A version of the straightforward plate-and-frame **filter press** (and probably the most common form), in which adjacent plates each have a cavity covering the plate, so that when the plates come together, a chamber is formed between them. **Filter medium** covers both plates, and is stretched into the appropriate cavity as **filtration** proceeds. The accumulated **cake** falls out of the chamber when the plates are opened.

Chamber press. [*See Chamber filter, Variable chamber press*]

Changeover plate. A special plate fitted in a **sheet filter**, to enable the filter to be operated in two parts, with coarse media in one part, and fine in the other.

Channel, Channelling. In **filtration** terms, a channel is a continuous space through a mass of porous material, such that the effective diameter of this space is significantly larger than the average **pore size** of the rest of the pores. Channelling may occur when a **cake** is being formed on a **filter medium**, and this is not a desirable occurrence because it allows flow of fine solids in suspension through the channel and out of the filter – until the channel is bridged.

Channelling is also a phenomenon observable in the **hindered settling** of particles out of a **suspension**. At certain concentrations, hydrodynamic instabilities can occur that create particle free zones in the direction of the settlement. Abnormal settling behaviour will then ensue until the channels are blocked.
[*See also Intermediate suspensions*]

Channel resistance. The resistance to flow of a gas in a **filter**, caused by the close proximity of two surfaces, especially of the two adjacent parts of a **pleat** in a pleated sheet filter.

Charcoal cloth. A material, such as a woven **fabric**, that has been charred to turn the carbonaceous material into charcoal, so that the cloth may be used for **adsorption**. Such material can also be incorporated with other materials into a **combination medium**.

Charge density. A term describing the electrical state of a polymeric material, defined as the percentage of monomer segments in the polymer that carry an electric charge.

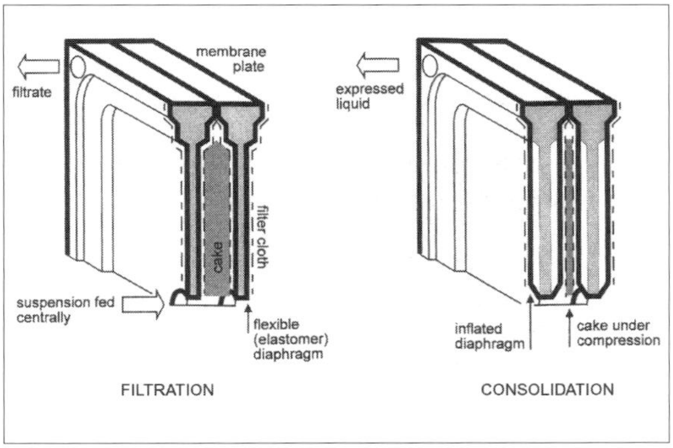

Chamber filter press

Charged medium. A **filter medium** made so as to carry a fixed electric charge. This will improve its **clarification** efficiency, by enabling it to **capture** particles carrying the opposite charge. A charged **membrane** material can also act as an **ion exchanger**.

Charge neutralisation. A process for improving the aggregation of charged particles (whose charges will normally repel one another), which involves neutralising the charges, either by pH adjustment or by **flocculation**.

Chatter. In a **centrifuge** that is separating solids from a liquid **suspension**, and continuously moving the separated solids through the centrifuge for discharge, the movement may not be smooth, but rather may occur in a series of small jerks. These jerks may be transferred into the **rotating assembly** as a series of torsional vibrations known as chatter. Too much chatter may lead to metal fatigue in parts of the rotating assembly and possible failure of the drive gear box. This is a particular problem with **decanter centrifuges**.

Cheesecloth. A very light, **plain weave** cotton material. Loosely woven and of open construction, it is used as a fine gauze for sifting light materials, or as a simple **strainer**.

Chelminsky method. A laboratory test to enable the estimation of the **torque** required to turn the **rake** used in a **thickener** for the collection of separated solids.

Chemical cleaning. [*See Cleaning*]

Chemical Oxygen Demand (COD). An important measure in pollution control, which shows the ability of a liquid **effluent** to deplete the **dissolved oxygen** content of a body of water (lake, river, etc). It is measured as the number of milligrams of oxygen that a litre of the effluent will take up from a hot acidic potassium dichromate solution. It is a faster test than a BOD determination (two hours, rather than five days), and gives a higher demand than that test.
[*See also Biological Oxygen Demand*]

Chemisorption. [*See Adsorption*]

Chitin. A structural polysaccharide, based on acetyl glucosamine, which forms the outer covering of crustaceans and insects. Extracted from shell fish, it is used as a **flocculating agent**.

Chitosan. A food-grade version of **chitin**.

Chlorine, Chlorination. A common procedure for the disinfection of water is the introduction of gaseous chlorine into it at low concentrations. What matters is the available (molecular) chlorine, not the quantity of chloride ion. It is necessary to ensure that there is enough chlorine present, so break-point chlorination is normally practised, wherein enough is added to ensure the presence of free chlorine, so that the water remains disinfected during its passage through pipelines. It is the problem of correct adjustment of chlorine levels that has led to the development of other forms of disinfection, including **micro-** and **ultrafiltration.**

Christmas tree. A process involving a number of stages of **cross-flow filtration**, in which the number of modules in each succeeding stage decreases because of the decreasing **permeate** or **concentrate** flows.

Chromatographic separation. The separation of components of a mixture by means of **chromatography**. Although chromatography began as an analytical technique – for which it is still excellently suited – an increase in operating scale enables the same techniques to be used to achieve the separation of several components out of a liquid or gaseous mixture.
[*See also Electrophoresis*]

Chromatography. A process that enables the differentiation among a large number of components in a liquid or gas, by means of their distribution (partition) between a flowing phase (liquid or gas) and a stationary phase (solid or liquid). In a typical process, a mixture of solutes is completely adsorbed from their original solution onto a solid **adsorbent.** This is then treated with fresh solvent and the various solutes distribute themselves between the solid and liquid phases in a manner specific to their chemical nature – enabling chromatography to be used as an exceedingly precise analytical process, or, at a larger scale, as a **separation** technique.

CIP. [*See Clean in place*]

Circular clarifier, Circular thickener. A gravity **sedimentation** device, in the form of a circular basin, which has a **rake** on the bottom, moving in a circle about the centre of the tank, to move settled solids towards the centre and the **underflow** outlet.
[*See also Gravity thickener*]

Circularity. A term used in characterising particles in a **suspension**. It is the ratio of the circumference of the circle having the same area as the projected area of the particle, to the actual perimeter of that projected area.

Circulating bed classifier. A simple cylindrical **hydrocyclone** with a flat (as opposed to conical) bottom, and the usual tangential inlet for the slurry used in the **classification** of solids in suspension. It can be used used for separation by particle size, particle shape or particle density.

Circulating tray filter. A **vacuum belt filter**, developed by Dorr-Oliver, in the form of a set of short rigid trays with raised side walls, and open leading and trailing sides, although the latter are fitted with seals to the adjoining trays. The filter operates continuously, with the trays passing over the end rollers to return to the start.

Clarification, Clarifying. The removal of fine particles and **colloid**s (and sometimes also dissolved colorants) from a suspension of them in a fluid, to produce an effectively clear fluid – liquid or gas. This is one of the main purposes of **filtration** and its related separations. It usually refers to the removal of a low concentration of suspended material, usually 1% or less, and possibly less than 100 ppm, and to situations where the removed solids are a waste material. The separation is normally expected to be highly efficient, and as low as possible in energy demand.

Because the concentration of the feed suspension is low, a **clarifying filter** will normally work by **depth filtration**, using thick **filter media**, such as a **bonded** fibre or **felt** for a gas, and a **deep bed** (granular) filter for liquids.
[*See also Absolute, Gas cleaning*]

Clarification enhancement. [*See Sigma enhancement*]

Clarifier. In principle, any device capable of achieving the **clarification** of a **suspension.** In filtration usage, it usually refers to the large settling basins employed for the clarification by **sedimentation** of large volumes of water, either fresh or waste, free of unwanted suspended matter. It is frequently necessary to coagulate the solid particles in some way, either outside (upstream of) the clarifier, or as part of the design of the clarifier, when it is termed a clari-flocculator.

Clarifying filter. A **filter** used to achieve **clarification** of a **suspension**. This will usually function by **depth filtration**, using a thick **filter medium** or a **deep-bed filter**.
[*See also Filtration*]

Clarifying length. In a **decanter centrifuge**, the distance between the middle of the **feed zone** exit into the **pond**, and the **front hub** (separated liquid discharge).

Clarifying separator. Any separating device that achieves **clarification**, but refers in particular to a special design of **disc-stack separator centrifuge**, which is intended to deal with a low-concentration **suspension** in a liquid (or possibly two immiscible liquids). The separated solids have to be removed from the centrifuge bowl, and this can be done by means of open **nozzles**, or **valved nozzles** (where the valves are opened automatically by virtue of the thickness of the layer of accumulated solids), or by the **opening bowl** form of discharge. The clarifying separator can remove particles of less than 1 μm.
[*See also Centrifuge*]

Clarity. The quality of a fluid, especially a liquid, which defines its freedom from suspended material. and, usually, from contamination.
[*See also Turbidity*]

Classification, Classifying. The separation of a mixture of **particles** of widely different sizes into two or more **fractions**, each with a much closer size distribution, often for the purpose of determining the particle size distribution in the original mixture, but also for process purposes, such as the preparation of a suspension of closely sized particles before subsequent treatment, for example in starch processing, or the production of china clay suspensions in the production of ceramic ware. Classification can be effected in the dry

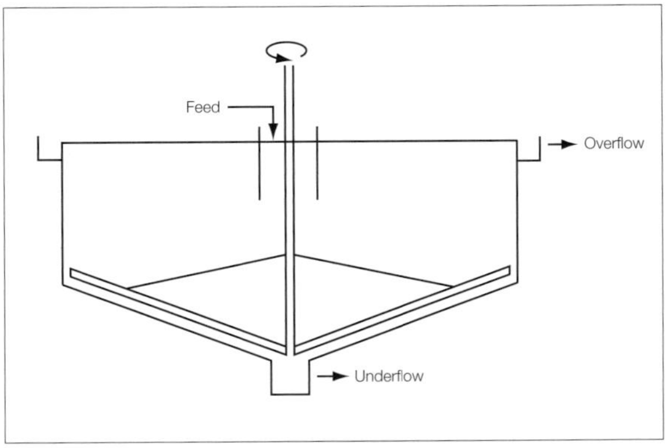

Clarifier

state, as in the use of a set of sieves, to produce as many fractions as there are **sieves** in the set, or in the wet state, as in a **hydrocyclone** or **centrifuge**, which can also be used in series, to produce a separate fraction for each processing unit.

The term is also, but less commonly, used for the separation of a mixture of two different kinds of solid by virtue of a density difference.
[*See also Cyclosizer, Elutriation*]

Classifier. Any equipment item intended for the **classification** of a mixture of solid particles, although mostly reserved for gravity **sedimentation**, with a cut size greater than about 50 μm. Two main types are used: **hydraulic classifiers** and **mechanical classifiers**.

Clay. A natural aluminosilicate mineral, occurring usually as a plastic paste, with particle size of 2 μm or less. A suspension of very finely divided clay particles will not settle very easily or quickly under the force of gravity alone. The prime function of clays in **filtration** is as **adsorbents**.
[*See also Bentonite*]

Cleanable filter. Usually refers to a **clarifying filter** (**strainer** or other **surface filter**), which can easily be cleaned for re-use.
[*See also Automatic filter*]

Clean air cabinet. [*See Cabinet*]

Clean cloth resistance. The resistance to fluid flow of a **filter medium** before it has been used – which should be the lowest value of its resistance during its useful lifetime. Different manufacturers of filter media define the resistance of their materials in different ways.
[*See also Filter medium*]

Clean filter factor. Term used for **deep bed filters**, using granular media, referring to the pressure drop across the clean bed measured under standard conditions.

Cleaning. The rendering of any object as fit for re-use, but particularly in reference to a **filter medium** it implies the removal from the surface, and more importantly from the interstices, of any accumulated solid particles. Cleaning can involve washing, blowing with gas, acid treatment, scrubbing, and so on. Media can be cleaned continually as part of the automatic **filtration** process, or **elements** can be removed from their **housing** for separate cleaning.

A different and more precise use refers to the cleaning of air or other gases free of dust.
[*See also Gas cleaning*]

Clean in place. CIP has become an important technique in the cleaning of process equipment of all types, where it is necessary both to clean to a high standard (e.g. sanitary, food, electronics), and to do so without having to dismantle the plant. The technique mainly uses appropriate cleaning fluids, with some mechanical systems such as scrapers or brushes, and usually finishes with a steam-sterilisation stage. It is also incumbent upon the equipment makers to design their plant to be as free of dead zones as possible. The process is especially relevant to **cross-flow filters** and some **centrifuges**.

Clean pressure drop. A term used to define the differential pressure across a **filter** and its **medium**, at its rated flow using the intended fluid, but before filtering any solids.

Clean room. A working space, for production, development or testing, where the ambient air should be as free of dust and aerosol particles as possible. The ventilation systems for such spaces are now normally provided with **"absolute" filters**, and a major industry exists to supply such facilities with associated **HVAC** systems. The air quality in a clean room would usually be of a very much higher standard than expected for a normal air-conditioned space.

Clear liquid zone. The zone of a **clarifier** (usually at its top) where settlement is complete, and the liquid is completely free of suspended particles. It is from this zone that the outflow from the clarifier is taken.

Clear run. In a **cake filtration** process, the cake has to be established by the **bridging** of the pores in the **filter medium**. Until that happens, fine particles will pass through the filter medium into the **filtrate**. When bridging is complete, then a clear filtrate appears – the start of the clear run phase.
[*See also Precoat*]

Clogging. This is a feature of **cake filtration** of a suspension containing particles of very different sizes. The larger particles will settle on to the **filter medium** first (and thus help to bridge the pores), and form the lowest layers of the cake. The smaller particles will settle together later, and form a much less permeable, clogging layer. This may increase the pressure differential across the cake, and also make **dewatering** and **washing** of the cake more difficult.

The term is also used to indicate the gradual build-up of captured solids in the interstices of a **depth filter medium**.
[*See also Cake*]

Clogging curve. The curve relating increase of pressure drop across a **clarification filter**, with time, as the build-up of captured solids increases.

Clogging indicator. A mechanical device, usually based on pressure drop indication, which tells when a **clarification filter element** is due for cleaning or replacement. The indication is usually visual, by means, for example, of a pointer, or a rotating coloured disc, which may be operated by the movement of a spring-loaded piston, or diaphragm, or bellows. This mechanism is contained in the **filter housing**, and has the up-stream pressure fed to one side, and the down-stream pressure fed to the other.

Closing pressure. The force necessary to close, and hence seal, a **filter press**, against the force exerted by the flow of the feed slurry into the press.

Cloth. The term "filter cloth" is a, by now, rather old-fashioned one referring to any **filter medium** made from a **fabric**-like material, such as **woven fabrics** and **felts**. It dates, of course, from the early days of the industry, when woven fabrics and simple felts were all that was available for filtration.

Cloth index. A characteristic of **woven filter media**, which is the ratio of the **permeability** of the clean (**multifilament**) cloth to the permeability of an equivalent clean **monofilament** medium, usually denoted by Β (capital β). It recognises that, in a spun **yarn** or multifilament material, fluid can flow through the structure of the yarn, as well as around it.
[*See also β ratio*]

Cloth resistance. A somewhat imprecise term used to define the clean resistance of the **filter medium**, and sometimes of the medium plus the initial layers of **cake**.

Cloth union. A means of sealing the **filter medium** around the central feed pipe in recessed **filter presses**. It is made from a tubular section of the medium, sewn with re-inforcing sections to the medium on either side of the plate, to make a double cloth. Alternatively, it can be a metal tube, flanged at one end, which passes through the feed aperture, and is threaded at the other end, to take a locking flange ring. This ring is then tightened to lock both flanges against the two filter cloths.

Cluster. [*See Floc*]

CMC. [*See Micelle*]

Coagulant. Any material, usually a specific chemical substance, which promotes the **coagulation** of a **suspension** of particles.

Coagulant aid. Colloidal inorganic material that assists a coagulant in its work.

Coagulation. The combining together of small particles into large ones, which, because of their larger **effective diameter**, will settle faster under conditions of **sedimentation**, or will be captured more easily on a **filter medium**, or will be more easily **dewatered**. Coagulation occurs by the destabilisation of the **suspension**, by neutralisation of the natural electrical charges on the particles, such that the weaker **van der Waals** forces of electrostatic attraction take over from ionic repulsion.

Lyophobic **colloids** (in which the particles are unsolvated) are relatively easily coagulated, but the result of coagulation may not be sufficient to enable the aggregated material to be separated by **sedimentation** or **centrifugation**, but it will be separable by **filtration**.

Coagulation is most often achieved by the addition of an **electrolyte** to the suspension, but can occur spontaneously (as happens with aerosols – as the particles get close enough, so they aggregate).
[*See also Flocculation, Solvation*]

Coal. The naturally occurring carbonaceous material is found in many different forms, graded from "soft" through "hard". Hard coals are the most efficient as sources of energy, but the hardest, also called anthracite, makes a very good **deep bed filter medium**, when crushed, washed and sized to the dimensions required. It is often used in a **multilayer filter**, as it has a lower density than sand.

Coalescer. A piece of equipment in which the droplets of one liquid phase suspended in another liquid, with which it is completely immiscible, are caused to combine to form larger drops, and so to become separated from the other liquid (the **continuous phase**). Most coalescers take the form of cylindrical vessels whose interiors are full of some kind of plastic or wire mesh, with which the dispersed droplets collide, and are captured, to merge with other neighbouring droplets, and so grow in size. The mesh should be preferentially wetted by the **dispersed phase**. Eventually the captured drops become so large that they fall (or float) from the mesh and form a separate layer.

Coalescing. The process by which finely divided liquid droplets dispersed in another liquid are made to merge together into larger droplets, and eventually into a continuous layer. The process is used, for example, in the recovery of waste oils from aqueous **emulsions**.

It can only be achieved if the two liquid phases are completely immiscible. If the two phases are widely different in density, then coalescing can normally be accomplished in a single step. As the densities approach one another, then two stages of separation may be necessary.
[*See also Demisting*]

Coalescing filter. A filter whose prime aim is the coalescing of a dispersed liquid phase, but may also effect some **separation** of solid particles as well.

Coarse capillary, Coarse capillary liquid, Coarse capillary system. These terms are occasionally used to refer to the moisture, held in the interconnected pores in a porous mass, which is easily removed by mechanical **dewatering** or gas blowing, as distinct from the **fine capillary** liquid that is difficult so to remove, and makes up the residual moisture content.

Coarse filtration. An imprecise term used to refer to the separation by filtration of the largest sized particles from a **suspension**. It is used to contrast with "fine filtration", the boundary between the two classes being generally accepted at around 1 mm.

Coarse fraction, coarse material. When a suspension of solid particles, or a dry mixture of granules, is separated into two parts by size, then that containing the larger particles is termed the coarse fraction, containing the coarse material.

Coarse grain yield. The efficiency of production of a higher size **fraction** in **classification**.

Coarse screen. Although in principle referring to any screen (perforated or woven), employed for the removal of coarse solids, the term is more particularly applied to those screens at the inlet to a cooling water system, or to a waste water treatment plant, whose function is to remove the really coarse (1 cm and upwards) objects carried by the liquid, especially including objects that would float in any subsequent **sedimentation** separator. The accumulated paper, rags, wood, dead animals, plastic bottles, and so on, are periodically or continuously removed from the screen, by brushes or scrapers, and then macerated before being added to other waste solids for disposal.
[*See also Screens and screening, Straining*]

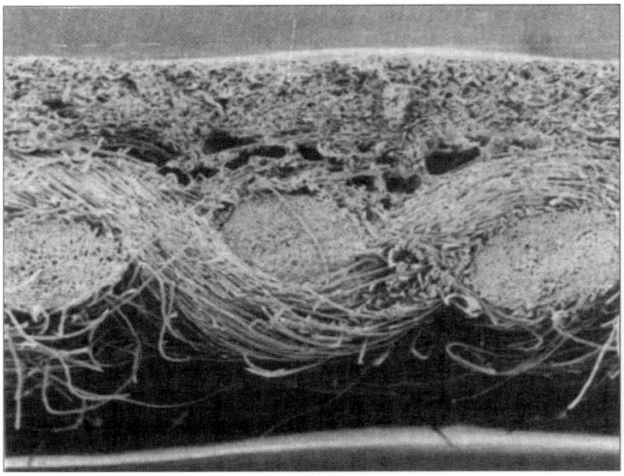

Coated fabric

Coating. In a general sense, a layer of one material applied to the surface of another, to change the properties of the latter (as in a coat of paint, or in the use of rubber-coated steel as a material of construction for filter vessels.

More specifically, the term is used in **filtration** to relate to the surface treatment of a **filter medium**, which renders that surface more suitable for use: such as making it less easily abraded, or more able to release a collected cake. Some coatings effectively convert a basic material into a quite different medium, such as Ravensworth's "Ravlex" or Madison's "Primapor" and "Azurtex".
[*See also Composite medium*]

Cocurrent. When two fluids or fluid suspensions flow in the same general direction, the flow is said to be cocurrent. The term has particular application to continuous separators that produce two **fractions** from a single feed suspension.
[*See also Countercurrent*]

Cocurrent flow decanter. A special design of **decanter centrifuge** that has the **feed zone** at the opposite end of the **bowl** from the solids discharge, so

that liquid and separated solids move together towards the conical end of the bowl. From there the **centrate** moves back through channels that are part of the **conveyor screw**, to the feed end of the bowl, and is there discharged. This design of decanter is intended to disturb the settling zone as little as possible, and so to give as clear a centrate as possible, being especially suitable for solids that are easily resuspended.
[*See also Centrifuge*]

Cocurrent washing. One version of the washing process for filter **cakes** or for solids separated in a **centrifuge**, in which the wash water is applied only once to the solids, and then removed having done its job. It uses more wash water than **counter current washing**, to achieve the same degree of wash efficiency, but there are some separation equipment situations where only cocurrent washing is possible. Thus only cocurrent washing may be applicable in centrifugal separations (because of the impossibility of segregating the centrate flows). Cocurrent washing can be carried out in either of the two main washing modes: **displacement washing** or **re-slurry washing**.

Coe and Clevenger method. A method, dating from 1916, that is used for the practical design of **gravity thickeners**, based on empirical correlations derived from experimental data. These are developed from a series of settling tests on **suspensions** of different **concentrations**, and the initial settling rate is determined for the feed concentration. The settling area necessary to give a required underflow concentration is then calculated.

Coefficient of permeability. A measure sometimes used to characterise the **permeability** of a medium in a particular **filtration** application. It includes fluid density and viscosity.

Coefficient of rugosity. [*See Angularity*]

Cohesion. A general term implying the forces that combine to hold particles together.

Cohesion tester. As part of the characterisation of a filter **cake**, the cohesion tester measures the cake's flow properties.

Coil (drum) filter. A type of drum filter in which the cake is formed around a set of coils, running across the drum, which carry the cake off the drum for discharge when the coil is momentarily bent or expanded.

Coiled candle. [*See Yarn-wound cartridge*]

Coil filter. A **metal edge filter**, from Cross Manufacturing, in the form of a stainless steel spiral coil with raised pips on the flat surface, which produce a precise **filtration** gap. Liquid can pass through the gaps in the coil but particles above that size are trapped by the coil. The element is cleaned by backwashing, reversing the flow across the coil, which forces it to open slightly and so allow complete flushing of any debris out of the system.

Co ion. An ion in solution that has the same electric charge as that on the surface of a particle, and so is repelled by it.
[*See also Counter ion*]

Collapse pressure. A term used in the characterisation of **cartridge** filter **elements**, and is that **pressure drop** across the medium of the cartridge at which it fails, as the pressure differential increases.

Collar. In a **filter press**, a collar is used to seal the external eyes of the plates and frames. Made of rubber or another elastomeric material, the collar is shaped to make the most effective seal.

Collecting, Collection. A fairly general term referring to the capture of suspended material by a **filter medium**, in the various ways that this is achieved. **Collection efficiency** is a more specific term, relating to the thoroughness with which the medium achieves its required separation.

The term is also used in a rather different sense for the manufacture of a material by the settling down of extruded **filaments** and **fibres** onto a flat (and usually moving) surface, such as in the manufacture of **spun bond** and **melt blown** media.

Collection efficiency. A term sometimes used for the total efficiency of a **filter**, and mainly used of gas/solid separations. In general, this refers to the fraction by number of particles, or by mass of material, that is captured by a filter from a **suspension**, and can be expressed as a fraction or as a percent figure. Usually determined by measuring the fraction penetrating the filter, so that:

$$\text{collection efficiency (as a fraction)} = 1 - \text{fraction penetrating}$$

Collector. The mechanical component of a collection device, which usually works through **impingement**, but can also work through electrostatic charge, as in an **electrostatic precipitator**.

Also refers to a chemical reagent that is added to mineral particles, and which adsorbs to the particle and renders it **hydrophobic**, and thus aids **flotation** separation processes.

In another sense, the collector in a **decanter centrifuge** is a shroud placed over the **cake** discharge from the beach, to direct the cake flow into, and so prevent the cake from sticking to, the **casing**.

Collision. A process in which a particle in motion in a fluid comes into physical contact with another particle, or with a solid surface. Depending on the kinetic energy of the incident particle, and of the electrical and other forces at the collision point, then the particle will either bounce away, or be captured and held in contact with the other particle or surface. If the solid surface is that of a **filter medium**, then the result of capture is that the required separation has been achieved. It is usually the aim of any **filtration** process, especially in the filtration of gases, to make the efficiency of capture as high as possible.

Collision efficiency. The fraction of actual collisions that result in capture of the incident particles. This is more a theoretical term than a practical one.

Colloid. A state of matter that is intermediate between true molecular (ionic) solutions and particulate **suspensions**. Thomas Graham (the "father" of colloid chemistry) invented the term as part of his work on **diffusion**. In 1860 he showed that finely divided substances could be divided into two classes: those that passed through a **dialysis membrane**, which he called crystalloids (common salt, acids, etc), and colloids, which did not (gelatin, proteins, etc). Colloids exist in the particle size range from 1 nm up to 1 μm, and they may be continuous (rubber and many proteins occur naturally as colloids), or dispersed into colloidal suspensions in a distinct continuous fluid medium, gas (as an aerosol) or liquid (a hydrosol in water).

Once dispersed into a fluid **suspension** (which is how many colloids are made), colloids are usually extremely difficult to separate from the fluid, especially from liquid suspension. Much of the colour, for example, in surface waters is colloidal in nature. **Membrane** separation processes are usually needed to effect removal of such dispersed material.

Colloids can be classed as **lyophobic** or **lyophilic**. Lyophobic colloids are unsolvated (i.e. they have not reacted with the suspending liquid to surround themselves with a layer of solvent molecules) and so are relatively easily coagulated. Lyophilic colloids, on the other hand, are solvated, and make a very stable suspension.

Colloidal particle. A particle (or droplet) in the colloidal size range, of an effective diameter in the range 10^{-9} to 10^{-6} m.
[*See also Colloid*]

Colloid flotation. An older term for what is better called **microflotation**.

Colour removal. An unwanted colour in a liquid may be of two types: a dissolved material or a colloidal suspension. Dissolved colour is usually removed by means of adsorption on **activated charcoal**. Colloidal material is more difficult to deal with, and usually requires **membrane** separation. Thus, the colour in surface water sources intended for drinking water is nowadays removed by **ultrafiltration**.

Comb. A device used in the processing of **staple fibres** to align fibres and improve the quality of the processed material.

In a quite different sense, a comb is a cleaning device for the surface of **edge filter cartridges**, or for **string discharge drum filters**.

Combination arrangement. An array of items of solid liquid **separation** equipment, arranged in series, with the prime aim of recovering and dewatering the suspended solids in the original suspension in as efficient and economical a manner as possible. The final item is usually a thermal dryer (to remove residual moisture), but, because thermal energy is more expensive than mechanical energy, the previous stages will try to remove as much liquid from the solids as possible.

The arrangement will thus normally include some kind of **thickener**, usually working by **sedimentation**, followed by a mechanical unit such as a press, and then the dryer. This kind of arrangement will normally be used where the solids are the required product of some prior production process, but can also be used for waste sludges.

Combination centrifuge. A term sometimes used for a **screen-bowl decanter centrifuge**.

Combination filter. A filter that achieves two (or possibly more) process purposes at the same time, one of which is solid liquid **separation**, and the other is the removal of unwanted colour or odour in the **filtrate** – the latter function usually being achieved by the use of activated charcoal.

Combination media. A **filter medium** capable of performing two distinct process tasks, which are usually particle separation and simultaneous removal of colour from a liquid **filtrate** or odour from a gas. Such a medium could be made from **charcoal cloth** that is then activated.

Combing. In the processing of **staple fibres**, when very fine **yarns** are desired, **carding** is followed by combing, a process that removes short fibres, leaving a **sliver** composed entirely of long fibres, all laid parallel, and both smoother and more lustrous than uncombed types. Slivers may be loosely twisted together, forming **roving**. Hackling, a process applied to straighten and separate flax, is similar to combing.

Combing also removes some residual dirt. It is a necessary stage in the production of fine yarns, and is used for the treatment of coarse yarns to give a high-quality material.

Compaction. A term that relates specifically to the **compression** of a **membrane** medium under high differential pressure, leading to a reduction in **permeate flux**.

Also used occasionally for the compression of a filter **cake**.

Composite, Composite medium. A **filter medium** that is made from a set of two or more layers of different materials, differing perhaps just in terms of varying effective pore size, or possibly made from quite different materials, with completely different structures. There are two main reasons why composite media are used: strength and filtration efficiency. The more delicate a filter medium is (usually to give it a low enough cut point), the weaker it is likely to be in terms, say, of tensile strength. Accordingly, fine filtration media are very often supported on a material whose purpose is entirely to support the finer material, and enable it to withstand a high pressure differential across the medium. The support layer (or layers) usually has no separating function at all.

This arrangement of a thin layer of fine medium supported on a much coarser (and usually thicker) support layer is normally intended to operate by **surface filtration** (or its variant, **cross-flow filtration**). Where **depth filtration** is the intended mechanism, then the medium will have a considerable thickness in the direction of fluid flow, so as to give it

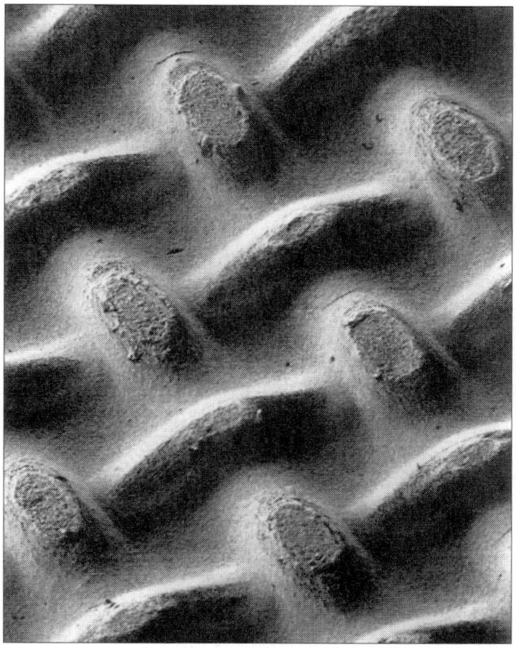

Composite: membrane on metal mesh

as high a dirt-holding capacity as possible. It makes sense, then, to provide this thickness in the form of successive layers of similar material, with each layer in the series having a finer pore size than the previous one.

[*See also Filter medium, Lamination*]

Composite membrane. Most **filtration** membranes are extremely thin materials, and therefore are usually quite weak in mechanical terms. Accordingly, they are almost always supported on a layer of stronger material, to give a composite membrane. This may be by means of creating the membrane as a surface layer on the substrate, as is done with **ceramic membranes**, or by impressing a plastic membrane onto a woven or non-woven fabric **substrate**, or by making the membrane as a thicker asymmetric plastic sheet, with very fine pores at one surface and much larger pores for the rest of the material.

Compost. A soil conditioner and low-nutrient content fertiliser produced by the decomposition of vegetable matter – usually agricultural and horticultural wastes. It should be free of the odours of sewage sludge, and of pathogens, and its fertiliser value comes from the fact that it releases its nitrogen content slowly to the soil.

Composting. The rotting process for waste vegetable matter that converts it to usable material, mainly for improving the condition of soil. It is usually undertaken in the open air, with the waste stacked in piles and turned over occasionally, to allow **aerobic** digestion of the waste.

Compressed air. Air at high pressure is a valuable operating means, either in pneumatic systems or as short pulses. The latter is the way in which it is usually met in **filtration**, but the need for purity, sometimes very high purity, in the air leaving the compressor provides an important use for filtration equipment, to remove dirt and oil.

Compressed air blowback. When a cake has formed on a **filter medium**, and after proper **washing** and **dewatering**, one of the ways whereby it can be released from the medium, for discharge from the filter, is to subject it to a sharp increase in pressure from the downstream side of the filter, and so to blow the cake off the surface (aided, where the filter medium is at all flexible, by an expansion of the material of the medium). The released cake then accumulates in the (now empty) feed zone of the filter – so it is important that all of the slurry has either passed through the filter, or been withdrawn from the feed zone.

Compressibility. The ability of a porous mass to decrease in volume, when subjected to an external mechanical squeezing force. The volume decrease occurs because the voids in the mass are made smaller, and any contained liquid would be squeezed out of the mass, as it is compressed.

Compressibility is thus a measure of the variation in **filter cake porosity** with increasing pressure, and therefore a measure of the ability of the **cake** to be dewatered by mechanical compression. For a filter cake of constant overall diameter, being compressed in a direction perpendicular to that diameter, the degree of compression is related to the cake thickness by:

$$U = (h_1 - h)/(h_1 - h_\infty)$$

where h_1 = original cake thickness
h = compressed cake thickness corresponding to U
h_∞ = cake thickness at equilibrium with the compression force

Compressibility coefficient. An experimentally determined factor (sometimes called compressibility factor), taken from the slope of the plot of **cake porosity** vs applied pressure, or of log (specific cake resistance) vs log (applied pressure), if either gives a straight line relationship.

Compressible cake. A filter cake is deemed compressible if its volume can be decreased by the imposition of an external compressive force.

Compression. In **filtration** technology, compression refers to the process whereby **capillary liquid** is squeezed out of a filter **cake**, which must, of course, be compressible, and not formed with a rigid structure. Compression is achieved by means of a mechanical force applied externally to the cake, by means of a belt and roller system, or in a **filter press** by the movement of the diaphragm. Compression of a cake can occur whether the particles comprising the cake are elastic or rigid.

Compression also occurs in the **sedimentation** of a suspension: as the layer of settled solids increases in height, the lower parts of the mass of solids are compressed by the weight of particles settling down above them.
[*See also Expression*]

Compression blanket. A term sometimes used for the upper belt in a **belt press**.

Compression cell. A piece of laboratory equipment used in the testing of a filter **cake** for its **compressibility**.

Also refers to a special design of **filter leaf** designed so that the **cake** forms, under pressure or vacuum, in a cylinder and can then be compressed by a piston.

Compression filter. Any filter in which **dewatering** of the separated cake is achieved by compression, such as a **filter press**, a **diaphragm press**, a **belt press** or a **tube press**.

Compression layer, Compression zone. As the particles in a **suspension** settle out in a **sedimentation tank**, such as a **thickener**, a layer forms at the base of the tank where the particles are sufficiently close that they can transfer mechanical force among them. This results in **compression** of the layer as more particles add to it, and the sedimented solids in any plane within the compression zone are compressed by the weight of solids above it.

Compression permeability cell. A term sometimes used as synonymous with the laboratory version of a **compression cell**.

Compression point. The point on a **batch settling curve** that corresponds with a sharp change in settling velocity, indicating a change to the compression stage.

Compression roller. A roller mounted on a **filter** in order to reduce the porosity of the filter **cake** (such as on a **drum filter** or **belt press**). The compression effect can be achieved by a dead weight roller, a spring or hydraulically or pneumatically loaded roller, or even a roller that is raised and lowered rapidly on to the cake.

Compression time test. Another term, but less frequently used, for the test used in a **sedimentation** process to determine the ultimate pulp density.

Concentrate. In **cross-flow filtration**, some of the suspending liquid flows through the **filter medium**, whilst the rest of the liquid, plus all of the material unable to pass through the medium pass out of the filter. The dissolved or suspended material is thus concentrated in the exit slurry, compared to the feed, and the exit stream is called the concentrate.

A rather different use occurs in the processing of minerals to extract metal ores. The as-dug material may have only a low content of the required ore, so the material is crushed and then treated to a **separation** process (which may be magnetic, or dependent on density difference), to produce a higher grade product – called the concentrate.

Concentration. Although more commonly used to describe the ratio of dissolved substance (solute) to the dissolving solvent, the term is also used to describe the amount of particulate solid held in a suspension. In this sense, concentration is a dimensionless number, usually expressed as a percentage (or in ppm, for example).

In a quite different sense, the word relates to any process by which a dilute stream of valuable material is treated to remove some of the diluent and so produce a concentrated stream of the valued material. [*See also Solids content*]

Concentration effect. Refers to the strong effect that the solids concentration in a **filter cake** has on the mechanical and filtration properties of that cake.

Concentration factor. In an **ultrafiltration** process, the concentration factor is the measure of the degree of concentration effected by the process. It is the ratio of the solute concentration in the **concentrate** (retentate) to that in the feed solution.

Concentration polarisation. One of the problems that affect the efficiency of **membrane separation** processes in **cross-flow** mode, and are generally termed **fouling** of the membrane. As the liquid moves up to and through the membrane, the dissolved material is left behind at the membrane surface. This increased concentration near the membrane acts to oppose the diffusion of liquid through the boundary layer, and so to reduce the **permeate flux**.

Similarly, if the filter is dealing with a particulate suspension, then there will be an increase in the concentration of suspended particles as the membrane is approached, with a similar resistance to liquid flow. [*See also Cross-flow filtration, Membrane*]

Concentration profile. A set of curves relating **cake** thickness to solids **concentration** at various times in a **batch test**, used especially in the assessment of a **rotary vacuum filter** design.

Concentrator. Any one of a number of items of equipment, used in the mineral processing industry, to increase the concentration of metal ore in a stream of material that has been extracted from the ground. The equipment may include some kind of **elutriation** process, or **magnetic separation**, or **froth flotation**.

Condensate front. When steam condenses in a **filter cake** that is being steamed, the condensate forms in approximately the same place all over the area of the cake, and the pressure of the steam forces this condensate through the cake, pushing **mother liquor** before it.

Condensation. Condensation of a **vapour** into a liquid normally occurs when the vapour is cooled below its boiling point. This is used in the **steaming** of **filter cakes** for the purposes of **washing** them free of **mother liquor**. Condensation also occurs in cool parts of **filters** and **centrifuges** that are processing hot slurries, and this may be an inconvenience, needing provision for condensate runoff to a discharge point.

Conditioner, Conditioning agent. One of a large range of chemicals used to change the nature of a **suspension**, in order that a subsequent **separation** process should be more efficient.

Conditioning. The treatment of a **suspension**, or a **sludge**, or a filter **cake** in order to change its physical and chemical properties, so as to make the next stage of an process more efficient. This subsequent process may be **separation** by **filtration** or **centrifugation**, or **dewatering**. A suspension may be conditioned by the addition of a **flocculant** in order that it can more easily be separated into a solid cake and **mother liquor**, while the cake can be conditioned by the use of an additive to increase the solids content (prior, for example, to disposal to landfill).

Cone. Sometimes used to refer to the conical **beach** section of the **bowl** of a **decanter centrifuge**.
[*See also Deep cone thickener*]

Cone screen centrifuge. A continuously operating **filtration centrifuge**, whose **filter medium** is some kind of screen arranged, not as a cylinder, but as the frustum of a cone, with its smaller diameter at the **feed zone** for the centrifuge. The separated solids move, under the effect of the **centrifugal force**, across the screen and discharge over the edge at the wider diameter. The angle of the cone is designed to match the intended application, so that, together with the rotational speed, the movement of the solids across the screen is sufficiently slow to allow good separation from the **mother liquor**, and even some washing if necessary. Mainly used for easily filtered materials such as coarse crystals.

There are several ingenious designs of the cone screen centrifuge designed to make it usable for a wider range of solids – according to whether the solids are to be held back on the screen, or made to go across it more rapidly. Some versions of the cone screen centrifuge have a small cone angle, such that the centrifugal force is insufficient to drive the separated solids out of the basket. This enables the solids to be held longer on the screen, and so improves dewatering and washing, but an additional device is now required to move the solids, and this is most often in the form of a screw conveyor running the length of the basket, scraping as close to the screen as possible. The screen can be vibrated or oscillated to achieve the same purpose.
[*See also Self transporting centrifuge*]

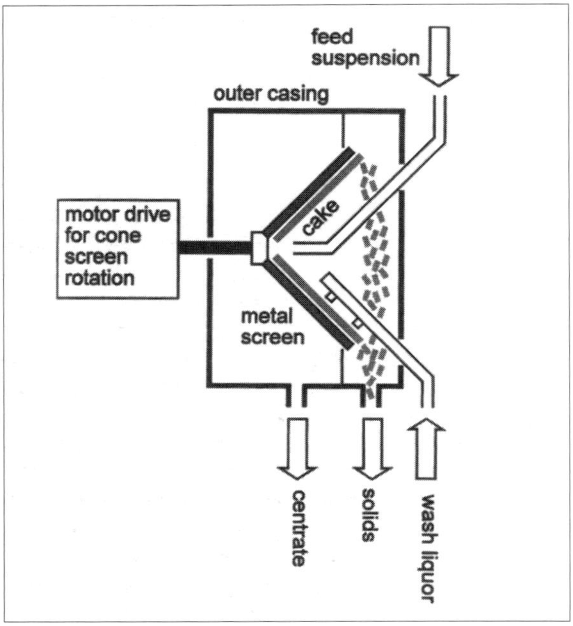

Cone screen centrifuge

Conical basket centrifuge. [*See Cone screen centrifuge*]

Conifuge. A sampling device for **suspensions** of solid particles in a gas, which deposits particles in a complete spectrum, according to their settling velocities. Typically, a conifuge can cover a range of 0.5 to 50 μm.

Consistency. In general terms, the consistency of a liquid or **suspension** refers to its mobility or stiffness, varying from thin (water-like) to thick (a syrup or paste). In the case of pure fluids, it is equivalent to the **viscosity**, but it is more properly used to define the viscosity of a suspension of fine solids in a liquid. It is also used as a term, similar to the viscosity of a liquid, to define the resistance to flow of a **non-Newtonian fluid**.

The term can be used in a different sense to refer to the percentage concentration of solids in a suspension (especially in the paper industry).

Consolidation. In principle, another term for **compression**, but relates in particular to masses of thick slurry (such as concrete) or accumulations of waste solids (such a landfill site) or regions of soil, which need to be made as firm as possible, either to support a load, or to take up as little space as possible. For wet slurries this requires some kind of agitation to expel air bubbles, for loose soil this may involve the injection of a binder, while on landfill sites this involves the application of pressure – by vehicle rollers or tamping.

A specific **filtration** use refers to the process whereby more slurry is forced into the filtration zone of a pressure filter, after it has already been filled with **cake** by the normal filtration process. The formed cake must be compressible under the pump's pressure for consolidation to occur.

Consolidation coefficient. A measure of the rate at which a filter **cake** will compress.

Consolidation tank. The term frequently used in the fresh water industry to refer to a **thickener**.

Constant pressure filtration. A method of conducting a **cake filtration** process in which the **differential pressure** across cake and **filter medium** is kept constant for the whole process. Consequently, the **filtrate flux** decreases with time, until it falls to so low a value as to stop the process.

Constant rate filtration. A method of conducting a **cake filtration** process in which the **differential pressure** across cake and **filter medium** is steadily increased, as the cake thickens, so as to maintain a constant feed of slurry to the filter. Eventually, the differential pressure reaches a level at which it is no longer economical to continue, and the process is stopped.

Constant solution underflow. A method of operating a countercurrent **decantation** process in which the **underflows** from all stages are equal. It follows that the **overflows** are also equal.

Contact angle. The angle formed by a drop of liquid on a flat solid surface at the perimeter of the drop, and measured in the liquid phase. It is a measure of **interfacial tension** between the two phases, and is a guide to the **wetting** characteristics of the liquid in relation to the solid. If the contact angle is less than 90°, then the liquid will wet the solid (and move through any pores in it without difficulty), whereas if the angle is greater than 90°, then the solid will not be wetted.

The nature of the contact angle, and its relation to interfacial tension is well illustrated in the behaviour of the liquid as seen through the wall of a capillary tube placed vertically in a pool of the liquid. If the tube is wetted, i.e. if the contact angle is less than 90°, then the liquid rises in the tube, and there is a concave meniscus at the boundary between liquid and air. If the tube is not wetted, then the level falls inside the tube, and the meniscus is convex.
[See also Capillary pressure]

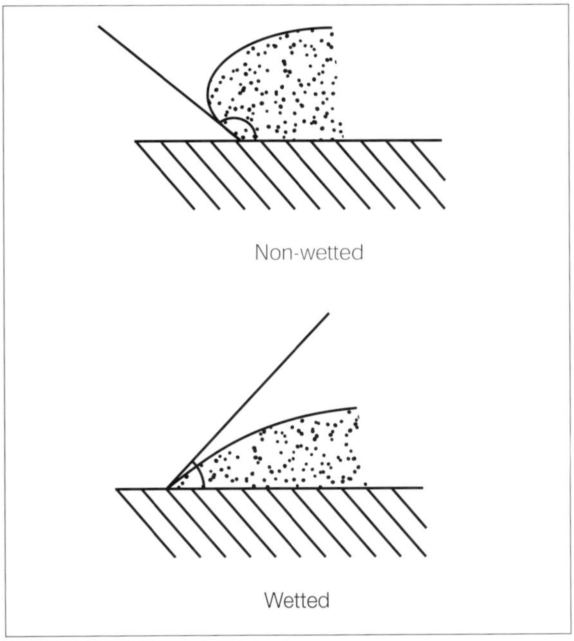

Contact angles

Contact electrification. The collision between a particle and a surface of a different material may create an electric charge on the particle. Thus, a stream of dust particles, blown against an obstacle, may acquire an electric charge on bouncing off the obstacle.

Contact flocculation. An **aggregation** process that occurs in a solid contact **clarifier**.

Contact point number. In an array of regular spherical particles, the degree of closeness of the packing of the particles can be described by the number of contact points, where one particle touches any of its neighbours. The most dense packing ("hexagonal close packed") has a contact point number of 12: 6 in the central plane, 3 below and 3 above. Any lower number than 12 represents a less well packed structure.

Although the completely spherical particle is a rarity, the CPN concept is a useful one for explaining some of the bulk properties of a porous mass.
[See also Packing density]

Contaminant. A general term referring to any unwanted component in any system, rendering it less pure than desired. Specifically refers to unwanted solids in a fluid.

Contamination classes. A system of defined numbers of particles of a defined size that are found in a unit volume of suspension, which are widely used to describe the cleanliness of hydraulic systems. NAS (National Aerospace Standard) NAS 1638 has 14 classes each specifying the particle number in 5 size ranges, from 5/15 μm to over 100 μm, to be found in a 100 ml sample.
[See also CETOP RP70 system]

Contamination profile. A method of specifying the solid particle content of a **suspension** by means of two numbers, the first indicating the number of solid particles greater than 15 μm in a 100 ml sample, and the second that number greater than 5 μm. The standard has 15 ranges of particle number (from 250 to 4 million), numbered from 0 to 14, and a suspension could thus be described as 3/7, indicating range 3 (1000 to 2000) above 15 μm, and range 7 (16,000 to 32,000) above 5 μm.
[*See also CETOP RP70 system*]

"Contibac". A proprietary design of semi-continuous **candle filter**, made by the Dr M company, for the thickening of difficult **suspensions** (such as catalyst slurries).
[*See also Fundabac filter*]

Continuous Area Press (CAP). The CAP is a proprietary design (Bokela) of **belt press**, for the final **dewatering** of a compressible **sludge**. The sludge, after initial dewatering, is spread to a uniform thickness on a **sieve belt**, and then compressed by another sieve belt above the sludge, by passage across a press roller. The CAP provides compressive pressures up to 30 bar, and one major application is in the preparation of paper slurries.

Continuous filament yarn. A **yarn** that is formed from a single endless length of material, usually of constant diameter. **Silk** is the only natural continuous filament, but any polymeric material can be extruded as a continuous filament, of a length limited only by the feed capacity of the extruder.

Continuous filter. Any **filter** operating in a fully continuous manner. Such a filter will need to have a method of continuous removal of accumulated solids, and there are many ingenious designs of filter that enable continuous operation.

Continuous phase. Where two immiscible phases co-exist, then one will usually be in the greater proportion. That in the smaller proportion will then be dispersed in the other, which is called the continuous phase. The continuous phase is completely interconnected, and encloses the other, **dispersed phase**. In a suspension of solid particles, for example, the liquid is the continuous phase, and the solids the dispersed phase, but when the solids have been separated from the liquid as a **filter cake**, then the solids now become the continuous phase, and the **residual moisture** is the dispersed phase.

Continuous pressure filter. This term should refer to a continuous filter that operates under pressure as the driving force, and which produces a dry **cake**. It thus does not include pressure filters that can be operated continuously, but which only produce a thickened slurry.

Continuous separation apparatus. Any item of separation equipment that runs without stopping, or any variation in operating speed. It will have arrangements for a constant supply of feed suspension, and for a steady removal of separated solids and fluid. Even if the equipment involves a series of process steps, these each continue to operate in co-ordination with the others. Such apparatus would normally be used for processes where the flows are large, and also continuous in quantity.

It should be noted that there are a number of items of separation equipment that run without stopping, but which do not maintain a continuous intake of feed or discharge of products. These are considered to be semi-continuous apparatus. A **decanter centrifuge**, or a **rotary vacuum filter** represent continuous items, while a **knife-discharge basket centrifuge** or an **"automatic" filter press** is semi-continuous.

Continuous vacuum filter. Any of a quite wide range of continuous filters that operate with suction as the driving force, and which produce a dry **cake**.

Control disc. The means whereby the various phases of a **rotary filter**'s operation are carried out. It is a disc fitted at the **control head** of a rotary filter (pressure or vacuum operated), into which the control zones are cut as slots. The control disc remains stationary, while the filtrate off-take pipes move from zone to zone.

Control head. The division between the stationary part of a **rotary filter** and the rotating part, to provide means for the control of **filtrate** discharge from the filter. The control head is divided into control zones: for **cake formation**, cake **dewatering**, cake removal (compressed air blowback) and medium cleaning (bubbling zone). This enables the right pressure differential to be applied at each stage of the filtration. There are two types of control head, axial and radial, of which the axial is the more often used. The axial type uses a **control disc**, which sits flat against the rotating set of filtrate pipes, with appropriate seals. The radial type has the rotating part inside a stationary cylinder.

Convective mixing regime. A regime in the **displacement washing** of **filter cake** where backmixing occurs mainly through the mixing of different liquid elements.

Convolution. Specifically refers to a single **pleat** in a fully pleated **filter medium**.

Also refers to the twisting or coiling of a **cotton** material, or the screwlike condition of mature cotton.

Conveying baffle. One of the developments of the simple **scroll** in a **decanter centrifuge** that is designed to improve the degree of separation. It consists of a section of a screw **flight**, of greater **pitch** than the main scroll, which runs from the front of one main flight to the rear of the next.
[*See also Cake baffle*]

Conveyor. One of a great variety of devices intended to move solid materials from place to place. They are used for piece goods after manufacture, such as a roller conveyor, or for bulk granular solids, such as a belt, pneumatic or screw conveyor.

A special meaning attaches to the **conveyor screw** (**scroll**) used in a **decanter centrifuge** to move separated solids along the cylindrical part of the **bowl** and up the **beach** for discharge.
[*See also Magnetic conveyor*]

Conveyor chute. An open channel from a **separation** device down which the separated solids move under the influence of gravity into an appropriate receiver (or into the feed zone of a **conveyor**). The solids must therefore be able to flow to a certain extent – or there has to be a **conveyor screw** in the chute.

Conveyor differential. In a normal conveyor, a screw rotates inside a stationary shell. The same effect is sought in a **decanter centrifuge**, namely movement of the **scroll** relative to the **bowl**, but now the bowl is rotating as well. To achieve the required relative motion, a complicated **back-drive** system is required to turn the scroll at a different rotational speed from the bowl. This differential speed is adjustable so as to give one of the process operating variants for the centrifuge, and can be positive or negative.

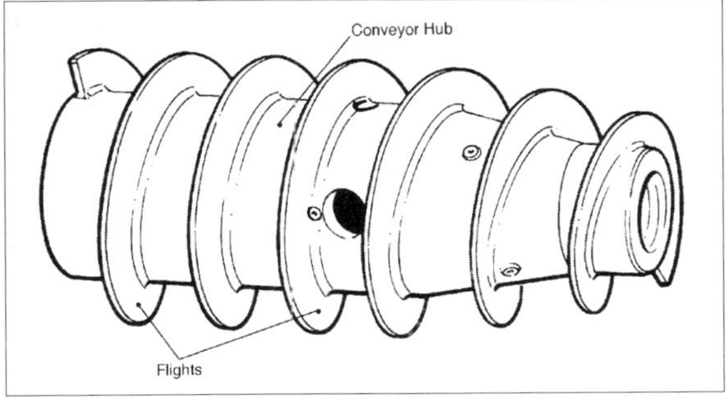

Decanter centrifuge conveyor

Conveyor hub. In a **decanter centrifuge**, the conveyor hub is the central tube that supports the **scroll**, and into which the feed tube is fitted, as far as the **feed zone** in the hub – where also are any **flocculant** addition and rinse chambers. The hub also carries the conveyor bearing assemblies.

Conveyor screw. A screw with one or more flights, rotating inside a cylindrical shell (which can be mounted horizontally or sloping up or down from the feed end), so as to move bulk granular solids or thick **sludges** for some distance. One of the reasons for the thorough **dewatering** of a filter **cake** is to enable it to be moved in a screw conveyor.

A special form of screw is used in a **decanter centrifuge** to move the separated solids, but conveyor screws are used in a variety of separation equipment for the discharge of separated solids, for example in **basket centrifuges**, **clarifiers** and **nutsche filters**.

Conveyor torque. In any screw **conveyor**, the force required to move the solids through the shell by means of the screw is equalled by the **torque** applied to the screw by the drive motor. In the particular case of the **decanter centrifuge**, this torque must be provided by the **back-drive** system.

Coolant filtration, Coolants. [*See Cutting fluids*]

Cooling centrifuge. The high speed of rotation of some **centrifuges**, especially the laboratory **beaker centrifuge**, can generate a considerable amount of heat. In order that this heat should not affect the samples being spun, some designs of laboratory centrifuge have integral cooling systems.

Copolymer. A normal **polymer** is made up of long connected chains of the original **monomer**, linked together in a regular way. In a copolymer, a different monomer is mixed with the basic one, either intermixed in the polymer chains, or as cross-linkages between the chains of the first monomer. The copolymer can have three or more monomers and is a true molecule, with properties that can be quite different from those of a physical mixture of the separate polymers.

Cores for cartridges

Copperas. Hydrated crystals of ferrous sulphate, green in colour, occasionally used as a **coagulant** in fresh water treatment, with pH in the range 8.5 to 11. The aqueous solution is acidic.

Core. The central part of a filter **cartridge**, on which the **filter medium** is supported, and through which the **filtrate** passes. It is normally cylindrical in shape, of a very open structure, but strong enough not only to give the necessary support to the medium during filtration, but also during the formation of the cartridge. It can be made of perforated sheet, of wire mesh, of metal rod, or of extruded plastic mesh (Netlon-type). Sometimes called a centre-tube.

Increasingly, where the used cartridge is to be sent for disposal, the material of the core and of the medium are being made from the same material, to make for easier recycling.
[*See also Air core, Cage*]

Corona. A feature of the **electrostatic precipitator** used for the removal of dusts from a gas flow. If the electric stress at a conductor surface is too high, then the surrounding air can glow, with a luminous discharge.

Corrosion. Unwanted chemical attack on a **filter medium** (or vessel) resulting in material loss, usually into solution, but corrosion by gases is by no means uncommon. The ability of a material to resist corrosion by the components of a fluid being processed is an important performance parameter.

Corrugation. The impressing into a sheet material of a wavy (sinusoidal) shape in one direction, to give added strength, in the form of resistance to bending in a direction across the corrugations. This can be done to paper, plastic or metal sheet. It may also be done to **filter media** to give a greater area-to-volume ratio in a particular **filter**.

Corrugations are often applied to a filter medium prior to pleating, in a direction at right angles to the pleating, in order to create a space between the pleats in use.
[*See also Creped paper, Pleating*]

Cotton. A soft, fibrous material, made almost entirely of cellulose, in the form of hairs attached to the seed of the cotton plant. Long staple fibres are from 2.5 to 6.5 cm in length and are the highest grade, medium staple is 1.3 to 3.3 cm, and short staple is 1 to 2.5 cm. The fibre diameter is usually in the range 10 to 25 μm. The individual fibres have a flattened cross-section and the fibres thus possess a natural twist.

Couette flocculator. Equipment used in the flocculation of suspensions that consists of coaxial rotating cylinders.

Coulter counter. A proprietary instrument used in the preparation of particle size analysis data for mixtures of granular solids. The counter scans a stream of particles, and reports particle numbers within a range of volumetric sizes.

Count. A term used in the definition of a **woven fabric**: either the size or number of a **yarn**, or (and more particularly in **filtration** terms) the number of **warp** or **weft** yarns per 25 mm of width or length, respectively.

Can also refer to a particle size distribution, describing the numbers and sizes of particles in different size ranges.

Countercurrent. Countercurrent flow occurs when two main process components move through the stages of a piece of equipment (or through several items of equipment in series) in opposite directions. The system is used for example in **leaching**, or in liquid/liquid **extraction**, or in some forms of **washing**, so that the most exhausted part of one process stream is contacted with the freshest of the other.

Countercurrent classifier. An inclined rotating drum used for continuous **classification**. A conveyor screw rotates inside the drum, both to break up any aggregates, and to carry the fines to the overflow.

Countercurrent (flow) decanter. The normal form of **decanter centrifuge** operation, with feed in the middle of the **bowl** length, liquid flow away from the conical bowl section and separated solids flow towards it.

Countercurrent decantation. A **reslurry washing** process carried out in a series of **gravity thickeners**, with **wash liquor** flow opposite to the flow of solids.

Countercurrent washing. The treatment of separated solids from an item of solid/liquid **separation** equipment with a **wash liquor** (usually fresh water), so as to remove any soluble materials from them, and also any residual **mother liquor** left after mechanical **dewatering**. The fresh wash liquor first contacts the almost fully washed solids flowing in the opposite direction, thereby removing the final traces of solute and mother liquor. As it moves through the washing process, the wash liquor meets progressively stronger residues, and so becomes enriched in these materials. This method of washing is especially useful if the enriched wash liquor can then be combined with the originally separated liquid.

Countercurrent washing is equally applicable to **displacement washing** and to **re-slurry washing**.

Counter ions. A charged particle in a solution will attract, from the solution onto its surface, a layer of ions of opposite charge. The concentration of these counter ions will decrease rapidly with distance from the surface. The effect of the counter ions is partially at least to neutralise the charge on the particle, and hence let the weaker **van der Waals** forces of attraction take effect.
[*See also DLVO theory*]

Countershaft back-drive. A **back-drive** system for a **decanter centrifuge**, whereby the pinion shaft is rotated by a belt connected to a lay shaft, which in turn is rotated by a pulley on the main rotating assembly. It achieves a fixed **conveyor differential**, but this can be changed by altering the sizes of the pulleys.

Counter top filter. An increasingly important component of the **filtration** business, the counter top filter provides on-demand filtration of domestic tap water for drinking purposes. It features a particle filter coupled with an **activated charcoal** bed for the removal of unpleasant tastes. It may also have other active materials such as a silver coating. The normal form is a jug with replaceable filter **cartridges**, working with just the small head of liquid in the jug as driving force.

Creep. A material's resistance to creep, or the elongation imposed on it by stretching forces during its use, is an important characteristic for many **filter media**, especially where the medium is expected to move during its operating cycle – such as a bag in a **reverse pulse fabric filter**, or the belt of several different types of **belt filter**. Media manufacturers will normally give the stress-strain properties for their materials.

The term also refers to a phenomenon in the **compression** phase of **sedimentation**, in which the **cake** continues to consolidate slowly over a long period, if left free to do so.

Creped paper. Creped paper has had a wrinkled finish impressed into it, which provides some elastic stretch to the material, in a direction at right angles to the creping.

Cresting. In a **decanter centrifuge**, the normal method for discharging the **centrate** from the **bowl** is to allow it to overflow from the **pond** by means of a **weir plate** mounted in the **front hub** of the rotating assembly. The height (distance from the bowl wall) of the weir nominally sets the pond level, but for any liquid to flow out then the pond level must be higher than the weir, and this difference is called cresting: the higher the liquid throughput, the higher the cresting.

Cricket filter. A proprietary design of **pressure filter** made by Amafilter. It is a semi-continuous filter, using a set of **candles** arranged in a circle about a central **filtrate** off-take system. Each candle is flattened on one side, roughly in the shape of a cricket-bat, hence its name. Several sets of candles may be enclosed in one pressure vessel.

The sets of candles are rotated about the central axis, each in turn being **indexed** through the slurry zone and then the other stages of drainage, **dewatering** and **cake discharge**. The cricket filter is a complex system, but has application for difficultly filterable slurries, and to **precoat filtration**.

Crimp, Crimping. The treatment of **monofilament** threads, especially metal wires, so that, after being woven together as **warp** and **weft**, they hold their positions with respect to one another. This can also be done to increase the bulking properties of the finished material.

Plastic monofilaments will naturally adopt a crimped shape when they are woven together and compressed, but it is usually necessary to prefix the crimped shape into metal wire.

Crimping can also refer to the treatment of woven fabrics, e.g. by soaking them in boiling water, to reduce the likelihood of shrinkage in use.

Critical coagulant concentration. The concentration of added coagulant that is just sufficient to achieve the level of **coagulation** required for the subsequent separation stage.

Critical concentration. A term specific to the **sedimentation** process, being the solids concentration at which the rate of settling of individual particles begins to be affected by the presence of other particles.
[*See also Hindered settling*]

Critical flux. A term used in the design of a **gravity thickener**, being the minimum point on a curve relating flux and concentration, from which can be calculated the dimensions of the thickener.

Critical speed. An important design feature of any **centrifuge**, being the rotational speed of the centrifuge that equals the natural frequency of vibration of the rotating parts. Operation at or near this speed would lead to eventual damage of the machine.

Critical thickener flux, Critical zone. A condition of a **gravity thickener** that exists when the feed flux of solids exceeds that which can be transmitted through some limiting concentration (the critical zone).

Crockett belt separator. A type of **magnetic separator**, in which a continuous belt dips into a trough containing the slurry. Behind the lower run of the belt is a series of fixed magnets (of alternating polarity) which attract magnetic materials to the belt and hold them onto it. The belt carries these materials out of the slurry, to a zone beyond the trough, where they are washed off the belt, leaving the non-magnetic solids to settle to the bottom of the trough.

Cross breaking strength. A measure of the mechanical strength of a material, in particular of ceramic material that is supported on two parallel knife edges, and loaded at the mid point between them, until it fractures.

Cross-flow electrofiltration. A **cross-flow filtration** process, in which **fouling** of the **membrane** surface is minimised by the use of an **electric field**.

Cross-flow (filtration). There are two distinct modes of **filtration**. In one, the **filter medium** sits athwart the fluid flow channel, all of the fluid passes through the medium, and any separated solids are held in or on the medium. This is called through-flow or dead-end filtration.

In the other mode, the flow of fluid is parallel to the medium, and some of the fluid flows through the medium by virtue of the pressure difference across it. The, now concentrated, slurry flows on and out of the filter. Very little of the suspended solid remains on the medium, and the intention is that the flow of fluid across the surface of the medium should keep it scoured free of any deposit. This is called cross-flow filtration (sometimes tangential-flow), and while it can give extremely clear permeates, it is only a **thickener** as far as the slurry is concerned.

Cross-flow filtration normally employs a **surface filtration** medium, and this is usually a **membrane**. The cross-flow principle began with the **hollow fibres** used in **reverse osmosis**, and has expanded to become one of the most important components of the filtration industry. In order to keep the surface free of deposit, then high-shear conditions are employed, and these can be created either by a high fluid velocity across the medium, or by some sort of movement (rotation, vibration, etc) of the medium with respect to the fluid flow.

Cross-flow filtration can be classed as one of a number of **thin layer filter** systems. It is one of the three main practical filtration processes (the others

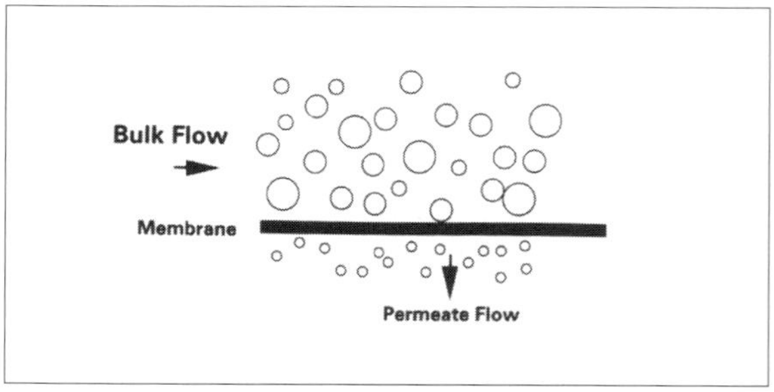

Cross-flow filtration

being **cake filtration** and **depth filtration**, with **surface straining** being a less-used fourth process).
[*See also Dynamic filter*]

Cross (machine) direction. In any woven material, it is easy to distinguish between the two basic dimensions – the **warp** threads run the length of the material, in the direction of production of the finished fabric, while the **weft** threads lie (usually) at right angles to the warp, across the width of the loom, and therefore across the material. Where other roll-goods are similarly made continuously on a machine (such as **paper** or **felt**) it is sometimes convenient to be able to refer to these same directions. The cross machine direction is the one at right angles to the main flow of the material.
[*See also Machine direction*]

Cross flow rotational filter. A type of **dynamic filter**, developed to treat waste water from the paper industry. It consists of a series of parallel circular cells, arranged as a long cylinder. To each cell is fed a suspension under pressure at one end of a diameter, with the thickened slurry leaving at the opposite end. The sides of each cell are made of polymeric or micro-metallic **membranes**, and the **permeate** flows through the membrane to a compartment between each cell, and so out of the filter. A central shaft carries a rotor for each cell to prevent cake formation within it.

Crystal, Crystalline. A crystal is a solid in which the component atoms or molecules are arranged in a regular manner, thereby dictating many of the properties of the solid, especially the shape of particles of it. The values of some solid characteristics are then dependent upon the direction in which they are measured.

Crystals are generally prepared by precipitation from a concentrated solution, and the crystallisation process is frequently followed by a **filtration** or **centrifugation** process to separate the crystals from their **mother liquor**.

CST. [*See Capillary suction test*]

Cumulative size distribution, Cumulative oversize. [*See Particle size distribution*]

Cumulative wash curve. A graphical plot of the fraction of solute left in the **cake** against the **wash ratio**, in a cake washing process.

Cunningham correction. In most **suspensions**, the size of the suspended particles is large relative to the distance that molecules of the fluid move at the ambient temperature. The suspending fluid can thus be considered to be continuous and isotropic. However, when the particles are very small, then they may be of the same size as, or even smaller than, the molecular mean free path of travel. Now the fluid can neither be regarded as continuous nor homogeneous, and, in particular, the standard equations, such as the **Stokes' law** relationship, for calculating particle velocities, do not apply exactly.

Under these conditions, particles can "slip between" the fluid molecules, and exhibit a velocity greater than that calculated. The corrected velocity, u_c, is greater than the calculated velocity, u, by the factor:

$$u_c = u(1 + 2.8\lambda/d_p)$$

where λ = mean free path of the molecules in the suspending fluid
d_p = particle diameter

In the case of air at 20 C and atmospheric pressure, the mean free path is about 65nm.
[*See also Brownian motion*]

Cup dam. A special design of **weir plate** in a **decanter centrifuge**, which arches out from the surface of the **front hub**, in order to give a longer weir rim over which the **centrate** can flow.

Curb. The raised edge of the carrying **belt** of a **belt filter** that retains the slurry on the belt. It may be formed as part of the belt, and hence move with the belt, or may be part of the filter machinery, and hence remain stationary.

Curb press. A type of **hydraulic press** used for expressing moisture from a **cake** or other mass of porous solids. It consists of a single vertical cylindrical vessel, whose walls are porous, filled with the material to be pressed, with a ram pressing on the top of the mass. It does not produce a very clear filtrate, and is used for fibrous non-oily materials like fruit, olives and fish.

Curved screen. [*See Sieve bend*]

Cushion module. A term describing a particular arrangement of **membrane** filter media, wherein two discs of membrane material are welded together around their perimeter, with some kind of spacer fitted between the discs to keep them apart, and so allow the **permeate** to drain away to an off-take pipe.

The cushion normally is circular in shape, but may also be rectangular. It functions in the same way as a **plate filter**.

Cut-off point. In principle, the lowest **particle size** in a mixture that is retained by a **filter** or other **separator**, and so should refer to 100% removal of particles above this size, but often refers to some lower degree of removal (eg 98%).
[*See also Absolute, Cut size*]

Cut point, Cut size. A key particle size that expresses the requirement for a particular degree of **separation** or defines the separation capability of a **filter** or other separating device. Because no filter is perfect in its separation capability, the cut size is frequently derived from a **grade efficiency curve**, and is variously chosen as that size at which the grade efficiency is 98%, 95% or more often 50%.

Cutter. Sometimes used to describe a device that removes a **cake** from a **filter** with a cutting action.
[*See also Knife, Heel*]

Cutting fluids. In the mechanical engineering industry, machine tools are used to cut, drill and otherwise shape pieces of metal. The process of machining generates heat, so the cutting site must be cooled, and also lubricated to ensure efficient machining. The cooling and lubrication duties are effected by a jet of cutting fluid, which also serves to carry away the fragments of metal, or **swarf**. The cutting fluids are often a complex mixture, water or mineral oil based, carrying rust inhibitors at least, and too costly to reject.

They are accordingly **filtered** or **centrifuged** to remove the swarf and then recycled to the machine tool. This may be done adjacent to the point of use, or, more likely, at a **central system**. It employs a wide range of filters or centrifuges, including some designs quite specific to the duty, such as the **paper-band filter**.

C-value. One of the terms used to describe the size of the force acting in the separating zone of any equipment that operates by centrifugal separation (**centrifuge**, **cyclone**). It expresses the centrifugal acceleration as a multiple of the gravitational acceleration, g. The maximum figure is that at the wall of the bowl of the equipment:

$$C = r\omega^2/g = n^2d/1800g$$

where r and d = radius and diameter respectively of the bowl
ω = angular velocity of bowl (or of rotating fluid)
n = number of revolutions (of bowl or rotating fluid) per unit time

The C-values of industrial centrifuges range from a few hundreds up to several tens of thousands. Also known as G-force.

Cyclamatic filter. A **filter** used for clarification of water that contains several spools of renewable **medium**, each being continually wound onto a central core.

Cycle. The chain of processes undertaken in a **filter** or **centrifuge** to complete the **separation** activity, which might include **pre-filtering**, **dewatering** and **washing**, as well as the main separation.

Cycle time. The time taken to perform any particular stage of a **separation** process that is continually repeated.

Cyclone. An item of **inertial separation** equipment that works by **sedimentation** under **centrifugal**

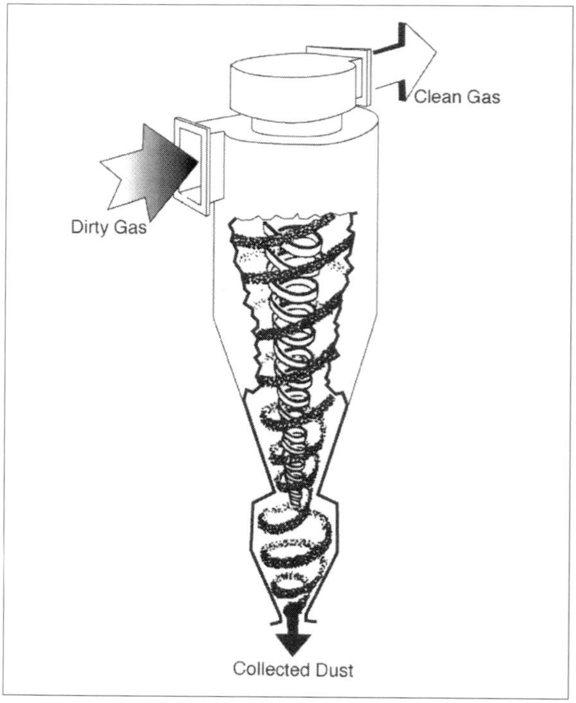

Cyclone dust collector

force. Although strictly applicable to any fluid suspension, the word is normally used for gas/solid separation. The suspension is introduced to a cylindrical vessel (that is usually mounted with its axis vertical) tangential to the inner wall at the top of the cylinder. The fluid and separating solids pass down inside the cylinder in a spiral path. The length of the cylinder dictates the residence time of the suspension, and hence the degree of separation achieved.

At the base of the cylinder the vessel becomes conical, leading to a solids off-take at the bottom. The fluid flow reverses at this point, and the cleaned fluid passes back up the centre of the cylinder, still spinning, to leave through a central off-take for the fluid at the top of the cyclone.

Cyclones are capable of handling quite high solids concentrations in their feed, and, being largely free of internals, are also capable of handling hot suspensions. They are normally followed by a **filter** that removes the trace of solids left in the gas as it leaves the cyclone.

Generally speaking, a smaller diameter cyclone will have a lower **cut point**, so, if a large gas flow is required to be efficiently cleaned, a number of small cyclones will be needed, connected in parallel.
[*See also Hydrocyclone*]

Cyclosizer. A series of **hydrocyclones** used for measuring **particle size distribution**, rather in the same way as a set of **test sieves**.

Cylindrical length. A dimension in a **decanter centrifuge**, which measures the length of the centrifuge **bowl** from the bowl/**beach** junction to the inside of the front (**centrate** discharge) **hub**. [NB Some decanter centrifuges do not have a cylindrical section of the bowl.]

D

Dalton. The unit of measurement of molecular weight (or, more correctly, mass), named after the physicist and chemist, John Dalton, who was a founder of atomic theory. The Dalton is used to denominate the molecular weights of very complex molecules, and so measures in kiloDaltons (kD) are quite common. The Dalton is defined as one twelfth of the weight of the atom of the C^{12} isotope of carbon. The efficiencies of **ultrafiltration membranes** are often expressed as Molecular Weight Cut-Off (**MWCO**) and measured in Daltons.

Dam, Dam plate. [*See Weir plate*]

Damper. A device installed in a duct or at the entrance to a fan, intended to control the flow of gas in the duct. This may operate as a simple on-off arrangement, as, for example, may be used to enable or prevent access to a side duct, or it may have a throttling action, controlling the rate of flow over a wide range of velocities.
[*See also Vibration damper*]

Darcy. The unit of **permeability**, with dimensions of $[L^2]$, named after the French hydraulic engineer, Henri Darcy. Working in Dijon on the flow of water through pipes, as part of a renewal of the city's municipal water system, he first showed that flow through pipes was heavily dependent on the roughness of the pipe wall, and then developed his work to look at flow through **packed beds**, deriving the relationship between flow and pressure drop that also bears his name.
[*See also Darcy's law*]

Darcy's law. The basic relationship between liquid (i.e. incompressible fluid) flow through a porous material and the pressure gradient in the fluid. Derived by Henri **Darcy** in 1856, specifically for the flow of groundwater through **packed beds** of granules, it has become a foundation for hydraulic engineering, enabling quantitative work on the movement of water and other fluids through aquifers and other porous rock strata.

In the case of water, whose viscosity is effectively unity, the relation shows that flow per unit area is directly proportional to the pressure gradient across the bed:

$$Q = kA\,(\Delta P/L)$$

where Q = volumetric flow of liquid
A = cross-sectional area of the porous mass
ΔP = pressure differential across whole bed
L = bed thickness

and k is an experimental constant, the **permeability coefficient**.

For a different liquid, of viscosity μ, the relationship becomes:

$$Q = (kA/\mu).(\Delta P/L)$$

The relationship can be expressed also in terms of the resistance to flow of the liquid through the porous mass:

$$v = Q/A = \Delta P/\mu L(r_c L + R_m)$$

where v = empty pipe velocity
r_c = specific **cake resistance** (i.e. resistance to flow per unit of bed thickness)
R_m = **filter medium** resistance

From these two main relationships it can be seen that the permeability coefficient is effectively the reciprocal of the filter resistance.

DDS vacuum disc filter. A vacuum disc filter in which the discs are immersed in turn into the slurry to be filtered. It consists of an array of nine discs set flat in a circle, which are attached to a central shaft, and rotate with it, **indexing** from position to position. **Cake** is formed on the disc when it is in its lowest position, totally submerged in the feed slurry. **Dewatering** by gas suction can occur when the discs emerge from the slurry. At the top of the circle, porous pistons move in from both sides of the disc to compress the cake. Vacuum is still applied at this stage. The cake is then discharged by gas blowback.

Dead area, Dead space, Dead zone. Any region in an item of **separation** equipment through which the **suspension** does not pass, and is thus either wasted space, or space in which material can settle, and, if putrescible, decay and cause the process to be less sanitary that it should be.

Dead-end (filtration). One of the two main modes of **filtration**, wherein all of the liquid flows through the **filter medium**, leaving any collected solids to build up in or on the medium. The build-up of solids eventually reaches a point at which the required driving pressure to force the liquid through the medium and accumulated **cake** becomes unacceptably high, whereupon the flow is stopped, and the accumulated solids are removed, by whatever means are suitable – by **backflushing**, or scraping away, etc. If the filter medium remains in the filtration zone, then this becomes a batch process. It can be made continuous if the medium is moved through the filtration zone, and then on to a cleaning or cake removal zone. Also known as through-flow filtration. [*See also Cross-flow filtration*]

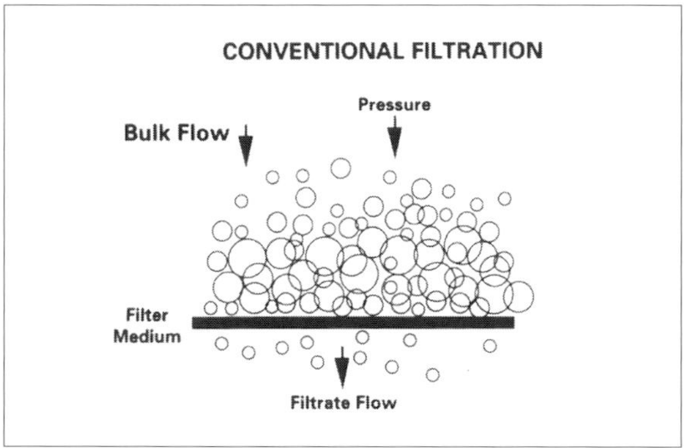

Dead-end filtration

Dead flux. That proportion of the fine particles in a slurry, smaller than the intended **cut-size**, which is discharged with the coarse **underflow** from a **hydrocyclone**. Applies similarly to any **classification** process with a high underflow rate.

Dead time. [*See Idle time*]

De-aeration. [*See Degassing*]

Debye length. In general physical terms, this is the greatest distance at which the electric fields of two charged particles can interact in a mixture of charged particles – basically intended for measurements in a plasma. Specifically it is used as a key term in the **DLVO theory**, which describes the behaviour of an electric boundary layer at a solid-liquid interface. It is then the characteristic penetration depth of the electric field caused by a surface charge on a particle suspended in a solution of electrolytes.

The Debye length:

$$I_D = (\varepsilon kT/2e^2nz^2)^{1/2}$$

where
- ε = dielectric constant
- k = Boltzmann constant
- T = absolute temperature
- e = electron charge
- n = density of charge carrier
- z = valence of the charge carrier

Decant, Decanting. To pour off, or to allow to overflow, the supernatant layer of clear liquid above a settled mass of solid particles or a dense **suspension** (or immiscible denser liquid).

Decantation. The separation of a layer of clear liquid from a settled mass of solid particles, either by pouring, or by allowing the liquid to build up to a point where it can overflow from the settling vessel. The term is more specifically used to refer to a multistage **separation** process, wherein a mixing stage is followed by settlement and then a **decanting** stage, after which fresh liquid is added, mixed, and decanted, and so on. This is a common means for the washing of a previously separated **cake** of solids. [*See also Countercurrent washing*]

Decanter. In principle, any device that allows the separation of a clear supernatant from a settled mass of solids, but, in common **separation** practice, now usually refers to the **decanter centrifuge**.

Decanter centrifuge. *See* pages 69–71.

Decanter centrifuge. The solid-bowl scroll-discharge **centrifuge** – now almost universally known as the decanter centrifuge (or even just as the decanter) – has become the workhorse of a wide range of solid/liquid **separation** activities. Its application to the **dewatering** of waste **sludges** has made it a most valuable tool in the combating of environmental pollution. This has made the decanter a well-known and widely appreciated piece of equipment.

The effective history of the decanter centrifuge occupies the second half of the twentieth century, although it originated at the start of that century. The **tubular bowl** and the **disc stack centrifuges** had, by then, been in use for some time, and there was, clearly, some incentive to produce a centrifuge that would enable the continuous removal of solids that accumulated during the processing of a liquid/liquid mixture. The first patent describing a device like the decanter was granted in 1902. Its resemblance to a vertical **basket centrifuge** is strong, but the design, which was intended to separate solids from two distinct liquids, contains all of the main features of a modern **three-phase decanter**.

Despite this development, nothing of consequence happened in the decanter centrifuge business until the sudden appearance on the market of simple designs of decanter, from both Sharples and Bird, in 1945/46. The success of these machines started a rush of development, leading to the sophisticated designs known today.

Although complicated in construction, the decanter centrifuge embodies a simple principle, that of the **screw conveyor**. In basic terms, the decanter comprises a solid cylindrical **bowl**, rotating at high speed. Inside the bowl is a **scroll** (screw conveyor) rotating at a slightly different speed. The differential speed between bowl and scroll provides the conveying motion to collect up and remove the separated solids. (The earliest versions had no cylindrical section, just a conical bowl – whose clarification performance was not all that good.)

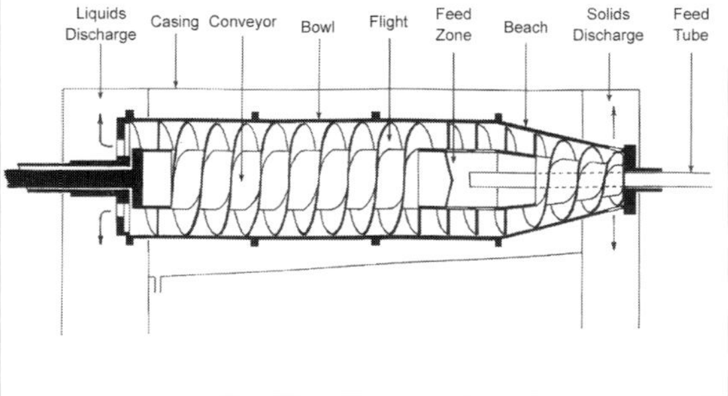

Decanter centrifuge components

A slurry of liquid and suspended solids is carried along the centre line of the bowl, to some fixed position within the bowl, and from there is accelerated outwards to join the **pond** of liquid held on the bowl wall by the **centrifugal force**. This same force then causes the suspended solids to settle, and accumulate at the bowl wall. The clarified liquid flows along the bowl, to leave at one end of it, over some kind of **weir** design, which sets the level of the liquid surface in the bowl.

The other end of the bowl is sloped inwards, towards the centre, thus providing a **beach**, up which the solids are conveyed, to be discharged from the bowl, at the top of the beach. Whilst the solids are conveyed up the beach, some, and hopefully most of the entrained liquid drains back into the pond, to join the liquid flow towards the far end.

The scroll usually is mounted on a hollow axial **hub**, through which the slurry **feed tube** passes to the **feed zone**. The diameter, the number, and the pitch of the conveyor **flights** are chosen to match the needs of the slurry being treated – as are the depth of the pond, the length of the bowl, the conveyor differential speed, and the angle of slope of the beach.

Most decanters operate with their axis horizontal, in which case they usually are mounted in substantial bearings at each end of the bowl. Vertical operation is possible, in which case the bowl is carried only on one set of bearings, at the top. If the decanter is short, then cantilevered horizontal operation is also possible, with bearings at one end only.

The rotating bowl is enclosed in a **casing**, which is divided to ensure that the discharged liquid (the **centrate**) and solids cannot remix after separation.

The basic decanter is completed with a drive motor, usually electrical, and a gearbox, which controls the differential speed of the conveyor.

The decanter operates almost entirely by **sedimentation**, so is capable of liquid/liquid separation, but its prime virtue is the ability to provide efficient solid/liquid separation, in a wide range of capacities, and to do this continuously, operating unattended for weeks or months at a time. The decanter centrifuge can be used for most types of liquid/solid separation, and its ability to handle a wide range of feed slurry concentrations adds to its general versatility.

Decanter bowl and frame

It can be used for the **classification** of solids in liquid suspension, where a single cut is required between two sizes of solid particle (or, less often, between solids of differing density). It is a very good device for this purpose, and its early history included development for the kaolin (china clay) industry.

The decanter *can* be used for the **clarification** of a liquid, and it is able to give a high degree of clarification, but it is not usually used to clarify a slurry that contains only a small amount of solids in suspension.

It can also be used in the **recovery** of a valuable solid from its suspension in a liquid, and following such recovery, it is capable of **washing** the recovered solid free of the original mother liquor, and of **dewatering** the washed solids to a high degree of dryness.

Where the slurry is a waste needing treatment prior to safe disposal, the decanter again can dewater such slurries to a high level of dryness.

Finally the decanter can be operated so as to act as a **thickener**, producing a clear liquid and a more concentrated slurry – either in a manufacturing process, or in waste treatment.

In the second half of the twentieth century, since that first really commercial entry onto the market place by Bird and Sharples, rapid developments in both machine design and application technology have occurred. These developments were largely market-led, with demands from end users quickly converted into machine improvements.

In relative terms, the greater changes occurred in the first 20 years or so of this period. This period saw the decanter expand into more than 100 applications within food and by-products processing, mining, energy materials and systems, chemical, petrochemical and pharmaceutical industries, and environmental engineering. The major manufacturers produced decanters suitable for a wide range of these applications, but there also grew smaller companies, specialising in just one industry, such as olive oil or starch.

Basic decanter performance (in terms of solids recovery, centrate quality and solids dryness) remained relatively little-changed in this period, but improving engineering and materials of construction enabled the use of longer bowls to give greater feed capacity.

Of the major design variants, the screen-bowl design was developed by Bird in the mid-1940s. Sharples produced the first vertical, pressurisable decanter in 1958, and three-phase operation came along in the early 1960s. The main gearbox choices were all in place before 1970, while the potentially wearing surfaces of the conveyor could be protected by a range of treatments or materials, including the range of tungsten carbide **tiles** developed by Sharples (tiles being small flat pieces of hard material affixed to the scrolling face of the conveyor flight).

As a separating device, dependent largely upon a combination of particle size and density to achieve good separation of solids from liquids, the decanter can show much better performance if the particles can be **agglomerated** before or during separation. The appearance onto the market of polymeric **coagulating agents** at the end of the 1960s quite revolutionised the use of the decanter for waste sludge dewatering, and offered great improvements in performance in other applications, as well.

In the subsequent 30 years, relatively few "new" applications have been developed, with the difficulty of achieving fully sanitary operation keeping the decanter from some of the food and biochemical operations. However, major changes have occurred in the refinement of machine design, and, therefore, performance, whilst modern **clean-in-place** technologies have largely dealt with the sanitary operation problem.

These changes have been most apparent in the way in which the decanter handles the discharging solids, and especially in the achievement of the maximum dryness (least moisture content) in those solids. Thus there are restrictions available at the base of the conical section (such as Sharples' "Centri-Seal" patent), operations with deep ponds, and improved control over conveyor speed and torque, leading to the "Dry Solids" decanter.

The availability of stronger stainless steels has enabled the production of very long bowls, as well as quite large diameter bowls. Very high rotational speeds are now available, giving separational forces of up to 10,000 times the force of gravity on smaller models.

Mechanical improvements include the ability to profile the protective tiles on the conveyor face, to use better bearings, and to use 3-stage planetary gearboxes. Improved machine driving mechanisms include inverter drives, and the use of back-drives with power regeneration.

Alternative means for liquid discharge have been developed, largely from other centrifuge types, such as **centripetal** and **skimmer pumps**. Decanters have been fitted with discs to improve clarification performance.

A very important system development has been the improvement in control methods, enabling the decanter to react automatically and quickly to changing feed conditions.

Decanting arm. A floating pipe at the top of a **settling tank**, which removes the supernatant liquid from a settled zone below it.

Decitex. A unit of measurement used by the textile industry for denoting the linear density of a fibre or filament, and hence, if the material density is known, a measure of fibre diameter. The decitex value of a fibre is the weight in grams of 10,000 m of the fibre.

Note that: 1 decitex = 10/9 denier
and that: 1 decitex = 10 **tex**

(the opposite of the normal meaning of the prefix "deci-").

Deck. Can refer to the **drainage** grids below a **filter medium** in a **vacuum filter**.

Also used of the **screen** surface in a **sieve**, or, more particularly, to one surface in a set of sieves ("the top deck").

Deckle strap. A strap on either side of an apron to confine the pulp flowing within it, and so to determine the width of the pulp flow into the **separation** equipment.

Decontaminate. To remove and dispose of undesirable matter, used especially of the treatment of chemical or radioactive wastes. Decontamination processes are becoming of increasing importance as environmental legislation requires the return of working sites to their original condition, or better, after use of them has ceased. **Separation** equipment of all kinds is used in these activities.

Decontamination factor. A term used in air and gas cleaning to indicate the efficiency of decontamination:

$$DF = 1/(1 - E/100)$$

where E = percentage efficiency of reduction in the concentration of the material being removed. (Also the mass of solids in the feed divided by the mass in the **filtrate**.)

Decontamination index. The logarithm of the **decontamination factor**:

$$DI = \log_{10} DF$$

De-emulsifier. When an emulsion has formed between two immiscible liquids, with one, the **discontinuous phase**, being found as droplets within the **continuous phase**, then the emulsion may be dilute, with the droplets widely dispersed in the continuous phase, or it may be concentrated, with the continuous phase found only in the interstices of the droplets.

A de-emulsifier can be intended either to concentrate the emulsion and then separate it from the bulk of the continuous phase, or to break the emulsion, by causing the droplets to coalesce. A tilting plate **sedimenter** could achieve the first task, a **coalescer** the second.

Deep bed filter, filtration. *See* pages 72–3.

Deep cone thickener. A continuous gravity **thickener**, usually conical in shape, characterised by large vessel heights, of the order of 30 m, which allow the development of high levels of **compression** in the settled solids, and thus high **solids concentrations** in the thickened **underflow**.

Defining. A term sometimes used to refer to the removal of the finest particles from a mixture of differently sized solids, by any one of a number of **classifiers**. The object may be to remove impurities,

Deep bed filter, filtration. Operation of a **filter** with a deep bed of granular material as its **medium,** usually with the fluid flowing downwards. This is normally cleaned by flow reversal, washing the dirt upwards and out of the vessel containing the bed (of sand, anthracite, coke, etc).

The deep bed filter (often called a **sand filter**, because that has for long been the filter medium used) has been the basic means of treating fresh water to render it safe to drink for over 100 years. The simple gravity sand filter is still its most common form, in terms of volumes of water treated.

The gravity filter exists in two main types:

- slow, characterised by a low water flow rate and a finer grade of sand, and
- rapid, with water flow rates 5–7 times higher, and using a coarser sand.

The main difference between the two types is, however, in their mode of operation. The slow sand filter works by a **straining** action, exercised by a shallow layer of material on the top of the bed, which contains biological matter. This "**schmutzdecke**" has both a filtering and biological destruction part to play in the water cleaning process. By contrast, the rapid sand filter aims for a truly deep bed action, with contaminant solids adsorbed onto the bed material for most of its depth. Both are capable of giving treated water that is free of solid particles above 0.5 μm, from raw water as high as 50 mg/l in solids concentration (or even 500 mg/l).

The slow version runs with water rates of about 0.1–0.2 m/h, downwards through a bed of sand in the particle size range of 0.35–0.5 mm (uniformity coefficient up to 1.75). It sits, as a layer about 0.6–1.0 m deep, at the bottom of a concrete tank full of the raw water. For a new bed, time must be allowed for the schmutzdecke to form, which can mean running for a while at about one third of full capacity, and then slowly bringing the rate up to its full value. Once established, however, the slow sand bed can operate satisfactorily for considerable periods of time (weeks or even months) before the flow rate drops too far. Then the top layer must be scraped off the bed, and removed to another container for cleaning.

The rapid filter is around 0.75 m deep and formed of sand in the size range 0.5–0.6 mm (uniformity coefficient up to 1.7). It accepts a downwards water flow in the region of 5–15 m/h. It must be cleaned much more frequently than the slow filter, perhaps as often as daily. It is cleaned by reversing the water flow (using clean **filtrate**), at a much faster rate than the processing flow, so as to expand and fluidise the bed completely. The higher water flow rate and the fluidising action together dislodge trapped solids into the wash water, which is removed to another **separator** for further treatment (usually a **gravity settler**, after **flocculation**). The **backwashing** flow is in the region of 30–35 m/h, and is usually augmented with air scouring at the base of the bed, or hydraulic jets on the surface. Backwash lasts only a few minutes and uses 1–5 % of the throughput (and so has about 50 times the solid concentration of the raw water).

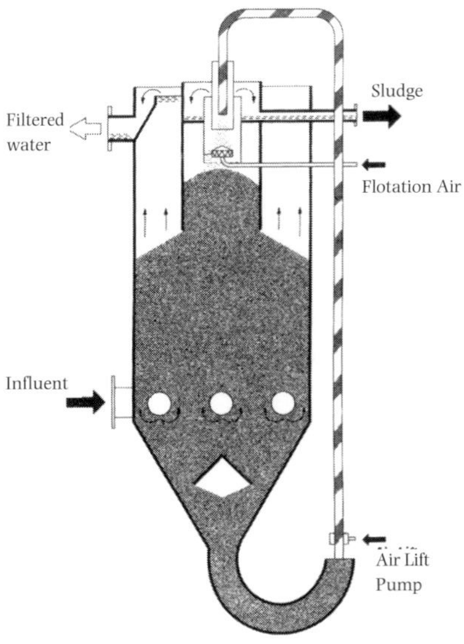

Moving deep bed filter

Whilst the backwash flow is a very effective means of cleaning the bed, it has also the feature that when the back flow is stopped, the fully expanded bed sinks back to its compact form, with all its constituent particles settling at velocities dictated by their size. The result is a stratified bed, with the coarsest particles at the bottom, and the finest at the top.

Unfortunately this is exactly the opposite of what is needed for a downflow depth filter, which should have the raw water meeting the coarsest particles first and the finest last. Nevertheless, the rapid sand filter was used in this manner for many years, because to have an upflow filter risked the expansion of the bed in the direction of the flow, and the consequent release of trapped solids into the filtrate. It was not until the development of the **Immedium** filter in the 1940s that upflow became possible, by virtue of an open grid of parallel bars just below the surface of the bed.

The upflow filter was one of the major changes in the rapid sand filter in the second half of the 20th century. Another was to operate it under pressure, which is done by building a cylindrical steel pressure vessel to contain the bed. Pressurised operation enables the filter to reach head loss figures 2 or 3 times those of the gravity filters. Pressurised filters offer the chance to operate at increasing pressures to counter the increase in head loss as the medium blocks up with trapped solids; however, declining rate operation is generally preferred, with the filter being backwashed when the flow falls below some predetermined value.

Another major development was the multi-media deep bed filter, which uses the stratification action of the return from the fluidised state to create the better size gradations – by using two or three different materials of markedly different density as well as different sizes. Materials such as anthracite, sand and garnet are graded such that the lightest (anthracite) has the coarsest grains and the densest (garnet) the finest. Then in the resettling, the density factor is greater than the size factor, and the finer particles sink to the bottom. A downwards flow of raw water then reaches the coarsest layer first, as it should.

These variants of the rapid sand filter are all suited also to what has come to be called "**direct filtration**" (or contact coagulation) in which a feed liquid is used to which coagulant has just been added, i.e. without time for **floc** formation prior to entry into the bed, so that the flocs form inside the bed. This process is used for the treatment of waters with very low concentrations of **colloidal** solids.

All of the types of sand filter so far discussed are discontinuous in their operation, even if the cycle is quite long in the case of the slow version. The most recent development of the rapid sand filter has been to make it continuously operating, by having the bed of sand or other materials move downwards through the filter. The dirty solids normally fall into a conical hopper at the base of the filter, whence they are carried by a jet of air to a wash zone above the filter. Here they are washed clean of trapped solids and then returned to the top of the bed in the filter. There can now be no stratification of the solids by size or density, but the flow of water, counter-current to the movement of the solids, ensures good solids removal.

to make a suspension more readily filterable, or to make it more acceptable as a product or in a subsequent processing stage.

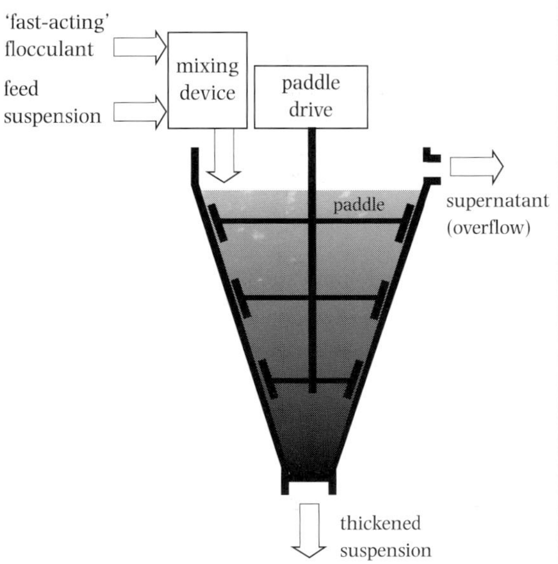

Deep cone thickener

Deflector plate. A plate fitted to the bowl of a **centrifuge** or to the body of a **filter**, to direct the flow of discharged solids.

Defoaming. In most solid/liquid **separation** processes, the presence of foam can create a serious limitation on process performance (as it can in many other liquid handling processes). It is thus necessary so to design the process that foams are not formed, but if their formation is unavoidable, then a defoaming device or system must be included to minimise or remove foams completely.

The effect of foaming is most pronounced with **vacuum filters**, and a range of methods is recommended to minimise foams. The most obvious is to add a defoaming agent if its presence can be accepted. Otherwise, the system should be designed to have as low a vapour velocity as possible, with fluid entry into vessels by means of tangential inlets, and internals designed not to add to the pressure drop. **Knitted mesh screens** can prevent the carry over of foam.

Degassing. The removal of dissolved or suspended gas from a liquid, especially where it might interfere with a subsequent **separation** process involving the

liquid. For example, the presence of the gas may cause particles to float rather than sink. Degassing (or de-aeration) may be done by evacuation of the gas layer above the liquid, by heating, or by mechanical agitation. Care must be taken to see that the degassing process does not result in unwanted foaming.
[*See also Defoaming, Flotation*]

Degree of consolidation. A parameter that measures the **compression** of a **filter cake**. Its value is zero for the freshly formed cake, and unity for a fully compressed cake. (Also called the degree of compression.)

Degree of saturation. [*See Saturation*]

Degree of separation. A measure of the efficiency of a **clarification** operation, given by the mass of solids retained in the **filter** divided by the mass passing through in the **filtrate**.

Degritting. The removal of a very coarse suspended solid, ahead of finer treatment. Achieved in **grit-traps** using either **sedimentation** or **screening**. Usually used for the first stage of municipal sewage treatment, where the foul sewage comes co-sewered with storm water, the latter having swept up a significant proportion of road grit.

Dehydration. A term occasionally used for the **dewatering** of a porous mass.

Delayed cake filtration. A term used occasionally to refer to the process better known as **cross-flow filtration**, the prime aim of which is to operate without the formation of any significant layer of solids on the surface of the **filter medium**.

Deliquoring. A term sometimes used instead of **dewatering**, especially when the liquid being removed is not aqueous.

Delta stack. A variant of the **lamella separator**, developed by Eimco, with specially shaped elements in the form of an inverted V, sitting one on top of another. Clarified liquid is collected at the apex of each element and is transported horizontally to a manifold, by which it is carried upwards to the **overflow** outlet. The solids fall to the base of the housing vessel and are compacted by vibration in the **underflow** hopper.

Demister pad

Demister. Any device used for the removal of a suspension of fine liquid droplets from a gas stream. This can either be done in a **depth filter medium**, where the droplets are captured as though they were solid particles, or in a pad of fine metal wire or plastic filaments, on which the droplets coalesce into larger droplets that then fall out of the collector, without being re-entrained.

Demisting. The process of removing liquid droplets from a gas stream, equivalent to the removal of dust particles in a **gas filter**, but with the additional mechanism of coalescence available.

Demixing. The separation of the components of a well-mixed mixture of solid particles, mixed by size or density (or any other parameter). Once a good mixture has been prepared, it is not usually desirable that the components should separate, and processes will need to be carefully designed to ensure that this does not happen. **Filtration** and **sedimentation** are, of course, frequently intended to achieve the demixing of differently sized particles. Unwanted demixing during a sedimentation process can be prevented by operating at as high a solids concentration as possible.
[*See also Segregation*]

Demoisturing. [*See Dewatering*]

Denier. A measure of the diameter of a fibre, expressed as a weight per unit length, from which the actual diameter is only determinable if the density of the fibre is also known.

The denier number is the mass in grams of 9000 metres of the fibre (and derives from an earlier measure of silk filaments, of a weight in drams for 1000 yards of filament). Thus, if a particular fibre

weighs 12 grams per 9000 metres, it is termed a 12-denier fibre. The higher the denier number, the thicker is the fibre.
[*See also Tex system*]

Dense media hydrocyclone. A form of **heavy media separator** in which the separating device is a **hydrocyclone**. Conventional large diameter cylindrical-conical hydrocyclones can be used. These have wide cone angles, large inlets and large underflow orifices. The feed and dense media are introduced tangentially either under gravity or through a pump. The dense solids (sinks) are discharged through the apex and the light solids (floats) through the **vortex finder**.

Dense media separation.
[*See Heavy media separation*]

Density. The mass of unit volume of any material is its density, and has units of $[ML^{-3}]$, such as gm/cc or, more correctly, $kg.m^{-3}$. This characteristic is often quoted as a ratio to the density of water, known as the relative density (a term used now instead of specific gravity).

The density of the fluid in a **separation** process is an important parameter of the process, as also is that of the solid being separated, especially in determining the settling velocity of a particle of the solid. Of equal importance, however, is the **bulk density** of the solid, which defines the bulk properties of the mass of particulate solid.

Deposition rate. A term used in **cross-flow filtration** to measure the rate at which **foulant** materials deposit in the filtering surface. Usually measured in $g/m^2.h$.

Depth filter. Any **filter** employing a **filter medium** that separates solid particles from a fluid by means of **depth filtration**, and not, in principle, by **surface filtration**. The range of depth filters includes the **deep bed filter**, and any filter using a thick filter medium, such as might be made by the **bonding** of a **mat** of fibres or granular particles.

Depth filtration. One of the two major divisions of **filtration**, according to whether the particles to be separated by a **filter medium** are trapped within the thickness of the medium, or on its upstream surface. In the case of depth filtration, the particle passes into a pore in the medium whose diameter at the surface is larger than that of the particle, which then moves along the pore until it becomes trapped by one of a number of mechanisms. It may simply reach a point where it is wider than the pore, and so become fixed (probably blocking the pore in the process), or it may adhere to a spot on the wall of the pore or to the surface of a particular granule or piece of fibre within the medium. The particle/wall collisions may not be very efficient in capture, but sufficient will occur to give an adequate retention efficiency.
[*See also Cake filtration, Surface filtration*]

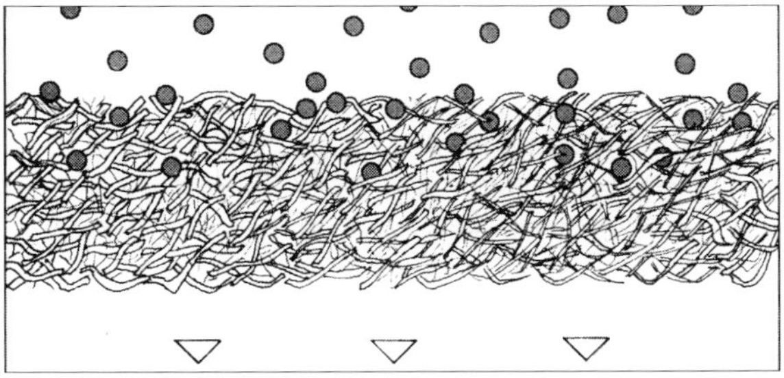

Depth filtration

Depth straining. A precise term referring to a subdivision of **depth filtration**, wherein the particles to be separated are trapped in the pores of the **filter medium** entirely by their size, and by no other mechanism.

Desalination. The removal of salts (mainly sodium chloride) from sea water or other salt (brackish) waters, to produce potable water. It can be achieved by a number of physical processes, including distillation and **electrodialysis**, and has important **separation** interest in the use of **membranes**, which use originated in the **reverse osmosis** of saline waters.

Desaturated cake. In the special sense of **saturation** as applied to **filter cakes**, namely that all voids are full of liquid, a desaturated cake is one in which the saturation is less than 100%, i.e. some of the voids contain gas rather than liquid. A desaturated cake may be prone to cracking.

Desaturation. A solution of a solute, or a mixture of a **vapour** with an inert gas, is said to be saturated when the concentration of solute or vapour is at its maximum, and any further increase in its concentration would cause the component to separate, the solute by precipitation as a solid, the vapour by condensing as a liquid. Either of these conditions may be deleterious to a **separation** process, and so the solution or gas mixture may have to be desaturated – either by increasing its temperature, or by increasing the content of solvent or inert gas.

Desiccant. A drying agent used to **adsorb** moisture from its surroundings, and so to dry a gas (or liquid), or to keep dry some valuable equipment. Materials commonly used include silica gel, activated alumina and molecular sieves.

Design factor. A term used for the multiplier that scales up the performance data in a laboratory test to predict the performance of a full scale piece of equipment. Also called the scale-up factor, but it should be differentiated from safety factor, which is a factor added on to the design factor to allow for unknown components of the process.

Desliming. A term mostly used in the minerals industry for a classification process in which a small proportion of **fines** is removed from the bulk of a mass of particulate solids.

Desorption. The separation of any substance that is adhering to the surface of a solid as a result of a previous **adsorption** process. It is achieved by overcoming the forces that hold the adsorbate onto the surface, by changing the conditions of the surrounding phase (temperature, pressure, etc). In principle, the term can also be used to refer to the recovery of material absorbed into a liquid bulk, but that is more likely to be called **degassing**.

Destabilisation. The change in bulk condition of an otherwise stable **suspension** (i.e. one in which the suspended particles stay separate), in order to enable the particles to be **agglomerated**. This requires a change in the electric properties of the suspension, so that the electric repulsive forces are reduced to the point where the van der Waal's forces of attraction become dominant. The change can be effected, for example, by a change in pH or by changing the ion concentration.

Detachment. A process by which a particle attached to a surface (say, the **medium** in a **filter**) becomes separated from that surface, as a result of which the particle re-enters **suspension**, and so lowers the efficiency of the **separation** process.

Detention efficiency. A correction factor used in the determination of dilute settling behaviour in a **long-tube test**. It is a function of the depth/diameter ratio, and usually has values between 20 and 70%.

Detention time. A term sometimes used instead of residence time – i.e. total volume divided by the volumetric flowrate.

Detention time test. A laboratory test used for sizing **gravity settlers**, **clarifiers** in particular.

Detergent. [*See Surfactant*]

Dewaterer. A term sometimes used for a **pre-thickener**, intended to create a thicker feed slurry to an item of **separation** equipment, which would not be able easily to treat a thin slurry. This is necessary ahead of a number of **basket centrifuges**, and also on some **drum filters**.

Dewatering. Literally the removal of water from any substance, but in **filtration** and **separation** usage, it refers to the removal of any liquid associated with a mass of solid being separated from that liquid – oil, mother liquor, liquid effluent, etc, as well as pure water. The word is used in two very specific cases: in the **dewatering of slurries**, where a thick suspension of solids is to be separated as far as possible from the suspending liquid, and in the **dewatering of filter cakes**, where the residual liquid is to be removed from the interstices of an already separated mass of solid. These two processes are vitally important to much of filtration and separation technology.

In cake dewatering, the process refers to the removal of liquid by mechanical means (**drainage**, **gas blowing** or **expression**), but not thermally (i.e. **drying**).

Dewatering aid. A surfactant added to the feed slurry to a **filter** to reduce the surface tension of the liquid in order to improve **drainage**, and so produce a cake with a lower moisture content.

Dewatering equilibrium. By whatever means **filter cake dewatering** is achieved – **draining** under the influence of gravity, suction, gas pressure over the cake, mechanical **expression** or **centrifugal force** – there comes a point at which the quantity of liquid in the cake reaches a steady minimum level, whose value depends partly upon the dewatering means and partly on the nature of the cake. The equilibrium occurs when the **capillary forces** balance the dewatering forces.

Dewatering filtrate. A term sometimes used of the liquid flowing from the **filter** after flow of feed has stopped.

Dewatering kinetics. Refers to the physical process of dewatering of any saturated porous mass, under the influence of one or more dewatering forces. When one of these forces is applied, the moisture content of the mass starts to decrease, quite rapidly at first, but more and more slowly as time passes, until the **dewatering equilibrium** is reached. A curve of moisture content, or **degree of saturation**, against time will thus fall at decreasing rates, towards the asymptote of the equilibrium value, and a set of curves can be generated to correspond to different values of the driving force in the dewatering process.

Dewatering of filter cakes. [*See Cake dewatering*]

Dewatering of slurries. This process is almost always a **thickening** one, where the product is still a flowing (or pumpable) **suspension**, but is now conditioned for some subsequent treatment. In the particular case of the dewatering of waste **sludges**, this is done either to make them suitable for landfill, or to reduce the volume so as to minimise transport cost.

Dewatering screen. A **screen** used for the **thickening** of a feed slurry to a **filter** or especially a **basket centrifuge**, which enables the processing of the slurry without its **washing out** of the equipment untreated.

Dewaxing of oils. Most oils, whether animal, vegetable or mineral in origin, are mixtures of organic liquids, with perhaps many components. A **wax** is of similar chemical composition to an oil, but has a melting point around or above the ambient temperature. Waxes can often be found as components of an oil mixture, and if the oil is cooled far enough, then the wax will precipitate as a cloudiness of the oil or as visible crystals. The formation of a wax precipitate in this way may be harmful to the acceptability of the product (a clear salad oil for example should not go cloudy in the refrigerator), and the oil should thus be dewaxed before being packaged.

Dewaxing usually involves the cooling of the oil to below any temperature likely to be met during its useful life, and the separation of the precipitated wax in a **filter** or **centrifuge**. The separating equipment may need to be properly housed to diminish the risks of fire.

Dewpoint. If a gas containing a **vapour** is cooled, then at some temperature the vapour condenses as droplets of liquid. The temperature at which the gas mixture first produces liquid droplets is its dewpoint. The **mist** formed by such condensation may behave differently from the gas mixture, especially with respect to corrosion by the condensed liquid, so it is often necessary to carry out gas processes at temperatures above the key dewpoint of any component of the gas.

Dewrinkling. Horizontal **belt filters** are prone to the wrinkling of their **filter media** as they pass through the filter and over the various rollers on their way back to the start. These rollers must be carefully adjusted to ensure the minimum of wrinkling, and an adequate level of tension to remove any wrinkles that might form.

Diafiltration. A means of operating any **filtration** process (but especially applicable to **ultrafiltration**) in which a diluent, normally the pure liquid of the **suspension** being processed, is added to the recycling slurry at the same rate as **filtrate** is being removed. It is most often used in **cross-flow** systems, using **membranes**, to enable a sharper division to be made in the **separation**, such as between salts and higher molecular weight solutes.

Dialysis. A **membrane separation** process, in which colloidal particles are separated from dissolved salts and small molecules, by **diffusion** through the pores of a semi-permeable membrane. The driving force is the **concentration** difference across the membrane (suspension/solution on one side, pure solvent on the other), and the separation occurs because the rates of diffusion are different – the colloids moving slowly if at all through the membrane.

First used by Thomas Graham in his research on **colloids**, dialysis is a very slow process, and can be accelerated by an increase in temperature, or, if the species are electrically charged, by **electrodialysis**.

Haemodialysis is the dialysis of blood by artificial means, to support or replace a failing kidney in the human body. The membrane allows impurities, such as salts and uric acid, to leave the blood, whilst retaining platelets, etc. Water diffuses into the blood during the dialysis process, and the dialysed flow needs **ultrafiltration** to return it to its proper concentration. This is a major use of membranes for separation of any kind.

Diamagnetic. A diamagnetic material is one that is repelled along magnetic lines of force to a region where the magnetic intensity is lowest.
[*See also Paramagnetic*]

Diaphragm filter press

Diaphragm. A barrier, usually made of elastomeric material, that effects the creation of two separate zones divided by the diaphragm (a word often used synonymously with **membrane**). Confusingly, there are two very different functions undertaken by diaphragms. In the first, the diaphragm is completely impervious, and the barrier is complete. This usage exists in the moving parts of some pumps, in the closing-off barrier of diaphragm valves, in the squeezing diaphragm of some **filter presses** or the **tube press**.

The other use has a **semi-permeable** diaphragm, capable of allowing passage of certain species through it, and preventing the passage of others – such as in the separators of batteries and fuel cells. In this usage, the word should really be replaced by membrane.

Diaphragm (filter) press. A development of the **chamber filter press**, in which a **diaphragm** plate is fitted between each pair of chamber plates. This plate holds two impervious sheets of elastomeric material, which can be expanded outwards from the plate, by pneumatic or hydraulic pressure between the diaphragms. As each diaphragm expands, it does so into the chamber of the filter plate, thereby squeezing the accumulated filter **cake**. This allows more **suspension** to be filtered, and then, at the end of the filtration cycle, the cake can be mechanically **dewatered**.

Diastar filter. A proprietary design (by Gaudfrin) of **pressure leaf filter**, in which the leaves are arrayed vertically in a star-like arrangement, around central feed and off take pipes, in a vertical cylindrical vessel.

Diatomaceous earth. [*See Diatomite*]

Diatomite. Also known as diatomaceous earth and kieselguhr, diatomite is the classic material for use either as a **precoat** or as a **filter aid**. Diatomaceous earth is the fossilised remains of microscopic algae, several million years old, of which over 10,000 varieties have been recorded. The name "diatom" is derived from the Greek word meaning "to cut through", referring to the way in which the individual cells are joined together into zigzag chains that are easily separated. But it also aptly describes the way in which single cells of algae reproduce or replicate by dividing into two almost equal parts.

The shell of the diatom is virtually pure silica, which is extracted from the water in which diatoms live. After the death of the diatom, the silica shell survives, and it is a vast multitude of these minute skeletons that goes to make up a bag full of diatomite for use in industry.

The diatomite deposits are worked by open cast methods, the amorphous rock being then subjected toa sequence of crushing, grinding, screening, washing, drying and calcining operations. The products are available in various grades, by chemical composition, by **bulk density**, and by **particle size distribution**.

Dielectric constant. A characteristic of a material in the context of electrostatic fields. It is the ratio of the permittivity of the material to that of free space, and is a measure of the degree to which the material can resist the flow through it of electric charge.

Dielectrophoresis. An electric field **separation** process in which neutral (i.e. without a net charge) but polarisable material is transported in a non-uniform (AC or DC) electric field. If the particles are more polarisable than the liquid, they will be attracted to a point where the field is strongest.

Differential pressure. The difference in absolute pressure between two points in a flow circuit, or two parts of a piece of equipment. Used almost inter-

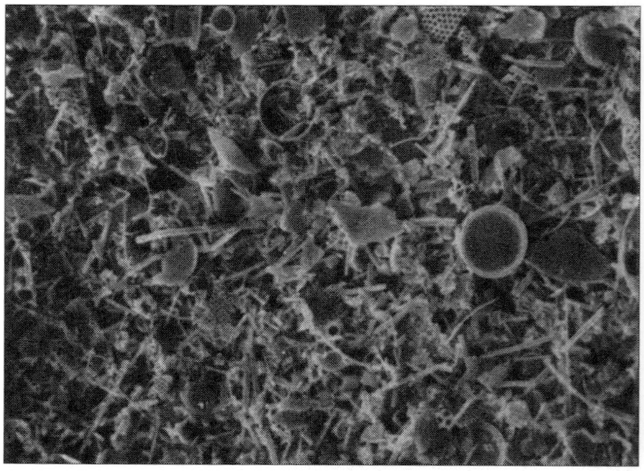

Diatomite filter aid

changeably with **pressure drop**. The differential pressure across a **filter medium** and its accumulated **cake** is an important process parameter in **cake filtration**. A filter may have a gauge to indicate this differential, and a switch to stop filtration when the differential pressure reaches a predetermined value.

Differential speed. Several types of continuous **centrifuge** discharge their separated solids by means of a **scroll**, or screw conveyor, rotating inside the **drum** of the centrifuge. Both scroll and drum rotate, and the conveying effect is achieved by having the scroll rotate at a different speed from the drum. This differential speed is usually effected by means of a gear box connecting the main drive to the conveyor drive.
[*See also Decanter centrifuge*]

Differential vacuum. It is frequently sensible to operate a **vacuum filter** at two or more levels of suction, such as for **cake formation**, **washing** and **dewatering**. This can best be achieved by having a main vacuum reservoir connected to the vacuum pump, maintained at the lowest pressure, and one or more intermediate reservoirs at somewhat higher pressure, with valved connection between them.

Diffuse double layer. [*See Electric double layer*]

Diffusion. The spontaneous movement of a molecule or ion through a continuous medium. The driving force for the movement is a difference in **concentration** of the moving species in adjoining regions of the medium, and the rate of diffusion depends markedly on the phase in which the medium exists – slow in solids, moderately quickly in liquids, and rapidly in the gas phase. The speed of diffusion also increases as the system temperature rises, and can be markedly effected by the presence of an electric field.

Diffusion of fine particles in a **suspension** can be effected by **Brownian motion**, and the term is also used of one of the mechanisms whereby particles are **captured** by a **filter medium**.

Diffusion battery. A term used occasionally for a liquid extraction (or **leaching**) process operating as a battery of stages, under pressure.

Diffusion controlled regime. The situation existing in the **displacement washing** of a **filter cake** in which **backmixing** of the solute occurs by molecular diffusion.

Digestion. The consumption of solid materials by bacterial action (and hence a variant of **fermentation**), which is intended both to reduce the quantity of solids and to sanitise them, so that they may be taken away for safe disposal in the ambient environment. Depending upon the nature of the bacteria (and the operating temperature), digestion can be either **aerobic** or **anaerobic**, and as well as processing the solids, digestion can produce valuable quantities of methane.

The term is used mainly of the destruction of **sludges** separated from the main liquid flow, but is sometimes also used of the secondary activated sludge process itself.

Digested sludge. Mainly refers to the **sludge** produced in waste water treatment works, from any of the processes, but mostly deriving from the **primary** and **secondary** separation stages, which are usually mixed before **digestion**, which, for these sludges, is an **anaerobic** process, and operates at temperatures of around 35 C. The energy to provide this heat comes from the methane generated by the digestion (with usually enough methane left over for other uses). Not only is the digested sludge smaller in quantity, but it is also much easier to **dewater** than the mixed primary and secondary sludge – a task increasingly being undertaken by the **decanter centrifuge**.

Dilatancy, Dilatant. The behaviour of some **non-Newtonian fluids**, which become more fluid on standing, and thicken markedly when stirred or otherwise subjected to shear; a dilatant is such a fluid.
[*See also Shear thickening, Thixotropy*]

Dilute systems. Refers generally to a thin **suspension** of solids in a gas or liquid, for which the **separation** task is much more likely to be one of **clarification** than **harvesting**.

Dilution. The addition of **solvent** to a **solution**, so as to reduce the **concentration** of the **solute** in the solution.

Dilution washing. [*See Re-slurry washing*]

Dilution water. Water used to reduce the strength of a **flocculant** solution, generally added in the flocculant line, ahead of the **feed zone** to a **clarifier** or **decanter centrifuge**.

Dimensional stability. The ability of any material to retain its original dimensions when in use, referring in particular to **filter media** that are constantly in motion.

Dimensionless groups. In many theoretical approaches to practical problems it can be a great help if a relation can be found among the relevant variables of the system under study that is independent of the dimensions of these variables. Such dimensionless groups can then be applied to empirical data, for scale-up purposes, or to define regimes for the system. Thus the **Reynolds Number** can be used to describe the nature of fluid flow through a pipe or any other channel: low values of Re occur under conditions of smooth, **laminar flow**, while high values refer to fully **turbulent flow**.

Dipole. An entity with two electric charges, of opposite sign, located a specific distance apart, is called a dipole. Most commonly this is a small molecule, or part of a larger molecule, whose dipole characteristics cause it to act in a way different from a fully neutral entity.
[*See also Hydrogen bridge*]

Directed flow (screen) centrifuge. A name used occasionally to describe a continuous **centrifugal filter**, with a conical **screen** widening towards the solids discharge end. The separated solids tumble across the inner surface of the screen under the influence of the **centrifugal force**, and the time taken to cross the screen may be too short to allow proper **dewatering** (let alone **washing**). One method of slowing down the flow of solids is to fix barriers to the inner surface of the screen. These barriers are angled slightly to the circumference of the screen, and only occupy an arc of that circumference, so that the solids slide along one barrier, than tumble past its end down to the next – the tumbling action providing adequate movement among the particles to mix them and improve dewatering.

Typically, such centrifuges would be used for quite dense slurries, with solid particle sizes of 0.5 mm or more.
[*See also Cone screen centrifuge*]

Direct filtration. A form of **deep-bed filtration** that is a mixture of **filtration** and **flocculation**. The suspension to be filtered enters the bed flocculated, but only for a short time, so that the **flocs** have not fully formed. The shear inside the bed provides good conditions for **orthokinetic flocculation**. It is used for low concentrations of colloidal matter, and is also known as contact flocculation.

Direct flotation. A term used for the most common form of **flotation**, in which the required product separates into the froth

Direct membrane cleaning. A term sometimes used for the electrolytic cleaning process for **membranes**.

Directional feed nozzle. A device fitted to a **decanter centrifuge** to improve the liquid flow. It consists of a nozzle fitted to each of the exit ports from the **feed zone**, and fitted radially. The exit of each nozzle faces axially or tangentially, such that the feed from the feed zone discharges onto the back or front face of the adjacent **flight** or tangentially into the helical channel between the flights, towards the rear or front of the **bowl**.

Dirt-holding capacity. In a **filter**, for gases or liquids, that is **clarifying** a fluid free from a small **concentration** of suspended solid, then the separated solids will be held on or, more likely, in the **filter medium**, until the latter reaches **saturation** and can remove no more particles. At this point, the filter must either be cleaned, or the filter **element** be removed for disposal. The amount of solid held in the filter at this point, its dirt holding capacity, is an important characteristic in the life of the filter and hence in the economics of its selection and operation.

Disc. In **separation** equipment terms, discs have two main functions – either as an **element** in a **filter**, which may be porous on one or both sides, and have fluid flow from the outside inwards, or as a conical divider in the space within a high-speed sedimenting **centrifuge**, to provide an optimum clarifying distance, for settling solids.

The word can also to refer to those individual flat discs that are stacked one above the other, perhaps around a centre core, to make a cylindrical **cartridge**, whose **filter medium** is the outer edges of these discs, with the narrow spaces between each pair of discs.

A meaning of increasing importance relates to the conical plates fitted around the **conveyor hub** in a **decanter centrifuge**, towards the **centrate** discharge end, to enhance **clarification** capacity. A ribbon **flight** is attached to the outside periphery of the discs.
[*See also Disc filter, Disc stack separator*]

Disc and spoke cartridge. A **clarification filter cartridge** made up of discs of paper stacked around a central former. The discs alternate between two shapes so as to give an internal honeycomb structure.

Disc centrifuge. [*See Disc stack separator*]

Disc filter. One of a range of **filters**, operating semi-continuously (i.e. with one or more **elements** each operating in a cycle of **filtration** operations), and usually under vacuum, in which **filter cake** is formed on both sides of a **disc**, which dips into the slurry to be separated

The most common form, normally called a **rotary vacuum disc filter**, has an array of large circular elements or discs with **filter medium** on both sides, mounted in a row on a common axle (which also contains a **filtrate** discharge line and the vacuum line). The discs rotate slowly, partially immersed in a trough holding the slurry to be filtered. The vacuum draws the filtrate through the immersed area, building a filter cake on its outside. As the disc turns, this cake will drain free of **mother liquor**, and it can then be **washed** and **dewatered**, before being cut off the filter medium on the descending side of the turn. By comparison with a **rotary vacuum drum filter** the disc format offers significantly larger filtration area in a given floor area.

A quite different type of disc filter has the discs in a circle at the end of a set of radial arms. Each disc in turn is then dipped in the slurry before **indexing** round to the other stations for the remaining steps in the filtration cycle.

The majority of disc filters in use are vacuum driven, but there are also versions of the rotary disc filter where the entire system is enclosed in a pressure vessel, and the filter can be operated under pressure as the driving force. The better filtration performance has to be paid for by a much higher initial cost.

Discharge (liquor). The liquid leaving a **filter** or **centrifuge** after passage through its separating zone, coming from any stage of the **cycle**, and so can include **wash water** as well as **filtrate**.

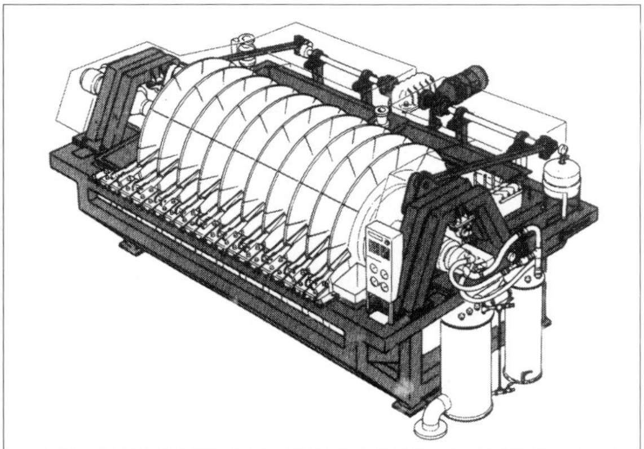

Disc filter

Discharge chute. A duct or large pipe attached to the **casing** of a **filter** or **centrifuge**, through which the separated solid is conveyed to the outside of the equipment, usually under the influence of gravity. The mass of solids normally keeps the chute free of blockage, but easy access is required, should a blockage occur, or a **screw conveyor** may need to be fitted in the chute to prevent the occurrence of a blockage.

Discharged solids. The solids separated in a **filter** or a **centrifuge** must be discharged from it, either intermittently or continuously. If the solids are valuable, then the discharge process must be such as not to damage the solids or cause any to get held up in the equipment casing.

Discharge mechanism. The mechanical means whereby the separated solids are removed from the **separation zone** of a **sedimenter**, **filter** or **centrifuge**, and passed on to whatever piece of equipment follows the separation process. This mechanism may operate batchwise, or semi-continuously, or fully continuously, according to the needs of the separation equipment (and, to a lesser extent, of the subsequent process). It involves the conveyance of the solids both out of the separation zone and through any **casing** around the equipment.

Discharge screw. A conveying screw used to move separated solids from the **separation zone** of a **filter** or **sedimenter** or **centrifuge**, through any casing structure, out of the equipment to a point where it can be transported to the next stage of the manufacturing process.

Discontinuous phase. Where two immiscible phases co-exist, then one will usually be in the greater proportion. That in the smaller proportion, called the discontinuous or dispersed phase, will then be distributed throughout the other, which is called the **continuous phase**. The continuous phase is completely interconnected, and encloses the other, dispersed phase. In a suspension of solid particles, for example, the liquid is the continuous phase, and the solids the dispersed phase, but when the solids have been separated from the liquid as a **filter cake**, then the solids now become the continuous phase, and the **residual moisture** is the dispersed phase.

Disc package. [*See Disc stack separator*]

Disc press. A moderate pressure device for the **dewatering** of pulps, in the form of two discs mounted at a slight angle one to the other, so that there is a wider gap between them at the top than at the bottom. When the discs rotate, feed at the top is compressed by the time it leaves at the bottom.

Disc spacing. The distance between two adjacent **discs** in a **disc stack**. The smaller this distance, the greater the number of discs that can be fitted into a given **centrifuge** shell, and hence the greater the **equivalent clarifying area**. However, there is a limit to how close they can be. Firstly, they must be far enough apart to achieve the required degree of **separation** of the finer solids in the **suspension**, and secondly, the separated solids moving outwards on the underside of one disc must have room to move past the clarifying slurry moving inwards over the upper surface of the adjoining disc, without causing re-entrainment of the separated solid by this differential movement.
[*See also Disc stack separator*]

Disc stack. The array of conical **discs** that provides the **separation** means in a high-speed **disc stack separator**. The array is held together by sets of rods running parallel to the **centrifuge** axis, and there may be holes in the discs to allow liquid flow other than at the centre or periphery of the stack. A given centrifuge shell can be fitted with a variety of disc stacks, in order to achieve different types of separation: solid/ liquid, liquid/liquid, or solid/liquid/liquid.

That is the most common use of the term, but it can also refer to the stack of discs of various types used as a filter **cartridge** and acting as an **edge filter**.

Disc stack centrifuge, Disc stack separator. *See* pages 83–85.

Disc strainer. The simplest type of **strainer** consisting of a disc of **wire mesh** mounted, usually between a pair of flanges, across the flow of fluid, so as to remove **tramp material**. It only makes sense to use such a device when the amounts of material that it will remove are very low, or it will need changing far too often.

Dispersed air flotation. A **flotation** process in which the gas bubbles are generated in the flotation cell by means of a perforated sparge tube or some other mechanical device.

Dispersed phase. Refers specifically to the **discontinuous phase** when it is fully dispersed through the **continuous phase** without any **aggregation**.

Dispersed phase leaching. A **leaching** process in which the particles being leached are freely suspended in the leaching liquid, or are in a fluidised bed, rather than in a **packed bed**.

Dispersion. A mixture of the **dispersed**, or **discontinuous, phase** in the **continuous phase**, usually homogeneously. Dispersions can be of solid particles suspended in a liquid or a gas, a mixture of gas bubbles in a liquid, or a **mist** of liquid droplets in a gas, and can be of three or more phases.

Dispersion coefficient. A parameter used in the **dispersion model** that is similar to an eddy diffusivity.

Dispersion model. A theoretical model used to describe the behaviour involved in the **displacement washing** of a **filter cake**. The movement of liquid through the cake, the hydrodynamic dispersion, is caused by different types of molecular movement within the pores of the cake, caused by: external forces acting on the liquid; the geometry of the pore system; molecular diffusion resulting from concentration gradients; variations in liquid properties (such as density and viscosity); and adsorptive interactions between solid and liquid.

Dispersion number. A factor involved in the theory of **cake washing**, equal to the product of wash velocity and **cake thickness**, divided by the axial dispersion coefficient, such that a high value of the dispersion number would be necessary for effective removal of the **solute** from the cake.

Dispersion parameter. A means of characterising a powder by two key elements of its **particle size distribution**, to give the width of the distribution. The dispersion parameter,

$$D = 100(x_{75} - x_{25})/\bar{x}$$

where x_{75} = particle size read from the graph of the cumulative percentage undersize distribution, at the 75 percentile
x_{25} = particle size read from the graph of the cumulative percentage undersize distribution, at the 25 percentile
$\bar{x}$ = mean particle size

Disc stack centrifuge, Disc stack separator. From simple beginnings as a cream separator for dairies, the disc centrifuge (as it is probably most commonly called) has become a very important processing tool, especially in the pharmaceutical and biochemical sector. The Swedish engineer, Gustaf de Laval, invented the continuous cream **separator**, his first patent, granted in 1878, being for a continuous **centrifuge**, in the form of a roughly spherical bowl, empty of any internal structure. The key patent covering the **disc stack** was granted to a German inventor, Clemens von Bechtolsheim in 1888, and was quickly bought by de Laval's company, which von Bechtolsheim soon joined. This gave de Laval the right to insert in his centrifuge's bowl a stack of conical discs, greatly to improve the **separation efficiency**, thereby creating the basic machine whose essential form has not changed since that time.

The key feature of the disc centrifuge is its stack of "discs", which are actually conical sheets of metal, mounted one above the other on a set of posts, with spacers between them, to give a rigid structure. The disc centrifuge separates by means of **sedimentation**, and uses the same principle as the **lamella separator**, namely to offer as short a settling distance as possible before the particle or droplet of the heavier phase is captured and so removed from suspension. The discs are normally mounted at a spacing of 0.5–2.0 mm apart, and have a cone angle of 35–50°.

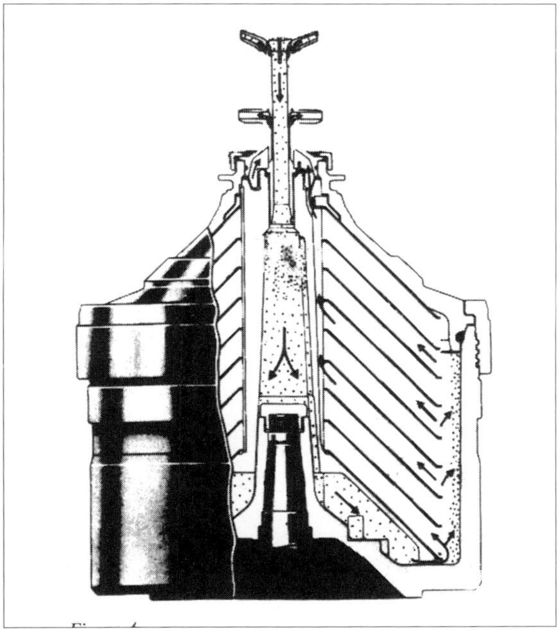

Disc stack in separator

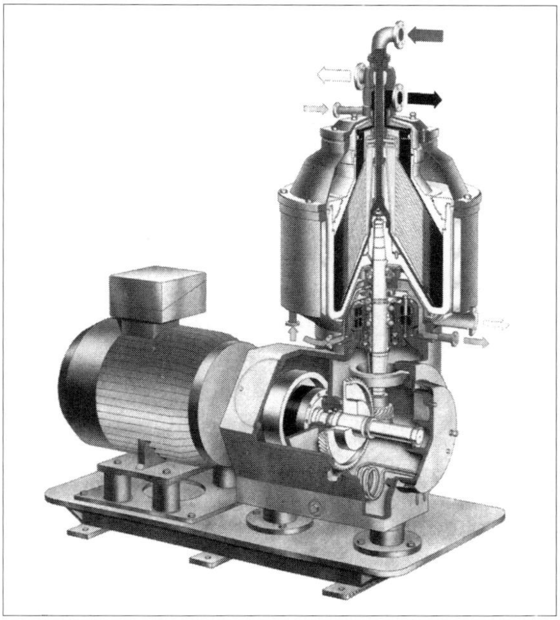

Disc stack separator assembly

The shape of the disc stack is thus a cylinder made up of the edges of all of the discs, surmounted by a conical top, the upper surface of the top disc. The bowl of the centrifuge, which is mounted so as to rotate about a vertical axis, is the same shape as the disc stack, the conical top its most visible characteristic. In the simplest form of the disc centrifuge, the feed mixture is injected at the axis of the bowl and at its bottom. It is deflected downwards outside the lower disc, into the annular space between the outer edge of the discs and the bowl wall. The feed flows up this gap, entering the space between each pair of discs as it does so.

Whilst the liquid flows between the discs, suspended solid particles or heavy liquid droplets move outwards horizontally until they hit the underside of a disc, down which they slide, under the centrifugal force, until they fall off the outer edge of the disc on to the wall of the bowl, where they accumulate. Meanwhile, the main liquid flow, now clarified, flows out of the centre of the discs and leaves the bowl by means of a central liquid discharge point at its top.

The original intention of de Laval was that the disc centrifuge should be a liquid/liquid separator (milk and cream), so the heavier liquid accumulates at the bowl wall to a point where it too will flow to a separate liquid outlet surrounding the lighter phase exit.

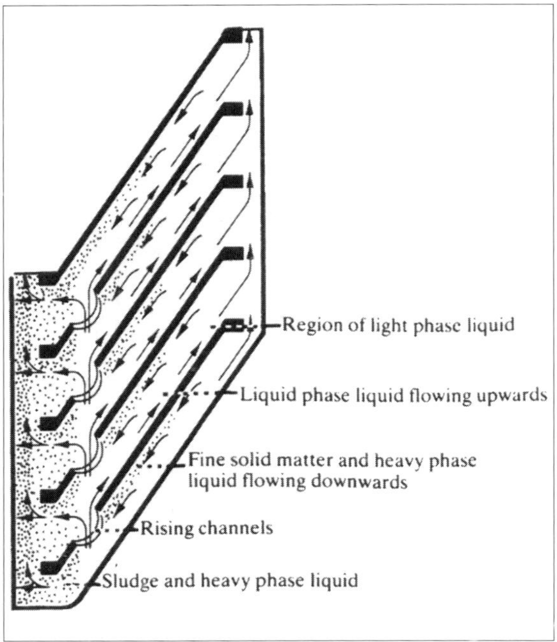

Action of disc stack

The term "disc separator" is most often used of this liquid/liquid separating centrifuge (sometimes referred to as a high speed separator). Other than increases in size and rotational speed (permitted by the development of stronger materials of construction), the separator of today looks very similar to de Laval's design of a century ago.

However, there have been three main areas of development of this basic design, which have turned it into the major processing tool of today. The first of these relates to the problem of suspended solids. Although the disc centrifuge is not a separator for slurries of any significant solids concentration, it is a very rare application that has a feed liquid completely free of solids. As already stated, those solids that do enter the bowl are collected at the bowl wall, and it is from this zone of the machine that they must be removed. The simplest process was to stop the machine and clean them out by hand, but automatic methods were a better idea. The first means developed was to angle the cylindrical part of the bowl to a < > shape, so as to concentrate the solids at the point of the angle. This zone was then extended further at points around the circumference leading to open nozzles, through which the thickened solids would flow continuously. This created the nozzle discharge disc centrifuge.

The nozzles, of course, would also allow some heavier fluid to escape, especially at start up. In order to prevent this, valves were fitted to the nozzles that could be opened by a hydraulic system, triggered either manually or by the height of the layer of accumulated solids in the bowl. This is the valved nozzle disc centrifuge.

The most sophisticated form of controlled solids discharge involves a redesign of the bowl of the centrifuge so that a sleeve on the outside, normally closed against a seal, could be made to retract intermittently and so discharge the collected solids. This is the opening bowl disc centrifuge. The opening is triggered by the height of collected solids in the bowl, and lasts a short time – only long enough to allow egress of most of the solids, and little or none of any heavy liquid.

The second main development in design concerned the means of liquid exit from the bowl. A liquid leaving the bowl has significant rotational energy and this is now converted into useful energy by means of a **centripetal pump**, able to deliver the leaving liquid some distance away from the centrifuge, or to some height above it. If two liquids are involved, then both leave through centripetal pumps. Another advantage of these pumps is that the liquid leaves through a contained channel and does not come into contact with the air.

The third development programme has been concerned with extending the flexibility of the disc centrifuge in handling a variety of feeds, by making changes in the flow paths of the liquid through the disc stack. These changes are achieved by drilling sets of holes through all of the discs in the stack, each disc having a ring of holes, the position of the holes determining the way in which the centrifuge

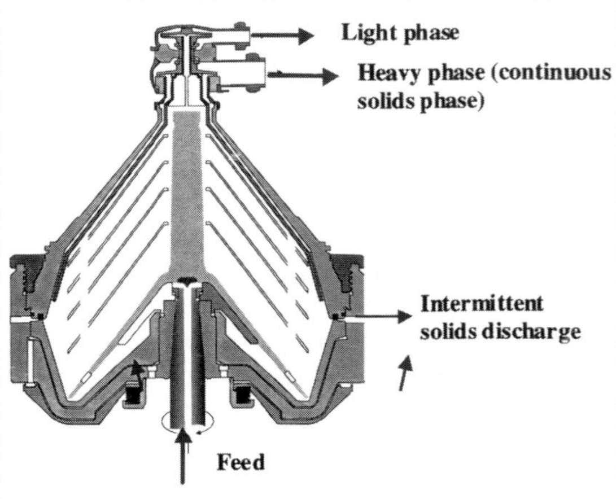

Discharging bowl

may be used. If it is to be used just for clarifying a liquid free from a small content of solids, then no holes are needed. If it is required to separate two liquids, then it is normal that one of these is more important than the other, and therefore needs to have the greater clarity. If the heavier liquid is to have the greater degree of clarification, then the holes will be nearer to the centre of the disc stack. Holes nearer to the edge of the discs give greater clarity to the lighter liquid.

The machines that eject accumulated sludge automatically through nozzles may not be able to create a thick enough sludge from the amount of solids in the feed. Under these circumstances, a special recycle bowl design may be used. In this bowl, part of the thickened sludge is recycled to an entry point at the base of the bowl and carried by a set of tubes over the base of the bowl to mix with the main slurry flow just before the nozzles.

Dispersivity. A general term describing the degree to which a mass of solid particles has been dispersed throughout a continuous fluid. However, it is also used for a specific measure of size distribution as measured by a range of properties of the **suspension**, such as the settling rate, or the scattering of a beam of light shone through the suspension.

Displacement washing. The process of washing **filter cakes** free of residual **mother liquor**, which is carried out in the **filter**, as part of the filter-wash-discharge sequence. A quantity of **wash liquor** is passed through the filter in the same manner as the original **slurry**, and so passes through the cake, intentionally sweeping the residual liquor in front of it in **plug-flow**. Not all the residual liquor is so easily removed, however, and some remains in fine **capillaries** or dead-end pores. This will need to be removed by **diffusion**, necessitating a further flow of wash liquor.

The efficiency of displacement washing is measured in terms of the volume of wash liquor required, and is measured by the **wash ratio**. The time required for effective washing is also important, as it can consume a significant part of the working time of the filter, which is out of productive use during the washing process.

Displacement washing can rarely be a highly efficient process because of the existence of **back mixing** and diffusional processes.
[*See also Re-slurry washing*]

Disposable element, Disposable filter. A **filter element** intended to be used once and then discarded. This will normally have been used in a **clarification** process, treating a relatively dilute suspension of solids (in a liquid or a gas). For this to be economical, the element must be of low cost, and its **dirt-holding capacity** should be high. An increasingly important feature of disposable elements is that they should be made of one material, such that material recycle is feasible, or that disposal is made more easy.

Disposable media. All **filter media** will ultimately need to be renewed, when they can no longer be returned to an adequate state of cleanliness and/or low working pressure drop. The term "disposable" implies that such media are quite quickly replaced, and are cheap enough for such replacement to be sensible.

Dissolved air flotation. A **flotation** process in which the air is introduced into the flotation cell in solution under pressure in the suspending liquid, and the bubbles are released by the resultant drop in pressure.

DLVO theory. A theory explaining the stability of suspended particles as a balance between **van der Waal's** forces of attraction, and electrostatic forces of repulsion. It was developed by Deryagin and Landau, working in Russia, and Verwey and Overbeek, working in the Netherlands, hence its name. The aim then was to develop a theoretical model to explain the stability of a **colloidal suspension** of particles in water. The particles remain dispersed because they are repelled from each other by similar (i.e. all negative or all positive) electrical charges on their surfaces. The source of these charges may be ionisation of soluble crystals, imperfections of the lattice structure of crystals, or absorption of ions from the liquid phase. The DLVO theory postulates that this results in the **electrical double layer** model.

The surface charge of the solid particle is balanced by a tightly held layer of ions of opposite charge. Beyond this is an outer layer through which the ionic concentration (and hence the charge) decays with increasing distance, until the equilibrium conditions of the bulk of the liquid are attained. As a particle moves, or as a liquid flows past it, it continues to retain the tightly held layer of ions, but leaves behind the outer layer, separating from the latter at the plane of shear. It is the potential at this plane that known as the **zeta potential** (ζ).

The magnitude of the zeta potential of a given **filter medium**, and therefore its adsorptive power, is not a fixed value but is dependent on a variety of related electrochemical phenomena, such as the nature and concentration of ions in the liquid being filtered.

Doctor blade, Doctor knife. A rather more precise term for the blade, scraper or **knife** used to remove an accumulated and dewatered **filter cake** from a **filter drum**, or the **basket** of a **centrifugal filter**. This knife may advance slowly into a cake, or move into the cake and back in each cycle of solids discharge.

DOP test. DOP is an abbreviation for dioctyl phthalate. This is used, as an aerosol, in the testing of high efficiency **air filters**. The test, standard practice in the USA, employs a DOP aerosol, of mean diameter 0.3 μm, produced by a thermal generator. The aerosol is passed through the filter, and particle concentrations upstream and downstream are estimated by a light scattering method.

Dorrco filter. The Dorrco filter is an internally fed **rotary vacuum drum filter**. The drum has an internal filtering surface and is open apart from an annular ring that forms a trough for the feed slurry. The drum is compartmented and vacuum is applied from 5 o'clock until 11 o'clock. The cake is discharged by release of the vacuum or by gas blowback, and falls into a chute that leads clear of the drum or into a troughed belt or screw conveyor. This type of filter is used for particles which are too coarse to pick up with bottom fed drum filters. However, maintenance costs are higher.

Dosing equipment. The equipment needed to apply a carefully controlled dose of some conditioning chemical, such as a **polyelectrolyte coagulant** to a slurry, usually ahead of a piece of **separation** equipment. It normally consists of a holding tank, a metering pump and appropriate valves and pipelines.

Double-acting pusher centrifuge. A **pusher centrifuge** in which a pusher plate mounted at the mid-point of the drum **screen** pushes the dewatering solids alternately one way across the screen, and then the other. The feed slurry is correspondingly fed in front of the pusher plate in either direction. This version of the pusher has a high solids capacity, but its washing capability is restricted.

Double belt press. A version of the continuous **belt press** in which the solids being **dewatered** are squeezed between two belts, as the belts are guided over a series of rollers, both pressing and deflecting the belt pair. The squeezing pressure is in the range of 2 to 4 bar, and the belts can be up to several metres long. The double belt press is used for difficultly filterable and very compressible slurries, which have been strongly **flocculated**, then gravity drained in order to be thick enough to be picked up by the belts. They have an important role in sewage sludge dewatering.

Double drum filter. A top fed **vacuum drum filter**, in which the slurry is fed between two counter-rotating drums, with the cake cut off each drum just beyond the top of its revolution.

Double filter. [*See Duplex filter*]

Double layer model. [*See DLVO theory, Electrical double layer*]

Double layer repression. The reduction in the thickness of the **electrical double layer** around a charged particle by the addition of an electrolyte to the suspension.
[*See also DLVO theory*]

Double lead. In any **screw conveyor**, but especially in the scroll of a **decanter centrifuge**, this means the provision of two equispaced continuous **flights** on the **conveyor hub**.

Double weave. A two-layered woven fabric used as a **filter medium**, with a fine tight **weave** on the upper side, facing the slurry, and a thicker coarser weave on the underside providing strength to the whole fabric. The two layers are strongly connected together.

Drag blanket. A layer of material over the surface of the **cake** on a **vacuum belt filter** whose purpose is to spread the cake over the filtering belt and minimise the channels through the forming cake.
[*See also Compression blanket*]

Drag classifier. A **gravity settler** used for **classification**, in which the **underflow** is removed from the base of the **settling tank** by means of an endless belt or chain having cross flights attached to it. Additional liquid can be injected through the belt to wash the solids, which can also be allowed to dewater on the **beach**.

Drag coefficient. When a particle and a fluid are in relative motion (for example, as a particle settles in a liquid), the particle experiences a drag force, D, parallel to the direction of this relative motion, but in the opposite direction to it. The drag force is given by:

$$D = k_{0.}\rho l^2 V^2$$

where ρ = fluid density,
l = a characteristic length of the particle,
V = relative velocity between particle and fluid,

and k_0 is the drag coefficient, a dimensionless number, related to **Reynolds Number**.

Drag effect. A term used of the phenomenon observed with **decanter centrifuges** in which already settled particles are picked up by the clear **centrate** and carried out with the liquid. This occurs because of the drag from the centrate flowing across the settled solids, and careful design is required at the liquid discharge end of the **bowl** to minimise this effect.

Drag tank. A term sometimes used for a **drag classifier**, or for a rectangular **clarifier** with a drag belt discharge.

Drainage. The gradual emptying from the pores of a porous mass of its contained liquid, under the force of **gravity** (or **centrifugal force**). The drainage rate will fall slowly with advancing time, to an equilibrium moisture level, decided by the capillary forces in the mass.

Drainage belt. [*See Gravity belt filter*]

Drainage deck, Drainage grid. A structure in a **vacuum filter** that both supports the **filter medium**, and provides a good drainage space below the medium for the **filtrate** to move on from the filtration zone. This is likely to be a large-aperture grid made of polypropylene. The grid should have as small a volume as possible consistent with adequate strength and minimum resistance to filtrate flow.

Drainage zone. The part of a **belt press** with no applied pressure, where the cake **dewaters** by drainage.

Drainlines. Pipes running from the **cells** on the periphery of a **vacuum filter** to the central **filtrate** offtake pipe. There may be several to each cell in order to provide an adequate drainage rate.

Drain plug. In a **decanter centrifuge** this is a spring loaded plug fitted to the **bowl** shell, usually in pairs fitted 180° apart. It is used for **CIP** work to allow the bowl to drain during cleaning, opening at low bowl speed and closing as speed increases.

Dredge. A scraper device for the bottom of a **sedimentation tank**, in which a set of continuous chains pull scraper bars across the base, to remove any accumulated sludge. Used in particular in settling tanks for machine tool coolants, and in the base of **wet scrubbers**.
[*See also Drag classifier*]

Drinking water filter. Literally, any **filter** used for the purification of water, in order to make it fit to drink. Nowadays, it has come mean specifically the point-of-use filters fitted close to the drinking water tap (just before it, or sometimes just after it), or those flasks of water sitting on the kitchen worktop containing a filter **cartridge**.

Drip pipes. Open ended pipes that carry **wash liquor** across a **rotary drum** or similar **filter** (thus obviating the use of sprays, which may have a tendency to block). The liquid flows from the drip pipe onto subsidiary systems that distribute the liquid over the area of the **cake**.

Driving force. The force needed to cause a suspending fluid to pass through a **filter**. In the simplest cases, gravity supplies sufficient force, with **vacuum filters** next highest (but limited to the maximum suction of 1 bar), and **pressure filters**, with, in principle, an upper value limited only by the strength of the filter **housing**. The highest pressures are those used on **reverse osmosis** processes.

Centrifugal force is another driving force used in **separation** processes, and this can also be very high.

Droplet. A spherical particle of liquid, many of which in a gas constitute a **mist**, and which corresponds in **demisting** to the solid particle in dust removal processes.

Droplet fractionation. The flotation process of **bubble fractionation** in which liquid droplets are used instead of gas bubbles.

Drum. A cylindrical structure, usually with quite a large diameter, which forms the basic filtering **element** of **drum filters** and **drum screens** (**trommels**). The drum normally rotates around a horizontal axis, with part of its surface immersed in the slurry being separated, and other parts available for **washing** and **dewatering**.

Drum filter. In its most common form, the **rotary vacuum drum filter**, the drum filter was, for a considerable period of time, one of the most obvious filters in the process industries, responsible for a significant proportion of solid **recovery** activities. It has lost some of that business to the **belt filter** and the **decanter centrifuge**, but is still a very important part of the **filtration** equipment spectrum.

The simplest form of drum filter is the gravity version, which consists of a horizontal cylindrical **drum** rotating slowly about its horizontal axis. The drum wall is perforated or slotted, and a **filter medium** is fitted to the inside of the drum wall. The filter medium may be a wire mesh or woven or non-woven fabric, and filtration normally takes place from inside outwards (although some designs operate in the reverse direction). Accumulated solids are removed by backwashing through the medium at the top of the drum's rotation. These filters are frequently used as intake filters on cooling systems.

The more complex version, the classic rotary drum vacuum filter, also has a horizontal drum, but there the resemblance ends. The drum is formed of a set of compartments or **cells**, each parallel to the drum's axis, carrying filter medium on the outside, with a vacuum chamber on the inside. The drum is immersed in a trough of the slurry, far enough so that at least one compartment is fully submerged. Vacuum is applied to this compartment, filtration starts and a **cake** builds on the filter medium. The slow rotation of the drum carries this compartment out of the slurry when sufficient cake has formed. Vacuum continues to be applied as the cake is successively **drained**, **washed** and **dewatered**, but is released as the compartment starts to fall back after the top of the rotation, and now the cake can be removed – by a wide variety of devices, either a knife, or a roller, or a set of chains or strings, etc.

A major alternative to this design puts the entire construction of the vacuum version inside a pressure vessel, and drives the filter by pressure rather than vacuum. This is a very complex mechanical construction, and so is much less common than the vacuum version.

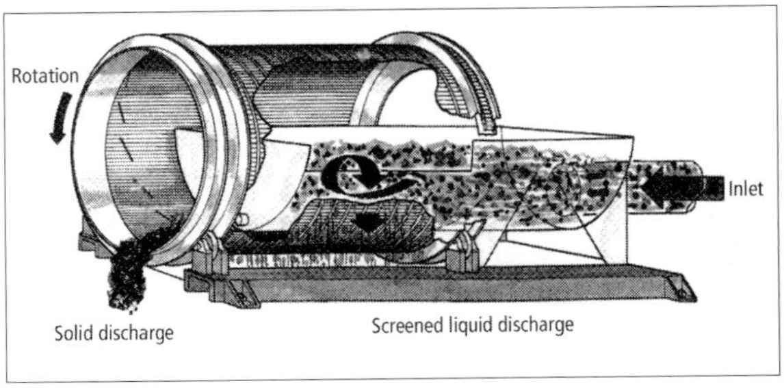

Drum screen

Drum precoat filter. Most **drum filters** are used for the separation of relatively thick slurries, where the solid forms a **cake** quite easily. A drum precoat filter has a consumable cake of **precoat** material, laid down on a fairly coarse wire mesh, to act as the filter medium. It is basically similar to the **rotary vacuum drum filter**, but now the filtration process is one of **clarification**, with the solids being trapped within the precoat cake. The precoat layer may be as much as 100 mm thick, and it is progressively removed as a thin layer, taking the trapped solids with it. Eventually the precoat cake has to be renewed, and the clarification continued.
[*See also Advancing knife, Precoat*]

Drum screen. The drum screen is an internally fed **drum filter**, rotating about a horizontal axis, which has a **wire mesh screen** as the **filter medium**, with a mesh size of about 50μm. Solids are washed off at the top of the drum's rotation into a channel that runs through the drum below the screen. The drum screen is much used as an inlet screen for fresh and waste water treatment.

Dry beach. The portion of the **beach** above the level of the **pond**, in a **decanter centrifuge** or **mechanical classifier**.

Dry bursting strength. The pressure at which a standard disc of **paper** or fabric medium bursts when subject to compressed air in a test rig.

Dry discharge. Any method of discharging a **filter cake** without the addition of liquid.

Dry filter. A term used in the cleaning of gases to differentiate the wide range of normal **filters** using dry filter media, from the wet or **viscous filters**.

Drying. The removal of residual moisture from a porous solid material, such as a **filter cake**. Final removal would normally have to be done thermally, in some sort of process dryer (or, if there is time, by exposure to a dry atmosphere). However, thermal energy is more expensive than mechanical energy, so it is important that, if a dry product is ultimately required, the maximum of mechanical **dewatering** be undertaken, before thermal drying is called upon.

Dry-laid. Fibres can be laid down to form a **mat** of random structure, which makes a very good **filter medium**. The earliest such material was **paper**, which was **wet-laid**, i.e. allowed to settle out of a suspension in water. Also fairly old was the parallel manufacture of **felt**, with fibres laid out on a moving belt, but laid mechanically. More modern processes generate the fibres as extruded **filaments**, which are then laid in random fashion through an airstream onto a continuous belt, or the filaments may be broken by horizontally directed air streams into shorter lengths, which also then fall through an air flow onto a belt. These dry-laid materials are the basis of **spun-bonded**, **melt blown** and **flash spun** materials.

Dry scrubber. Gases carrying suspended solid particles, or liquid droplets, or gaseous impurities, may be purified by passage through a mass of dry solid granules, usually able to react chemically with the gaseous impurity, and to remove the suspended material by a **depth filtration** process. Such dry scrubbers, usually in the form of vertical towers filled with a **packed bed** of the solid scrubbing material, act in a manner similar to that of a **deep-bed filter** for liquid suspensions, but with the additional benefit of the removal of the gaseous impurity.

Because their primary function is usually that of gaseous impurity removal, these units are not normally considered to be filters, but they are an important tool in any system required to clean process exhaust gases.

Dry solids, Dry substance. The solids content of a **suspension** in a liquid can be calculated by determining the weight of a sample of suspension, separating the bulk of the liquid, and drying the residual solids to constant weight at 100 C. The results are then usually expressed as a weight %DS. This may not be a meaningful figure if samples of solids of very different densities are being compared.

DS mode. The design and operation of a **decanter centrifuge** for DS, i.e. dry solids, production: the formation of an extra dry **cake** of solids

Dual angle beach. A **decanter centrifuge** whose **beach** is made in two sections, each with a distinct semi-included angle, and arranged with the steeper angle first, followed by the less steep, or vice versa.

Dual beach. A **decanter centrifuge** with a **beach** at both ends. Used for separating light and heavy solids from a **suspension**.

Dual strainer. [*See Duplex filter*]

Duct. A conveying channel for a gas, as distinct from a pipe, which is normally considered a flow channel for a liquid. In an air distribution system, for example, a main or trunk duct will carry the air from the **HVAC** system through the centre of the building, with side ducts taken off wherever necessary to the individual air registers in each room or working area.

Duplex filter. Almost every industry has need of small, **cartridge**-like **filters** to undertake miscellaneous liquid cleaning processes – lubricating oils, for example, or machine tool coolants. The **filtration** needs here range from coarse straining to submicrometre level.

These filters will need cleaning every so often, and so would require the liquid flow to be stopped while the cleaning is done. This loss of flow can be overcome by mounting such filters in pairs, side by side, and transferring liquid flow from one unit to the other, while the

Duplex filter

first is cleaned. These duplex filters are very economic to install, even if the flow switching and cleaning is done manually – but are even more of an advantage where the change can be automated.
[*See also Automatic filters*]

Duplex system. Any operating system or process that is supplied in parallel, duplicated format, such that one half is always functioning whilst the other is being cleaned or otherwise made ready for use. Changeover between the two parts is rapid.

Dust. Fine, unwanted solid **particles** suspended in a gas flow are usually referred to as dust, and their removal is usually referred to as gas cleaning or dust collection. This can be done by any of a number of **dust collectors**. Particles capable of forming a dust are relatively easily swept up into a passing gas stream – it would necessary to keep a gas flow at below 3 m/s to avoid re-entrainment of a settled dust.

A rather more particular usage of the word occurs in the manufacture of samples of fine solid particles to be used for testing **filter** and **filter media** performance by **challenge tests**.

Dust is normally classified as made of solid particles of 75μm or less, while a material with particles larger than 75μm would be called a **grit**.
[*See also Test dusts*]

Dust collecter. Any of a number of types of equipment designed to clean a gas stream free from a small concentration of unwanted solid particles. The most common **filters** for this purpose are the **fabric filters** or **bag houses**. Small gas flows will be cleaned by **cartridge filters**, while gas cleaning can also be achieved by means of an **electrostatic precipitator**, or by a **scrubber**, either **wet** or **dry**. Some of the simplest types of dust collector involve flow of gas through a device which imposes sharp changes in flow direction on the gas flow.

Dust control. The generation of dusts in industrial processes can cause considerable difficulties in meeting environment pollution legislation. Dust control has become a sophisticated system of prevention, minimisation and **filtration**, all to ensure that dust emissions are kept below legal limits.

Dust spot efficiency. A measure of the efficiency of a piece of **filter medium** in the **clarification** of an airstream. It is part of a system of tests under an **ASHRAE** procedure, and involves the comparison of the darkening (by collected dust) of two sample papers, one upstream and the other downstream of the test piece

Dutch weave. Weaves used mainly for **wire mesh**, in which the **warp** and **weft** threads are of different sizes, thereby giving a mesh of adequate strength, yet with a relatively fine degree of **separation**. The weft is composed of a larger number of fine wires, whereas the warp is made from a smaller number of thicker wires. A simple over-and-under weave then gives a plain Dutch weave, while twilled Dutch is a denser mesh having the weft cross two warp strands at a time, as with **twill weave**. (Also known as Hollander weave.)

Dynamic cross-flow filter. A **cross-flow filter** in which the **membrane** forming the **filter medium** is agitated in some way, so as to increase the **shear** at the surface of the membrane, and so decrease the problem of **fouling** or **concentration polarisation**. The motion of the membrane can be achieved by rotating the assembly holding the membrane, or vibrating it.

Dynamic filter. Any **thin layer filter** in which **cake** thickness is minimised by shear forces caused by rotating surfaces close to the **filter medium**, or by movement of the filter medium.
[*See also Cross flow filtration, Delayed cake formation*]

Dynamic membrane. A term sometimes used for a **cross-flow membrane filter** in which a layer of deposited solids acts as the **membrane**.

Dynamic viscosity. The property of any fluid whereby it resists attempts to make it flow. The viscosity of a liquid, under **laminar flow** conditions (low **Reynolds Numbers**) is defined by Newton's law of viscous flow:

$$F = \eta A(dv/dx)$$

where F = tangential force between two parallel layers of liquid of area A distance dx apart moving with a relative velocity dv

and η is the coefficient of viscosity, usually just called the viscosity (and here called the dynamic viscosity, to distinguish it from the **kinematic viscosity**). Viscosity is measured in units of Nsm^{-2}, and for liquids usually decreases as temperature rises (the reverse being true for gases).

The key feature of fluids described in this manner is that viscosity is independent of velocity gradient (dv/dx). Such fluids are termed **Newtonian fluids**, as opposed to **non-Newtonian fluids**, for which viscosity is dependent also upon **shear rates**.

DynaSand filter. A **deep bed filter** used for the **clarification** of liquids, in which the filter bed material flows continuously between a **filtration** zone and a **washing** zone. The movement is caused by an internal air lift, with the flow of liquid upwards and against the flow of sand.

DYNO-Filter. A proprietary design of **dynamic cross-flow filter**, made by the Bokela company, using parallel discs of **membrane** media, mounted on a central shaft, and rotated within a cylindrical housing.

E

Ecart probable. A term sometimes used to describe the sharpness of the **cut** by a piece of **classification** equipment. It is defined as half of the difference in the sizes corresponding to 25% and 75% on a **grade efficiency curve**.

Eccentric dam. A feature of a **decanter centrifuge**, which controls the depth of liquid (**pond** height) in the **bowl**. It takes the form of a circular **weir plate** with an eccentric hole, mounted in the end wall of the bowl. Normal practice is to have several such plates, in even numbers, with 4 or more to a set. If the plates are rotated in their mountings, different pond heights may be obtained. (Sometimes called a universal dam.)

Ecosuc filter. A **vacuum disc filter** in which the **filter medium** is a porous ceramic material, which does not allow any gas flow, and so operates by **capillary suction**.

Edge filter. A **filter** using a **filter medium** built up in such a way that the channels through the medium are formed by the adjacent edges of the structure, which is usually made of metal (and hence the filters are as frequently known as metal-edge filters). The array of active edges can be created from a wire or ribbon wound into a helical structure, from a stack of flat discs, from a series of parallel bars (such as **wedge-wire**) and so on, each having spacer lumps on their flat surfaces to keep them apart.

Edge filters are used as clarifying filters and **strainers**, with a precise figure for the **cut-size**. They are easily cleaned mechanically and so often feature in **automatic filters**.

Eductor. A water jet pump (or aspirator) used for creating a **vacuum**, in which the water jet drags air with it into the exhaust funnel of diameter only a little larger than that of the water jet. It is only capable of creating a vacuum equivalent to that of the water vapour pressure, but since water is the commonest fluid being filtered, this is usually sufficient of a pressure differential to drive a **vacuum filter**.

The term also refers to a water jet system used to dissolve solid **flocculants**. The turbulence in the region of the suction point is sufficient to agitate the resultant **suspension** and cause dissolution.

Effective area. The total area of a **filter medium** exposed to the slurry flow and actually working to achieve the purpose of the **filter**.

Effective size. A term sometimes used for the nominal minimum size of the **particles** in a mass of granular solids, although it is usually defined as the 10% passing size.
[*See also Equivalent diameter, Equivalent pore size, Hazen effective size*]

Efficiency curve. A graph showing the performance of a **filter** or sample of **filter medium** when challenged with a **standard test dust**. The curve can show **penetration** or **separation efficiency**, plotted against particle size, for a given **face velocity**.
[*See also Grade efficiency curve*]

Efficiency of collection, of separation.
[*See Collection efficiency, Separation efficiency*]

Efficiency factor. A term sometimes applied to the operation of a **sedimentation centrifuge** to account for the fact that the real performance is less good than predicted by the **sigma factor**. (Note that other definitions of efficiency are given under the appropriate headings, such as wash efficiency.)

Effluent. A term occasionally used for **filtrate**, but much more commonly used for any fluid leaving a process, especially if it is contaminated, and hence must be treated as a waste product.

Effluent treatment. [*See Waste water treatment*]

Elasticity. The ability of a material to deform under stress, and return to its original shape when the stress has been removed.

Elastic limit. The maximum stress that can be applied to a material without causing permanent deformation.

Electret. A **fibre** made in such a way that it has an intrinsic electrostatic charge, and can thus be used, when made up into a **filter medium**, to capture particles more effectively if they too carry a charge. This is an important material in the cleaning of gases, when almost all suspended particles are charged.

Electrical double layer. Most particles suspended in a liquid carry an electric charge at their surface. To this surface there then adheres a layer of oppositely charged entities, the **Stern layer**, probably of ions. This adhered layer almost compensates for the charge on the particle, but the balance is not exact. Complete neutralisation is achieved through a second, diffuse, liquid layer, outside the first, which has a slight excess of the counter ions. These two form the electrical double layer, and only outside them are the charges in the suspension fully balanced.
[*See also DLVO theory, Zeta potential*]

Electric fields. Electrically assisted **separation** processes are now quite common – the principle being that the particles or droplets to be separated carry an electric charge and will therefore move in the appropriate direction, according to that charge, under the influence of the electric field. Usually this is a supplementary effect, but there are some processes for which the electric field is the prime separating influence, as in **electrophoresis** or **electrostatic precipitation**.

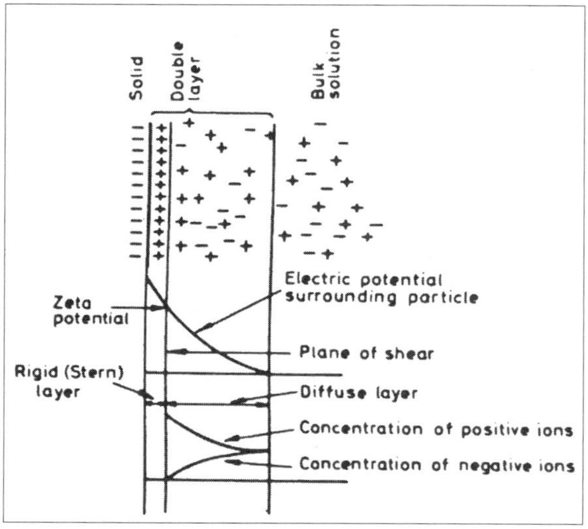

Electrical double layer model

Electroacoustic belt press. A **belt press** marketed by Simon Hartley, and used to **dewater** wastewater sludge. The feed **sludge** is fed between two polymer belts, which pass around a drum and pass close to an ultrasonic transducer. The drum forms an anode, and outside the outer polymer belt is a **wire mesh** belt that forms a **cathode**, which is earthed. The belts move slowly and electro-osmotic **dewatering** occurs in the early stages, but this becomes less effective when the **cake** dries out, at which stage the ultrasonic field takes over.

Electrocoagulation. Fine **particles** carrying a surface charge form a stable **suspension** and are difficult to **agglomerate**, and so settle out of suspension. If the suspension is subjected to an electric field between two electrodes, then the particles will move to the appropriately charged electrode, and the charge on the particle is there neutralised. The particles are now able to agglomerate by the attractive effects of **van der Waal's forces**. This method of **destabilisation** is of value when the addition of chemical **flocculation** agents is not feasible.

The term is also used for systems in which ions are produced by electrolysis with a suitable electrode, so as to neutralise the charges.

Electrodecantation. In electrodecantation, a feed slurry containing **colloidal particles** is fed between vertical **membranes** in a tank that has electrodes at each end. Negatively charged particles are attracted to the **anode** and form a layer on the intervening membrane; similarly for positively charged particles. When the polarity of the electrodes is reversed, the solids sink to the bottom and can be withdrawn as a thickened slurry, leaving clear liquid to be collected from the top. There can be several membranes in a tank, which can be separated by as little as 1 mm. The system is used for concentrating latex, but is difficult to control.

Electrodialysis. Dialysis under the influence of an electric field. If a **colloidal suspension** containing unwanted salts is to be separated by dialysis, then the application of the electric field will cause the dissociated ions from the electrolytes to move rapidly to the electrodes, and so help to purify the suspension.

Electrofiltration. The application of an electric field to a **suspension** of **particles** in a liquid causes a number of electro-kinetic effects to occur. The charged particles in the suspension will move towards the appropriately charged electrode, and a separation is thus achieved between positively and negatively charged species (or between charged and uncharged).

If the charged particles move through a suspension, then the process is **electrophoresis**. If components of a liquid mixture move through a solid by diffusion in an electric field, then the process is **electro-osmosis**. Electrically aided **filtration** and **separation** processes are currently the subject of considerable research and development effort.
[*See also Electric fields*]

Electroflotation. A version of **flotation** differing only in the method of formation of the air bubbles. In **dissolved air flotation**, for example, the bubbles are generated by a reduction in pressure of a solution of air in water. In electroflotation, the bubbles are generated electrolytically, by the passage of a direct current through the water. This produces very small bubbles of hydrogen and oxygen, which attach easily to small particles in suspension in the water.

Electrokinetic filtration. A term sometimes used for a **filtration** process enhanced by **electrophoresis**.

Electrokinetic potential. [*See Zeta potential*]

Electrolysis. The decomposition of an electrically conducting liquid by the passage through it of an electric current. If an electric field is applied to such a liquid, the charged ions move to the electrodes (**cations** to **cathode**, **anions** to **anode**) and are there discharged to form neutral entities, the current being carried by the moving ions. The best known electrolytic processes are the splitting of water into hydrogen and oxygen, and the production of chlorine and caustic soda by the electrolysis of sodium chloride brine.

Electrolyte. Any liquid that is electrically conductive, by virtue of its chemical nature, or because it has ionisable solutes dissolved in it.

Electrolytic cleaning. A method of cleaning conducting **cross-flow filter elements** by generating gas bubbles by electrolysis.

Electro-osmosis. A term most commonly used to describe the movement of a liquid through a **membrane** under the influence of an electric field. However, it also refers to the process of dehydration of porous solids under the influence of an electric field, in which the ions caused to move by the electric field carry molecules of water with them as hydrate shells around the ion. In this way, water is removed from the porous solid and deposited around the electrodes (a process sometimes called electro-osmotic **dewatering**).

Electrophoresis. The migration of charged particles, colloidal particles or ions, through a solution, under the influence of an electric field. The process is pH sensitive, as movement will cease at the **isoelectric point**. Electrophoresis can be used as an analytical tool, and also for **separation**: if the product goes to the **cathode**, the process is called cataphoresis, and if to the **anode**, anophoresis.

Electrosedimentation. An **electrophoretic** process for enhancing **sedimentation**. The performance of a **gravity settler** can be improved by making the bottom an **anode** and putting a perforated **cathode** near the feedwell. The process can be used for **clarification** or **thickening**. In the clarification zone, sedimentation is accelerated by electrophoretic forces. In the thickening zone, the concentration of the thickened **sludge** is increased by **electro-osmotic** forces.

Electrostatic filter. A **filter** employing a **filter medium** whose **fibres** carry a permanent electrostatic charge. This enables it to be very effective in capturing dust particles, which themselves normally carry a charge. Such filters are much used for **respirators**.

The term is also used of a particular filter used for **clarification** and operating by **dielectrophoresis**.

Electrostatic precipitator. A major class of equipment used to separate dust from **suspension** in a gas, although the mechanism by which this is done is very different from the equipment covered elsewhere in this book. An ESP works by creating an electric charge on the dust particles, which are then attracted to the appropriately charged electrode of the separator, where they are neutralised, and fall into a collection chamber.

Electrostatic repulsion. The stable **electrostatic charges** built up on solid particles stay with them when they enter a **suspension** in a liquid, the more

Electrostatic precipitator

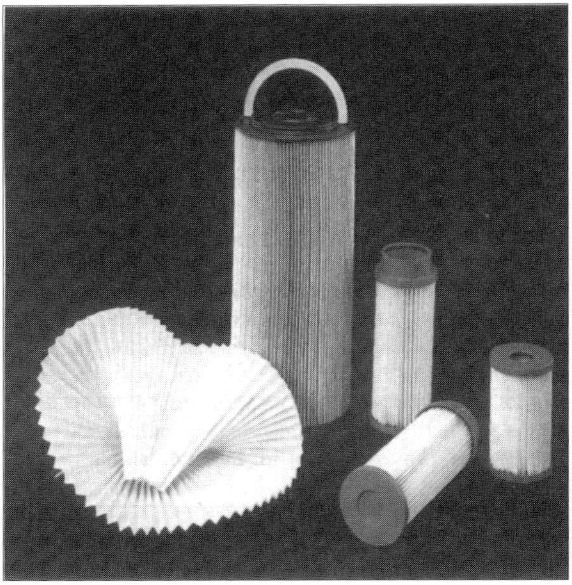

Pleated filter elements

so the less conductive that liquid. If the particles are of the same chemical nature, then the charges will be of similar sign, and the particles will therefore repel one another. The electrostatic forces are greater than the naturally attractive **van der Waal's** forces, and so the suspension is made stable (i.e. the particles are prevented from **agglomerating**), until some **destabilising** action is taken.

Element. In **separation** terminology, the filter element is the unit containing the **filter medium**, through or in which the separation occurs. Most usually, it refers to the easily removable elements such as **bags, cartridges** or **panels**, which can quickly be changed when fully loaded with separated solids. The element may have any one of a number of shapes or structures, designed to fit in a (usually standardised) **housing**, from which it is removable for cleaning (or disposal).

Elongation. A term employed in the characterisation of materials, involving the determination of tensile strength, given by the strain on a material at failure.

Elution. The removal of an **absorbed** material from a **packed** bed by a fresh **solvent** stream, as done in **chromatographic separations** or **gel filtration**.

Elutriation. A dry **classification** process in which the smallest and/or least dense particles in a mixture of particulate solids are carried away from the bulk by a stream of air, leaving the larger and/or most dense particles behind. The air flow should be sufficient to fluidise the bulk, to ensure that no particles are trapped, and then a figure higher again than that to carry away the lighter fraction. This would normally be a continuous process, with the solids carried on a perforated belt, which enable fluidising and elutriating air to enter the mass from underneath it. It may also be carried out in a series of elutriators of decreasing diameter, which, with a constant flow of gas, will result in **separation** of larger **fractions** in each successive stage.

Although predominately a dry process, elutriation can also be effected in a liquid flow, which may be necessary for dense particles. The **hydraulic classifier** works basically by elutriation.

Empty pipe velocity. A great deal of mechanical **separation** takes place in vessels filled or divided by the **filter medium**. Whilst it is usually very important to know the fluid velocity through the **pores** of the filter medium, it is often equally important to know the fluid velocity just before the medium, the empty pipe velocity – and this is usually more easily determined, experimentally or from the known system design parameters. Upstream of the filter, the empty pipe velocity gives the approach velocity at the surface of the medium, and this may need to be limited for practical reasons.

The average velocity in the pores of the medium can be calculated by dividing the empty pipe velocity by the porosity of the medium.

Emulsion. A **dispersion** of one liquid, as droplets, in a continuous mass of another liquid, with which it is immiscible (such as oil and water). The dispersion is stable because of surface effects at each droplet, although the emulsion can be broken by passage through a **coalescing filter**.

Many commercial products are stable emulsions, and their composition is such as to ensure the continuing stability of the emulsion, because an emulsion stabilizer has been added.

Emulsion breaking, Emulsion separation. If the two liquids forming an **emulsion** are truly immiscible, then the emulsion can be broken by passage over a mass of material of large surface area, usually in the form of a **knitted wire** or **wire mesh filter medium**, thick in the direction of emulsion flow. The droplets attach to the solid surface, where they coalesce to form larger drops, which eventually grow large enough to drop off the packing and thus separate from the continuous phase.

An emulsion can also be broken chemically, by the addition of a material that has the opposite effect from an emulsion stabilizer. Metal salts are commonly used for **destabilisation**.
[*See also Coalescer*]

Emulsion fractionation. The **foam fractionation** process of **separation** by **flotation**, using liquid droplets instead of gas.

Enclosed filter press. [*See Pressure plate filter*]

Enclosure. A cover fitted to a **gravity** or **vacuum filter** that is leak tight and so isolates the contents from the atmosphere.

Endcap. The fitting at one or both ends of a filter **cartridge**. Refers in particular to the shallow annular dish fitted over the ends of a cylinder of **pleated medium**, and bonded to the pleats with adhesive, so as to form a pleated cartridge. As also with the material of the core, so it is now good that the material of the cap should be the same as that of the **filter medium**, to make disposal of used cartridges easier.

End-flow filter. A particular design of **rotary drum filter**, which does not have internal drainlines. The liquid flows, instead, to the end of the drum under the drainage deck. The end of each **cell** or compartment has a collecting channel, which is connected externally to the rotary valve. This is advantageous in corrosive liquid applications, because it allows the drum to be completely sealed.

End-load rating. It is normal in some **filter cartridge housings** for a replacement **element** to be screwed on to a connecting thread. The end-load rating describes the maximum axial force (torque) that can be applied to the end of the element, without causing permanent deformation of the cartridge, or failure of its seals.

End-of-line strainer. A **strainer** fitted at the delivery end of a pipe, in order to remove **tramp material** from a liquid product.

Enhanced settling area. A term used to describe the increase in **effective area** of a **sedimentation** process when it is carried out over an inclined surface – such as in a **lamella separator**.

En masse settling. [*See Zone settling*]

Entrainment. The picking up of dust particles or liquid droplets by a moving fluid, followed by their transport to another part of a system. This is exactly what is done in pneumatic conveying, for example, but the normal objective of **separation** equipment is its opposite, so it is something to be avoided as far as possible, in the design of separation systems.

Entry cross section. The area of the **feed zone** of any item of **separation** equipment, measured perpendicularly to the direction of flow of the suspension.

Epicyclic gear box. Probably the most common form of gear box, used to drive the **conveyor** of a **decanter centrifuge** at a slightly different speed from the **bowl**. It has two or three stages, each consisting of a "sun" gearwheel surrounded by two or three "planet" gearwheels, running on a toothed track on the inner wall of the gearbox. The planet wheels are mounted on a carrier plate to which is centrally mounted the pinion shaft for the sun wheel of the next stage or to engage the conveyor itself.

Equalizing line. A pipe connecting two parts of any item of equipment, so as to equalize the pressure (or vacuum) at the two points.

Equilibrium flux. A feature of **membrane** separations, in which system design and operation are organised to achieve a steady flow of **permeate**, rather than a state of continuing decline. Various equilibrium flux models are in use to describe membrane filtration processes.

Equilibrium loading. The maximum **dirt holding** capacity of a **filter** or piece of **filter medium**, for a given temperature, pressure and **suspension** concentration.

Equilibrium moisture. Once a **filter cake** is fully formed then the next stage in a **cake filtration** process is to empty the pores in the cake of residual liquid as far as possible, by gravity or centrifugal **drainage**, by blowing with gas, or by **expression**. The cake is **dewatered** in this way until a point is reached when the dewatering forces are balanced by the **capillary forces** retaining the liquid in the **pores**. The residual liquid in the pores at this point is termed the equilibrium moisture (content), or equilibrium **saturation**.

The removal of residual moisture will slow down considerably as the equilibrium moisture level is approached, and achievement of the equilibrium moisture content may take far too long to be economic in the filtration process, especially if the cake is to be washed after dewatering.

A similar principle of equilibrium moisture relates to **sedimenting** separations, but here the minimum moisture content is that corresponding to the maximum **compression** of the sediment, with the minimal pore volume still being full of liquid.

Equilibrium porosity. The **porosity** achieved in a **filter cake**, after compressing it for a long time.

Equilibrium saturation. [*See Equilibrium moisture*]

Equiprobable cut size. The **cut size** in a **classification** process defined as that particle size at which the **grade efficiency** is 50%.

Equivalent cake thickness. The ideal **cake filtration** process would lay the separated solid down on the surface of an infinitely thin **filter medium**, which would present a vanishingly small resistance to fluid flow. However, all real filter media have a finite thickness, with a finite value of flow resistance, which is increased by the blockage of the pores at the upstream surface by the first particles of separated solid formed on it.

The flow resistance of the filter medium can be converted to an equivalent cake thickness, from consideration of relative permeabilities or resistances:

$$h_m = p_c R_m = R_m / r_c$$

where h_m = equivalent cake thickness
p_c = specific cake permeability
R_m = filter medium resistance
r_c = specific cake resistance.

Equivalent clarifying area. The major separational parameter of an item of **gravity sedimentation** equipment is its area in a horizontal plane, which determines its flow capacity. In a **sedimenting centrifuge**, the equivalent separating force to gravity is the **centrifugal force**, which may be many times that of gravity. It has become the practice, therefore, to express the sedimenting characteristics of a centrifuge in terms of its equivalent clarifying area, defined by the factor **Sigma**, derived from the application of **Stokes' law**. One of the definitions of Sigma is that it is the ratio of the volumetric feed flow rate to the centrifuge to the Stokes settling velocity of the typical particle in the slurry:

$$\Sigma = Q_f / v_s$$

From this relation it can be seen that for different sizes of centrifuge treating the same slurry, the Sigma value of each centrifuge gives the possible feed rate, since v_s is constant.

In area terms, Sigma is the settling area of the centrifuge multiplied by its **C-value**.
[*See also Sigma theory*]

Equivalent diameter. It is easy to define the properties of a **suspension** if every particle in that suspension is a perfect sphere. However, this is very rarely the case with real particles, except for specially prepared experimental suspensions of glass **ballotini**, for example. Nevertheless, it is of great benefit for many separational problems if some term can be derived that describes the behaviour of a mass of particles as if they were spheres, and that is their equivalent diameter. This term usually has to be determined experimentally, for example by allowing the particles to settle under the influence of gravity, such that their **settling velocity** can be measured, **Stokes' law** applied and the diameter determined for the sphere that would have the same settling velocity.

Equivalent pore size. As with the shape of real particles, so with real **pores** in a porous mass: very few of them will have a circular cross-section, and some will be widely divergent from a circle in shape. Nevertheless it is often convenient to treat all pores as though they are circular, and so to calculate their equivalent pore size as equal to the internal diameter from this shape. Indeed, most methods of determining pore size, such as the bubble point test, can only give a pore size on this basis.

Equivalent sphere. [*See Equivalent diameter*]

Etched filter medium. A **filter medium** made by the chemical or electrochemical erosion of apertures through a sheet of metal (usually) or plastic.
[*See also Veco screens*]

Euler number. A **dimensionless number** used to characterise the resistance to flow in a **hydrocyclone**. It is a pressure loss factor, based on the static pressure drop across the cyclone.

EUROVENT. A system of classification of **air filters** designed for living and working space ventilation. It has the same 17 classes as does CEN EN779.
[*See also CEN*]

Evaporation. The removal of a liquid from a mixture with a solid or another liquid, by converting it to a **vapour** and carrying that vapour away in a stream of incondensable gas. This is a major **separation** process in industry, and has its relevance in **filtration** in the removal of a liquid from a porous mass, such as a **filter cake**. It is then the basis for the **dewatering** of the cake by gas blowing, when the gas is hot, or when the blowing time is long.

Exhaust gas filtration. A major **filtration** application, strongly influenced by the fact that gases leaving as exhausts (from engines or industrial processes) are usually both dusty and hot. In many cases these are hot enough to justify an attempt at heat recovery – a process made much more difficult by the dust content of the gases. However, any attempt to filter the gases prior to heat exchange is frustrated by the high temperature of the gas. This vicious cycle is broken by the availability of **ceramic filter elements**, able to resist the high temperatures and achieve a high filtration efficiency.

Exhaust hood. In industrial processes that generate dusts or emit fumes, it is sensible to fit a collection hood over them, with a flow of air into the hood, which can carry the dusts or fumes away to subsequent processing, usually by **filtration**. The hood reduces the flow of air to be treated (since otherwise the whole room air would have to be treated) and protects the workers from the effects of the contaminants.

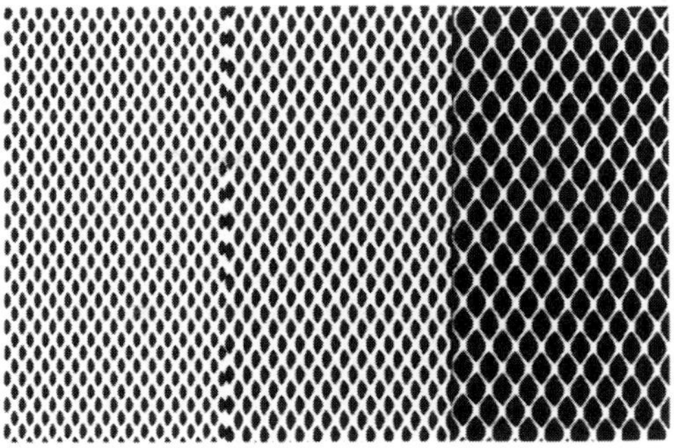

Expanded metal mesh

Expanded metal. A perforated material made by pressing a regular set of slits in a metal sheet, which are then opened to the required extent by pulling the metal sheet in a direction perpendicular to that of the slits. The expanded sheet is then suitable as a coarse **screen**, of regular sized apertures, or as a support material for other finer **filter media**. The sheet may be left in its extended form, which has a three dimensional structure, or it may be rolled flat.

Expanded perlite. [*See Perlite*]

Expression. The squeezing of a **porous** mass, by means of mechanical pressure, in order to displace the liquid held in its interior. This may be the major **separation** process, using, for example, a **screw press**, or it may be a final stage in **cake filtration**, wherein the residual moisture in the cake is removed. It is the main **dewatering** process in **variable chamber presses** and in the **tube press**.

Where expression is the main process, then this term is used (rather than **compression**) – the feed is solid (e.g. seeds or nuts) and there is no **cake formation** stage.

Expressor press. A **belt press**, made by Eimco, in which a pair of belts, carrying an already partially dewatered **cake** between them, passes around a large diameter drum. Outside the belts is a series of rollers, pressed against the belts, with the ability to increase the pressing force of each roller as required.

External discharge. A term sometimes used for any method of discharge of accumulated **cake** that requires the filtering **element** to be removed from the **filter**.

Extractables. The perfect **filter medium** would, among other valuable properties, have no impact whatsoever on the fluid system being filtered. One possible impact would be the dissolution from the material of the medium of some component of it that is soluble in the liquid being filtered. This would be particularly deleterious if the purity of the **filtrate** is an important requirement. Accordingly, manufacturers of filter media take great pains to ensure that the extractables content of their material is vanishingly small for as many liquids as possible, and otherwise state for which liquids their media are unsuitable.

Extraction. A term used sometimes for the **leaching** of soluble materials out of a mass of solid, particularly for the recovery of oils and juices from seeds, nuts, fruit and vegetables.
[*See also Liquid extraction*]

Extraction battery. A series of **percolators** operating with **countercurrent** flow of **leaching** agent.

Extractor. A term sometimes used in reference to a horizontal **belt vacuum filter** which, because of its horizontal orientation, is particularly suited to **cake washing** or even **extraction**.

Can also refer to any piece of equipment in which **leaching** or **liquid extraction** is taking place.

Extruded mesh. A continuous plastic **mesh** made by extrusion from special rotating dies in which the extruding filament is successively brought together with the adjoining filament to form a node in the mesh, and then separated to form the aperture. The system can produce filaments, and hence meshes, of very different sizes, and can be made as a rigid cylinder or a delicate sheet. It has very wide uses, in a range of applications, its filtration uses including **cores** for **cartridges**, and as a support layer for sheet **filter media**.

Extruded plastic mesh

Extrusion. The forcing of a molten solid through a hole, the die, of specific shape, so as to form a continuous filament whose cross-sectional shape is that of the die. This technology forms the basis of the production of major components of the **filter media** industry, extruding synthetic polymers, glass, metal and even carbon.

It is, of course, vital that the extrusion process be smoothly continuous, with no occluded gas bubbles, and especially with no solid contaminants that could block the die. The **filtration** of molten solids, especially synthetic polymers, ahead of extrusion processes is thus an important application for fine filtration, making severe demands on the medium used.

Eye. An aperture in the **plate** or **frame** of a **chamber filter** for entry of feed slurry or wash liquid or exit of filtrate.

F

Fabric. A generic term for a relatively soft, relatively flexible sheet material used as a major component of **filter media**. Fabrics are made from **fibres** or **filaments**, and are either interlaced (**woven**, **knitted**, lace, braid) or non-interlaced (**felt** and **needlefelt**).

Fabric collector. [*See Fabric filter*]

Fabric filter. Despite the general meaning of "**fabric**", the term "fabric filter" is employed specifically to refer to the large **filter** housings used for the cleaning of exhaust gases, in places like power stations or industrial furnaces, also known as bag houses. These are fitted with large numbers of filter **elements**, such as **filter bags** or **cartridges**, and are used to remove dust from gases before the latter are discharged to the atmosphere. The flow of dirty gas may be from inside outwards (when the dust collects on the inside of the bag or **sleeve**, and may be shaken or blown off by a reverse flow) or towards the inside (when the cake forms outside the element it is usually blown off by a **reverse pulse** of clean gas).

Fabric weight. A bulk property of any **fabric**, and, in particular, of fabrics used as **filter media**, expressed as the weight of the material per unit area (e.g. g/m^2).

Face plate. The end plate of a **filter**, especially a **filter press**, machined true to ensure the alignment of the other plates in the filter.

Face velocity. The velocity of a fluid as it approaches a **filter medium**, but before entering the medium. A common term in **gas cleaning**, but also used of liquids.

Fanning friction factor. A **dimensionless group** concerned in the study of incompressible fluid flow in pipes and channels. The friction factor relates the frictional pressure loss per unit length of pipe ($\Delta P/L$) to the pipe diameter, the fluid density, and the average fluid velocity:

$$\Delta P/L = 2f\rho V^2/D$$

where f = Fanning friction factor
ρ = fluid density
V = average fluid velocity through the pipe
D = pipe diameter

Fatigue. The progressive decay of a property of a material in use, especially when that use involves a cyclic operation.

Fatigue test. A test, used particularly for **filter media** to be used for **clarification**, in which the medium is subjected to repeated start-stop cycles or pressure fluctuations to determine whether it will retain its integrity in operation.

F curve. [*See Cumulative wash curve*]

Feed, Feed material. General terms for the **suspension** being fed to any piece of **separation** equipment, prior to its treatment.

Feed and bleed. A term descriptive of a method of operating **cross-flow filters** by recirculating the bulk of the **retentate** and bleeding off only a small fraction as **permeate**, which is then made up with fresh **feed**.

Feed chamber. [*See Feed zone*]

Feed cross-section. The open area of the **feed zone** of any item of **separation** equipment, in a direction perpendicular to the main **feed** flow direction. This cross-section may be the narrowest part of the flow system, and therefore may be the critical point in deciding the capacity of the equipment.

Feed nozzle. In a **decanter centrifuge** a short tube, the feed nozzle, often lined with erosion protection material, may be fitted to each exit from the **feed zone**.

Feed systems. The mechanical system for introducing **feed** into an item of **separation** equipment, which may include the holding tank for the suspension, the feed pump and the appropriate pipelines, valves and flow control instrumentation.

Feed tube. Any pipe that carries feed from the outside to the **separation** zone of the equipment. Specifically refers to a tube running from outside the **bowl**, along the axis of the **conveyor hub** of a **decanter centrifuge**, which discharges the feed into the **feed zone** of the decanter.

Feed well. The central part of a **gravity settler** through which the **feed** is introduced.

Feed zone. The region in any item of **separation** equipment that receives the **feed** from outside the equipment and then distributes it to the active areas of the equipment. Specifically, in a sedimentation **centrifuge** (such as a **disc stack separator** or a **decanter centrifuge**) it is the region in which the feed slurry is accelerated up to the rotational speed of the centrifuge **bowl**.

Felt. Probably the earliest fabric made by man, felts are produced from natural and synthetic **fibres**. The natural fibres, especially wool, but including cotton, jute and linen, are rough on their surface and so are able to hold together when placed in contact, without any bonding agent. Synthetic fibres on the other hand, having been made by **extrusion**, will be smooth surfaced and so will not easily hold together, and need to be bonded, thermally or by use of a bonding adhesive.

The manufacture begins with a bundle of fibres that is opened up and fed to a **carding** process, which pulls the mass of fibre into thin layers (**webs**) of fibres that are generally oriented in the direction of the carding action. These webs are usually then piled one on top of another, with the fibre orientations at right angles in alternate webs. The required felt thickness is achieved by adding enough webs to the pile, and then compressing the resulting **batt**, through heated rollers, after addition of any necessary adhesive. The area of felt produced is dictated by the size of the original carded web.

The main problem of a simple felt as a **filter medium** is that its tensile strength is low, and that fibres easily become detached from the felt and are **shed** into the **filtrate**. The tensile strength is increased by interlaying a very coarse net (a **scrim**) among the webs, while the shedding problem is reduced by imparting mechanical strength to the felt by entangling the fibres, usually by needling.
[*See also Hydroentanglement, Needlefelt*]

Feret diameter. A statistical dimension used in **particle size analysis**. It is measured from a profile of the particle, as the distance between two tangents on opposite sides of the particle, parallel to some given direction (usually vertical).

Ferric chloride. An inorganic **flocculant**, of increasing importance in the treatment of waste waters, by comparison with alum or polymeric flocculants, being more effective both as a **coagulant** and for the removal of odour-inducing sulphides.

Ferromagnetic. Materials that are strongly **paramagnetic**, and which retain their magnetism when removed from the magnetic field, are said to be ferromagnetic. They are used in permanent magnets.

Ferrous sulphate. [*See Copperas*]

Fibre. A particle that is very long in relation to its diameter, the diameter being very small (measured in hundredths of a metre, if not micrometers). It is the basic component of many textiles, and comes either from natural sources (**staples** such as **wool** or **cotton**), or from the cutting up of a continuous **extrusion**. Fibres can be used *en masse* as a **felt**, or spun into **yarn** – for which latter use the fibre should have adequate strength, pliability and cohesiveness. Among natural fibres, **cellulose** from softwood trees is the shortest, and some wools are the longest.
[*See also Filament*]

Fibre migration. The escape of loose **fibres** from a fibrous **filter medium** into the **filtrate** downstream of the medium, thereby contaminating the filtrate. A situation to be avoided as far as possible by careful manufacture and selection of filter media.
[*See also Shedding*]

Fibrillated fibre, filament. Any natural fibre (wool, cotton, etc) whose microstructure includes the protrusion from the surface of the fibre of **fibrils** that enable it bond naturally in **felts**. Also refers to any synthetic fibre or filament that has been processed to raise fibrils on its surface.

Fibrillated film. A perforated sheet material made by embossing a pattern on an impervious sheet and then stretching the film to cause the indentations to rupture.

Fibrillated yarn. A **multifilament** (or sometimes **monofilament**) **yarn** in which the surface of the **filaments** has been roughened to produce short **fibrils**. This gives the yarn some of the properties of a **staple** yarn.

Fibrillation. The processing of fibrous, filamentous or flat sheet material to create a very fine structure of open area and microfibrous protrusions, to give an effectively much smaller diameter material.

Fibrils. The very short, and extremely fine "hairs" on a natural **fibre**, or produced artificially on a synthetic fibre. The interlocking of fibrils of wool or cotton enables these materials to form coherent **felts** with little processing other than laying them in a pile and pressing them together.

Fibrotex filter. A novel **clarification filter**, developed within the research arm of BP, which consists of a cylindrical bundle of long **fibres** or **yarns**, held at each end in a solid plate, which can be rotated about the axis of the bundle. When the plates are fully turned (in opposite directions) to squeeze or tighten the bundle, the latter forms an excellent **depth filtration medium**, with the liquid flow perpendicular to the threads. When the bundle is fully loaded with dirt, the plates are twisted in the opposite direction, to open the spaces between the threads, so that the bundle can be washed free of accumulated solids.

Fibrous. Any material made from **fibres**, natural or synthetic, whether they be packed together in random fashion or woven into a **yarn**, is described as fibrous.

Ficks law. A relationship used in the study of concentration-driven **membrane processes**. It states that **permeate flux** is proportional to diffusivity times the concentration gradient.

Filament. A single continuous strand of indefinite length. **Silk** is the only natural filament of interest, but any **thermoplastic** material can be extruded into a filament, which can then be used unbroken, or broken into smaller lengths. Most synthetic filaments are smooth surfaced, and lack the natural cohesiveness of the **staple fibres**. This definition also covers extrusions of molten metal, i.e. **wires**, but the term is rarely used of metallic filaments.

Filling thread. [*See Weft*]

Fill wires. The **wires** in a **woven wire cloth** that run the short way across the loom, equivalent to **weft** threads in a woven cloth.

Film flow model. A model of the flow in the **dewatering** stage of a **cake** in a **centrifugal filter**. It applies especially to thick cakes, and cakes formed from coarse particles. The model assumes an initial **plug-flow** regime in the **drainage** process, followed by film flow across the surfaces of the particles.

Film theory model. A model used to describe the behaviour close to the upstream side of a **membrane** in any membrane **separation** process, consideration being given to the concentration gradient formed at the surface.

Filter. *See* pages 104–6

Filterability. In essence, the filterability of a **suspension** is a measure of how easy or difficult it is to filter, by whatever type of **filter** is being considered. It can be measured in a number of ways, a key method being the determination of the specific filter **cake resistance**, by, for example, the **t/V=f(V) method**.

Filterability index. [*See Boucher filterability index*]

Filterability number. A parameter used in **deep bed filtration** to characterise a **granular medium filter**. It is defined as head loss times average **filtrate** quality, divided by **feed** quality times velocity multiplied by time.

Filter aid. A specific and very important type of **filter powder** that is used to improve the **filterability** of a **suspension**. Sometimes included under this heading are the materials used to provide a **precoat** layer directly on the **filter medium**, at the beginning of the **filtration** process, to start the formation of the filter cake, without any loss of suspended solids through the **filter** (which might mean loss of valuable solids, but more likely prevents contamination of the **filtrate**). The term should not be used for those chemicals added to the suspension to **condition** it

More correctly, the term refers to those body aids (or body feed) that are added continuously to the **feed zone** of a filter (or in a prior mixing stage) to provide extra bulk in the **filter cake**, which makes the cake more permeable, and also helps it to trap any fine particles in the suspension. Materials commonly used

Filter. A filter is basically a device for holding a piece of **filter medium** across a fluid flow, well enough sealed such that all of the fluid must flow through the medium, in so doing depositing most or all of any suspended solid **particles** or liquid **droplets** on the medium. In this way, the fluid is cleaned of its **contaminants**, or a valuable fraction of the suspended material is recovered.

In its simplest form, a filter need be little more than a frame holding the filter medium across a flow line for a fluid. At the other extreme, a filter may be an exceedingly complicated machine, with many functions other than just **filtration**, whose complexity largely lies in the way the filter is designed to enable the removal of collected solids, especially when the filter is operating under super-atmospheric pressure.

Essentially, then, a filter is a device for allowing filtration to occur, in a number of different ways, and is quite distinct from the types of equipment used for that other major mode of **separation**, **sedimentation**.

There are several ways in which filters may be classified: by purpose or function, by nature of the fluid filtered or by concentration of the suspension to be filtered, or by the type of machine or mechanism of filtration involved. The discussion that follows is largely based on machine type, with some input from the nature of the filter medium and the filtration mechanism.

Filter cartridges

The first group to be considered is that which is perhaps the closest to real **surface filtration** in practice, namely the range of **screens**, which can operate wet or dry, and in the latter case actually have no fluid passage through the filter medium at all. Screens have two main functions: to remove larger particles, of **tramp materials**, from a liquid flow, and to separate a mixture of particles into different fractions according to their mean particle size (it being this second purpose which can be operated with or without fluid flow through the screen). On the whole, screening is a fairly coarse process, and the equipment used may be stationary or moving. Stationary screens include:

- in-line **strainers,**
- horizontal or slightly inclined,
- curved ("sieve-bend"), and
- vertical,

while moving screens include:

- continuous (vertical, or rotating drum/trommel), and
- oscillating (**vibrating** or **gyratory**).

Sensibly included with the screen group are the **demisters**, which operate in a similar manner, but also have an element of **depth filtration** to them, in the removal of liquid droplets from a gas flow, by entrapment of the droplets in a mass of wire followed by coalescence of the liquid.

A very different mechanism is involved with the **separators** that work by mechanical pressure, even though some of them use a screen as the filter medium. The group includes the **band press**, either horizontal or vertical, and the batch operated **expression presses** and the continuous **screw press**.

Depth filtration characterises the next group of filters, used to clean a fluid from a relatively low concentration of suspended solid particles. It is formed of two very distinct types of equipment, one using a mass of granular solid as a filter medium, the other using a thick felt or similar material. The **deep bed** filter is mainly used in the cleaning of raw water for drinking purposes, although there is increasing interest in it as a cleaning device for hot gases, and consists of a **packed** or **moving bed** of granules, in which the dirt particles are trapped. The thick media filters are mainly used for gas cleaning, although

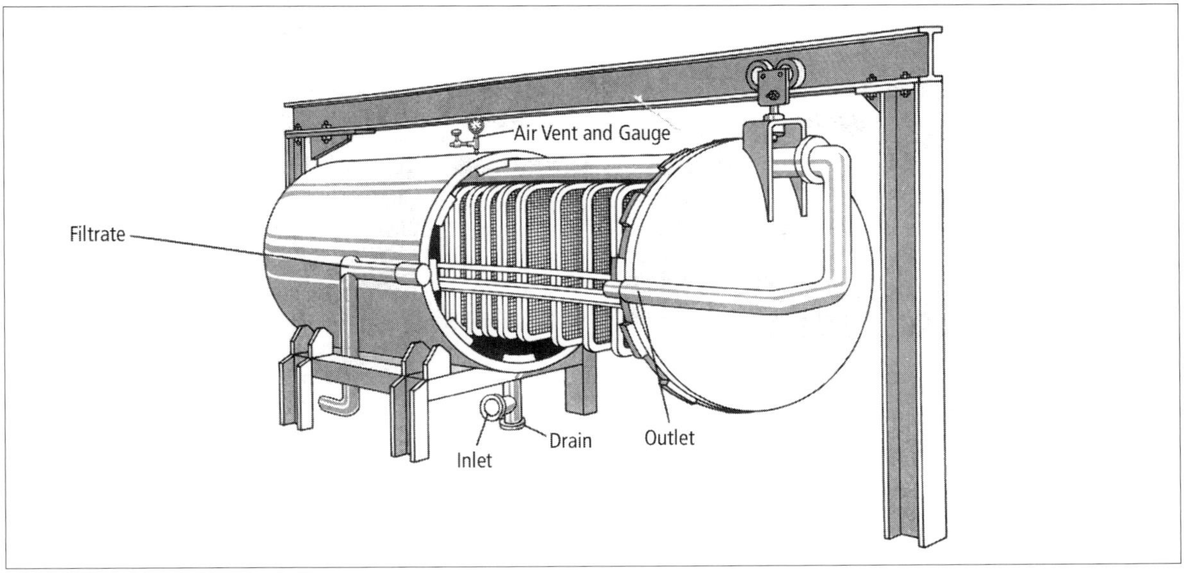

Vertical leaf filter

with a sizeable application to the removal of dirt from liquid streams as well. These include:

- **pads** and **panels** (cassettes) for air cleaning,
- **bags** and **pockets** (including bag houses),
- thick **cartridges**, mainly for liquid filtration, and
- **sheet filters** for liquids.

In the case of cartridges, inclusion here is a factor of the filter media used, while bags, especially those used in large bag houses (or **fabric filters**), can build quite thick cakes on the media surfaces.

There is next a significant group of filters operating by virtue of the hydraulic head of liquid above the filter medium, mainly where this head is greatly amplified by the use of **centrifugal force**. The simplest such filter, operating just by gravity, is the **flat-bed (roll)** filter, although even this now is often operated under vacuum. Most of the rest of the filters that are called gravity filters are actually versions of the rotating drum screen.

Centrifugal filters are mostly quite complex items of machinery, trading their higher cost against a significantly higher filtration efficiency. They are of two types, the first being batch or semi-continuous in operation, with either:

- manual discharge (by scraping or using a removable bag), or
- automatic discharge (vertical, horizontal, inclined axis, **inverting basket**).

The other type has a bed of **dewatering** solids moving across the screen, either as a

- cylindrical screen (horizontal axis – **pusher**), or a
- conical screen (horizontal or vertical axis), with free or controlled discharge.

Finally, the equipment items perhaps most commonly identifiable with the word filter are those machines that work with a differential fluid pressure head across the filter medium and growing cake, for they mainly work by **cake filtration**. The first group operate with vacuum as the driving force, and are therefore limited to a driving differential of less than 1 bar – but have the advantage that the filter cake remains available at ambient pressure for removal from the filter by relatively uncomplicated means. Vacuum filters include both batch and continuous (or effectively continuous) designs. Batch vacuum filters include:

- **nutsche** (manual or scroll discharge),
- **rotary table**,
- **tilting pan** (single or rotating/indexing), and
- **leaf**, **plate** or **tubular element**,

while the continuous types include:

- **rotary drum** (with a range of discharge mechanisms),
- **rotary disc**,
- **indexing disc**, and
- **belt** (single or multiple chambers).

Filters operating by means of above-ambient pressure include some of the most complicated filter

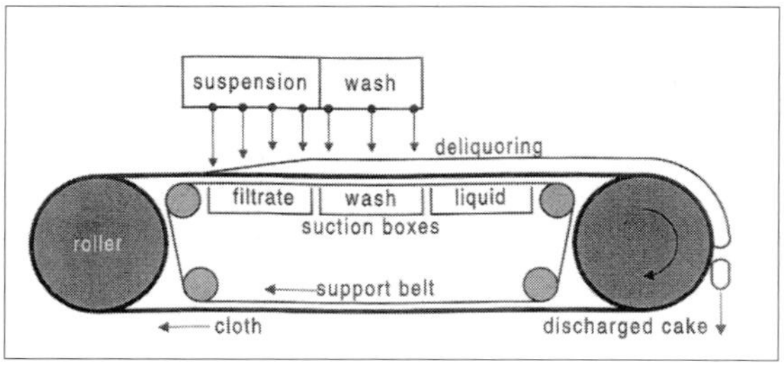

Horizontal belt filter

designs, as well as some of the simplest. They all require the relaxing of the pressure before the dirty element can be cleaned or replaced, or before the accumulated cake can be recovered (unless a very complicated solids recovery mechanism is built in). The simplest types are those filters which operate largely in-line, and which have a single filter element in a pressurised housing:

- **basket strainers**,
- **capsules**,
- bags and **sleeves**,
- cartridges (with a very wide range of filter element, including **membranes**), and
- other membrane filters (**hollow fibre**, **spiral wound**, tubular, etc).

The remaining pressure filters either have a multiplicity of **filter elements** inside a pressure vessel, or a more complex structure still. The pressure vessel group includes:

- tubular element (bag or candle filter),
- **filter coalescers**,
- flat elements (**plate** or **leaf**),
- **thickeners**, and
- enclosed rotary (drum, disc, etc).

The filter press group is a major component of the filter business, covering:

- simple **plate and frame**,
- **recessed plate**,
- **chamber press** (**diaphragm plate**, plate press),

and the developments of the filter press such as the **tower press** (with a continuous medium). Pneumatic compression of the filter cake is also a feature of the variable volume presses, such as the **tube press**.

By far the largest number of these types of filter are used for filtration from liquids, mainly because much of liquid filtration involves handling high concentration feed slurries. Gas filtration, on the other hand, largely involves the cleaning of gas streams free of relatively low solids content, for which depth filtration is mainly used. Of the types of filter listed above, gas filtration uses in-line strainers, pads and panels, bags, capsules and pockets, and thick cartridge. Fabric filters and demisters are entirely devoted to gas filtration, which is also taking some deep bed filters to itself.

The choice of filtration equipment involves selection both of the right medium, and of the best type of filter in which to mount it. The **harvesting** processes are more limited in equipment choice for a particular application (although representing the widest range of equipment types), because of the need to remove accumulated solids in relatively large quantity, often leading to the need for complex types of filter. **Clarification** duties are much less concerned about solid removal, and so can be satisfied by simpler (and, therefore, cheaper) filters, with correspondingly simpler media formats.

for this purpose are **diatomite**, **perlite** and **cellulose flour**. The filter aid may be granular in shape or fibrous.

The use of filter aids in a filtration process will almost certainly completely change the composition of the cake. They can only be used, therefore, when the composition of the cake is of little or no importance – as in the dewatering of sewage sludge, or the clarification of beverages free of organic solids. It is theoretically possible that the filter aids could be separated from the original solids, but this is unlikely to be an economic process.

Activated carbon granules can be used as a filter aid, as well as giving an **adsorptive** purifying action to the filter cake.

Filter area. The active area of **filter medium** in any **filter**, through which separation takes place, followed by any necessary **drainage** and **washing** stages. Thus the entire surface of a **drum filter** is active, with some part filtering, some draining and another washing, but a **vacuum belt filter** has more than half of its area inactive at any time, because it is being returned, below the filter, to the start of the filtration zone.

Filter bag. A very common filter **element** has the **filter medium** made into a bag-like shape, a cylinder of porous material, closed at one end with the same material, and fully open at the other. The bag is attached to its **housing** by a flange device at the open

Filter bags and supports

end, and hangs down from that fixing. It may be mounted singly, or (and much more commonly) as one of many attached to a **tube-plate**. Filter bags are used for the **filtration** of liquids and gases, although in numerical terms, many more are used for gas cleaning.

The suspension to be filtered may be fed to the inside of the bag, with the **filtrate** leaving from the outside, and the separated solids building up inside it. When the bag is full, filtration is stopped, the bag is removed and emptied, washed and replaced. In this mode the bag does not need any extra support.

Alternatively, the flow is directed from outside the bag towards its interior, in which case the bag is placed on a **cage** to support it and prevent its being squeezed shut by the feed pressure. The filtrate leaves from inside the bag, and the separated solids build up as a **cake** on the outside of the bag. When the cake has reached its optimum thickness, it is removed from the bag by a sharp reversal of the gas or liquid flow through it, so as to blow or flush the cake away, which falls to the bottom of the housing. The slightly loose fit of the bag on its cage, and the flexibility of the bag itself result in a slight expansion of the bag under the reverse flow conditions, which cracks the cake and helps in its removal.

The cleaning flow of fluid is rarely fully effective, and a little more residual solid is left on (or more correctly, in the material of) the bag. Eventually the bag has to be removed and cleaned (or possibly discarded). The reverse flow can be applied to all the bags at once, but where the application allows it, then each bag in turn can be **back-flushed**, without stopping the main filtration flow. The cake from this one bag falls to the bottom of the housing, where separated solids collect, until the whole filter is shut down and the waste solids are removed.
[*See also Candle, Pocket*]

Filter cake. The layer of separated solids that forms on the upstream surface of the **filter medium** in any **filtration** of thick **suspensions**. The first layer adjacent to the medium surface **bridges** the holes in the medium, and the rest of the filtration is through the increasing thickness of the cake layer. As the thickness of the cake increases, so does its **resistance to flow** of the fluid; this means that the flow rate will decrease if the **feed** delivery remains constant, or the pressure will have to increase to maintain the flow rate.

Depending upon the design of the **filter**, the cake will either be allowed to build up to an optimum thickness, whereupon suspension feed is stopped, the cake is drained and washed, and is then removed, or the cake is continuously formed on fresh medium in the filtration zone, and moved on to other zones in the filter for draining, washing and discharge.
[*See also Cake formation, Dewatering, Washing*]

Filter candle. [*See Candle, Candle filter*]

Filter cartridge. [*See Cartridge, Cartridge filter*]

Filter cell. Large, continuous **rotary drum** and **disc filters** usually have their filtering unit divided into a number of cells or compartments, each with its own **filter medium**, and **filtrate** drain connection. In the case of a drum, the cell is a flat rectangular pan, with several mounted in a circular array. The pan may have a plastic inlay to which the filter medium is attached.

The disc filter has cells in the shape of a segment of a circle, mounted to form the complete disc. Each segment will have a perforated layer, probably of metal, on both sides, over which the filter medium, usually in the form of a bag, will be pulled and tied at the neck. The filtrate leaves the cell through the **cell foot**.

The term is also used of a laboratory scale filter for predicting the performance of a **filter press**.

Filter cloth. A general, and now somewhat old-fashioned, term for a **filter medium**. Used usually to refer to **woven** fabrics (and possibly **nonwovens**),

which are bought as **roll goods** by the filter maker or end-user, from which the particular shape of medium is cut and fitted to the **filter**. As a filter medium, a woven cloth is able to separate in the spaces between the yarns, but also in the interstices of the yarns themselves.
[*See also Filter medium*]

Filter cloth clogging. [*See Clogging, Filter medium clogging*]

Filter coefficient. A parameter used in **deep bed filter** studies to define the performance of a **granular medium** filter.

Filter dryer. A batch **filter** set up to be able to carry out a number of processes in the one vessel. This is normally a **nutsche** filter, perhaps with a stirrer in the zone above the **filter medium**, with the capability of **washing** the filter **cake**, and then drying it thermally to produce a dry product.

The same type of equipment may also be designed to undertake a chemical reaction prior to the filtration step, such as a precipitation process, or the production of crystals. In these cases, the equipment may be referred to as a filter reactor. The more complex the processes, the more complex the equipment is likely to be, sometimes with the ability to be tilted.

The benefit of such a filter is its ability to undertake a series of processes without needing to handle the materials in between treatment stages, thus avoiding the risks of spillage, or contact with the air.

Filter housing. A working **filter** has two main components: the **filter medium** and the rest of the structure, which provides the **feed system**, and the **filtrate** and **cake** discharge mechanisms, all but the medium effectively being the filter housing. The term, however, is usually used of the simple cylindrical case for a **cartridge**, or other single **element** filter unit, which holds the element in place and has connections for feed suspension and filtrate discharge. Such a housing is easily opened for **element** change, and will normally be of a standard size and therefore able to accept elements from a variety of manufacturers.

Filtering centrifuge. [*See Centrifugal filter*]

Filter integrity. A term generally used of the mechanical correctness of a **filter** or **filter element** – such as the absence of holes, especially at seams, and the avoidance of **shedding** of the medium material.

Filter leaf. A filter leaf is formed by fixing two (normally rigid) pieces of **filter medium** close together (but not touching), and sealing their periphery. Two or more leaves are then mounted one next to another and sealed onto a central collecting tube. The whole assembly is placed in a vessel full of slurry under pressure, whereupon the **filtrate** goes through the space between the pieces of media, and into the central tube. The leaves may be held horizontally or vertically (with cake removal easier in the vertically mounted case), and they may be washed or blown free of **cake** *in situ* or lifted out of the vessel for cleaning in another place. (A vertically mounted leaf is able to collect cake on both sides, whereas a horizontal leaf will normally only accumulate solids on its upper surface – in which case a one sided leaf or **plate** is used, sometimes with a raised rim as a **tray.**)
[*See also Leaf filter*]

Filter mat. A material for use as a **filter medium**, that is usually a **non-woven felt**, in gas cleaning applications. Usually made as a long roll, cut to the appropriate shape for the **filter housing**.

Filter medium. *See* pages 109–10

Filter medium clogging. Any **filter medium** will slowly accumulate deposited solids when in use for **filtration**. If these solids collect entirely on the upstream surface, then they can, in principle, be easily removed and the medium returned to use.

Filter bag in its housing

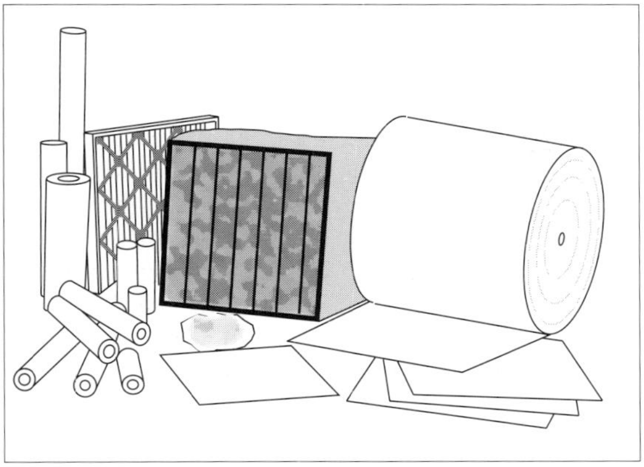

Filter medium

Filter medium. The filter medium is the most important component of any **filtration** process. If filtration is defined as the physical **separation** of one or more components from a **suspension** or **solution** in a fluid, by passage of that fluid through or across a barrier that is **permeable** only to some of these components, then the key element in this description is the barrier, permeable only to part of the feed suspension or solution. This barrier is the filter medium, and a **filter** is then any mechanical structure that holds the medium in the most efficient way.

A filter medium thus is any material that, under the operating conditions of the filter containing it, is permeable to one or more components of a mixture, solution or suspension, and is impermeable to the remaining components.

The retained components may be particles of solid, droplets of liquid, colloidal material, or molecular or ionic species in solution, while the **permeate** (or **filtrate**) will normally be the suspending fluid or solvent, possibly together with some of the other components.

The number and variety of materials embraced by this definition of filter medium are truly vast, ranging from metal plates with holes measured in centimetres, to microporous **membranes**, and from sheets of **woven cloth** to beds of sand. Filter media may be made from any material that can be rendered permeable or made into a permeable form, including organic ones, such as natural and synthetic fibres, and synthetic sheet material. Inorganic materials are also widely used, including natural minerals, carbon and charcoal, glass, metals and metal oxides, and other fired ceramic materials.

These materials can then be made up into filter media in a variety of forms: rod or bar, sheet, loosely packed or bonded fibres or granules, **wire** or **monofilament**, and so on. These in turn are used as **yarn**, fabric or **mesh**.

The range of actual filter media types includes:

- woven and knitted fabrics,
- non-woven fabrics, including felts, needlefelts and spun-laced materials,
- wet-laid materials (papers made from cellulose and other fibres),
- screens and meshes, made from metal wire and plastics, including sintered mesh,
- coarse porous sheets and tubes, made from plastic, metal, ceramics and glass,
- membranes, made from synthetic polymers and inorganic materials,
- fabricated cartridges (yarn-wound, bonded layer and metal edge), and
- deep beds of granular solids.

Clearly, not all permeable materials are necessarily usable as filter media, but certain of their inherent properties potentially enable them to be applied in this way, if they are combined with a compatible filter in an appropriate operating environment.

Neither is it necessarily true that a particular form of a material would be suitable as a filter medium in its original shape or format. For example, a loosely formed yarn of fibrous material is of little use as a filter medium, if it remains as a single yarn. However, once it is wound around a supporting **core**, with successive layers of yarn wound at an angle to the previous layer, then it becomes an excellent form of filter medium.

A successful filter medium is likely to be required to combine many different properties, ranging from its filtration characteristics and its chemical resistance to its mechanical strength, the dimensions in which it is available, and its wettability. These properties may be conveniently divided into three major categories:

- *machine-orientated* properties, such as its rigidity, strength, or fabricability, which may restrict the use of the medium to specific types of filter;
- *application-orientated* properties, such as its chemical and thermal stability, which control the compatibility of the medium with the process environment; and
- *filtration-specific* properties, which determine the ability of the medium to achieve a specified filtration task.

It is those properties of a filter medium that control its ability to undertake a filtration process that are the most important of its characteristics.

These include:

- smallest particle retained, which is an obvious and simple measure of the ability of a filter to remove particles from suspension, but is less precise than it might appear to be because of problems with particle shape;
- **retention efficiency** versus particle size, which is a relationship generally expressed in the form of a **grade efficiency curve**, from which a **cut-off point** may be determined, and the shape of which tells a lot about the nature of filtration performance, and also something about the structure of the filter medium;
- **particle shape**, which is normally assumed to be spherical, but in fact can diverge considerably from this ideal, requiring the use of **shape factors**;
- resistance to flow (of the clean medium), which depends both upon the size of the individual **pores** and on the number of pores per unit area, i.e. upon the **porosity** of the medium;
- **dirt-holding capacity**, which is an important characteristic for **depth filter** media, as it determines their operating life before cleaning or disposal;
- tendency to blind, which shows how easy it will be to free the medium from trapped particles in any cleaning process; and
- **cake discharge** characteristics, which are important for media used in cake filtration, especially those used on a continuously operating filter.

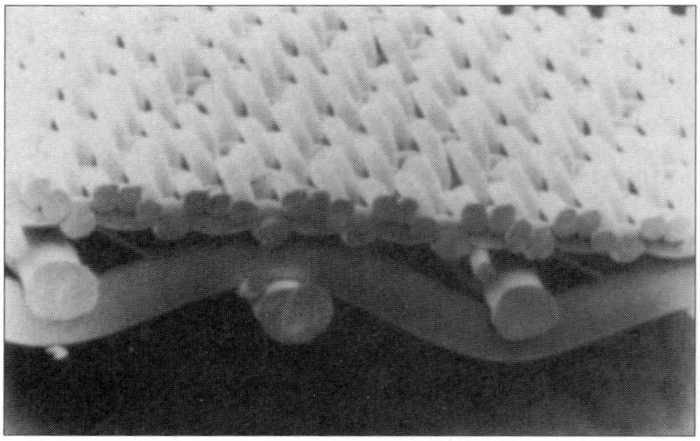

Double layer filter belt

However, any real filter medium will slowly accumulate solids within its thickness, and these may be less easy to remove, leading to slow clogging of the material, perhaps ultimately rendering it useless, so that it has to be thrown away.

Filter medium resistance. The resistance to flow through a piece of clean **filter medium**, which should then be additive with the resistance of any **cake** formed on the medium. If the medium is used for **depth filtration**, then the resistance will rise as the dirt load increases.

Filter paper. A simple **filter medium**, one of the earliest in use, made from a porous **paper**. In use every day in many households as part of the coffee making process (relying on gravity filtration), the filter paper is also a valuable item of laboratory equipment, either as a flat disc, or folded to fit into a conical funnel, and vacuum driven (usually from a water-jet pump).

Much more complex forms of paper media also exist – as **edge filter cartridges**, or in **pleated** form – for industrial and automotive applications. The quantity of paper used for filtration makes it one of the largest segments of the filter media industry – in area terms, if not so large in sales volume.

Filter papers are made from **cellulose** or **glass** fibres, with the production process and the nature of the basic fibre leading to a wide variation in **permeability** for a given thickness and weight per unit area. The strength is much less when wet, so some grades are impregnated with resin to improve the **wet strength** characteristics. A further variation involves the **creping** of the paper to give it an increased filtration area for a given diameter of filter.
[*See also Filter medium*]

Filter plate. [*See Plate*]

Filter pocket. [*See Pocket*]

Filter powders. A generic term for the various particles or powders used to assist the different **separation** processes, such as **coagulation**, **flocculation**, **filter aids**, **precoat**.

Filter press. *See* pages 111–12.

Filter rating. The various ways in which **filter** performance is measured.

Filter reactor. [*See Filter dryer*]

Filter press. The filter press is probably the most common of all of the **process filters**, and has, indeed, been called the workhorse of the filtration world.

In its simplest and earliest form, the filter press consists of a series of chambers, each formed by a pair of flush **plates** and a distance spacer, with each plate being double sided and therefore being part of two chambers (the plate and frame filter press). The plate and frame are suspended, by arms fitted integrally as part of the top of the plate and frame, from an overhead suspension beam, running on roller bearings. Alternatively, each has side arms fitted to it that run on bars along each side of the filter press.

The plates and frames are closed up together, along the support bars, between the end plates (**King** and **Queen**, or head and follower plates). They are held close enough together to provide seals between adjacent components of the filter press (although leakage from the joints is a common problem with the filter press, because the material forming the sealing gasket is part of the **filter medium**, and therefore intrinsically porous). The closure of the plates is effected by a hydraulic ram, or mechanically, or by electrically powered screws.

A sheet of suitable filter medium is hung or fitted over the plate, so as to be supported across the surface of the plate by the ridges surrounding the drainage channels on that surface. The spacer frame thus has filter media on either side of it. The slurry to be filtered is fed to the filter through appropriate channels in the fixed King (or head) plate, and flows along corresponding channels through every plate in the array. As it passes each spacer chamber it enters the space and fills it with slurry. The fluid pressure in the slurry causes **filtrate** to flow through the filter medium into the drainage channels in the flush plate, then along these channels to the filtrate exit port, which is usually at the bottom of the plate. The solids left behind accumulate in the **cake** space and gradually fill it, starting at the surface of the filter medium, and growing in thickness towards the centre of the chamber.

When the cake space is nearly full, the delivery pressure of the slurry starts to rise (or the filtrate rate to fall), giving the signal to stop filtration. If the cake is to be washed, then **wash liquor** can be introduced from the same channel as the feed (simple washing), and can then be discharged with the filtrate or separately. Alternatively, special wash ports may be used, to give flow across the thickness of the cake (**through washing**).

When the cake is washed and adequately drained, the filter press is opened from the Queen plate inwards, and the cake in each compartment is discharged into a collection trough or screw conveyor below the filter. Once every cake is discharged, and the filter media washed, then the filter is closed and the cycle is repeated.

It is possible that the filtration cycle should be stopped short of maximum cake thickness, since the pressure starts to increase well before that point, and filtrate rate may fall to an unacceptable level. The question of optimum cake thickness is an important parameter in filter press design.

The first main development from the basic plate and frame filter press saw the alternating plate and frame replaced by a pair of recessed plates. Now, each plate had a definite depth along the axis of the filter, formed by a rim protruding from its base all around the plate, so that when adjacent plates are closed up, the chamber is formed from the two recesses (the recessed plate press). The filter media are fixed to the face of the plates, so that when they are closed up, the pieces of medium butt up to one another. Now the slurry is fed into the small gap

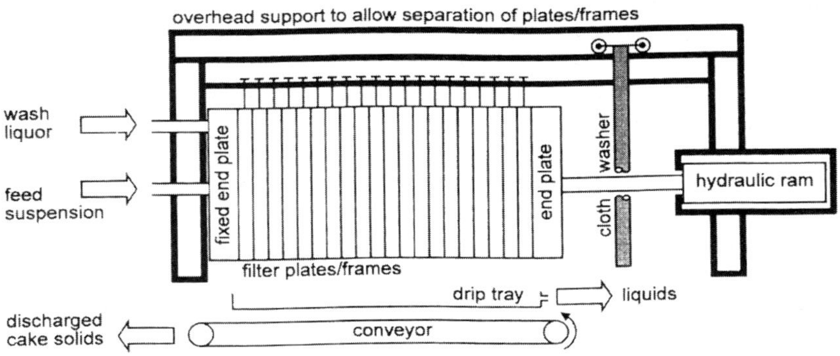

Filter press components

Simple plate-and-frame press

between the sheets of medium, and then filtrate flows through the media into the drainage spaces on the surface of the plates, and so away to the filtrate discharge port.

The filter cake is left in the space between the sheets of filter media, which slowly expands, as it fills with cake, into the recesses on either side, until it sits on the raised parts of the plates. The resultant straining of the filter media during each cycle of filtration gives a shorter medium life for the recessed plate filter, to offset against its simpler design and lower first cost.

Some degree of mechanisation of the solid discharge process has been achieved, with these versions, leading to a measure of automation in the operation of a filter press.

Operating pressures for filter presses are upwards of 7 bar. Plates, originally of wood, are now made from a variety of materials, including rubber-lined steel and moulded plastic. They are available in a range of sizes, from 15 cm square, up to 2 m square, and they are installed up to 100 or more in one filter press.

The plates are normally square in shape, with delivery and offtake ports in the centre, and/or at the corners. This gives the chance to have the feed point at the centre or at the top, as well as at the base – from which the slurry will rise to fill each compartment (there usually being some problems with cake distribution across the area of the plate). Feed from two points is also possible to alleviate the distribution problem.

Not only does the filter medium act in its separative function, but it also provides the seal between plates – not just at the outer edges, but also around every port. The sheets of media must therefore be made to quite strict tolerances, from material that will not change in shape with continued use.

The main development from the simpler versions just described has been to include an impervious elastomeric sheet in each plate compartment, which can be inflated pneumatically so as to press down on the drained cake and so dewater it by **compression** (the **diaphragm filter press**). Initially this sheet was referred to as a membrane, and the membrane filter press became quite confusingly well known. The alternative meaning of membrane as a semi-permeable material required the change to the use of the word diaphragm.

There have also been greater strides towards the fully automatic filter press, including the use of compressed air for cake dewatering.

The idea of the filter press has led to several other designs of filter employing the plate principle in this way, most notably the **tower press**, which is basically a filter press stood on one end with a continuous band of filter medium passing progressively over each plate in turn.

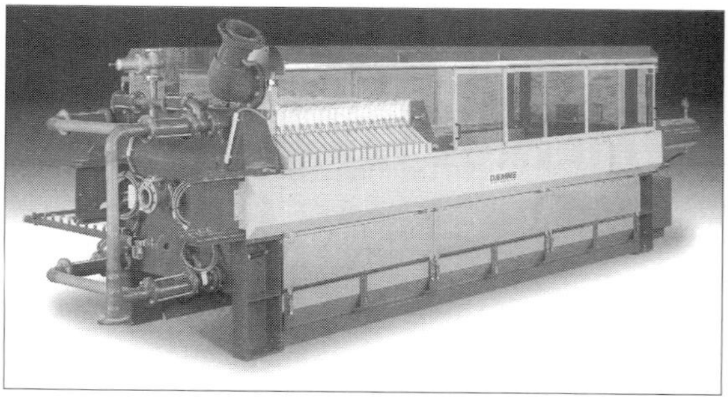

Automated filter press

Filter segment. [*See Filter cell*]

Filter sheet. [*See Sheet, Sheet filter*]

Filter system. A great many **filtration** tasks are simple, and are simply achieved. However, there are also many that are more complicated, usually because of the complexity of the **feed suspension**, but also because of the delicacy of the **separation** that is needed. Accordingly, the answer to many filtration problems requires the provision of a filter system, containing more than one item of separation equipment. These would normally be set up in series, with the coarser or more simple task undertaken first, and the more complex or more demanding stage to follow.

Filter system optimisation. Wherever there are equipment items connected together to perform a process engineering task, then there is scope for careful design in the first place, or re-engineering later, to create the optimum combination of items. Very often the development in the design or performance of a type of equipment, or the availability of a new type, makes an old equipment combination less efficient than it might be. Thus, complexity of system will always provide scope for a company's own engineers, or for external consultants, to improve on the system design, and to optimise it in terms of cost or performance.

Filter thickener. A **filter** whose purpose is to produce a more concentrated **suspension** from a less concentrated one by the removal of some of the suspending fluid. The clarity of the **filtrate** is not usually an important parameter in such a system, but the turbidity should not be too high, or solids value will be lost.

Cross-flow filtration is essentially a thickening process, because the **retentate** has to be fluid enough to flow out of the process.
[*See also Gravity thickener*]

Filter trough. The container in a **rotary filter** holding the **feed suspension** and into which the **cells** of the **drum** or **discs** dip in sequence to suck up the slurry through the **filter medium**. If there is sufficient space, then a stirrer will normally be fitted to the trough, to maintain the uniformity of the suspension. In the case of disc filters, individual troughs may be provided for each disc, the movement of which through the trough may provide adequate agitation of the slurry.

Filtrate. The liquid leaving a **filtration** process, after removal from it of some or all of its suspended solids, by passage through the **filter medium**. In theory the term applies to gas filtration as well as that of liquids, but is rarely so used. It is used specifically for the liquid leaving a **filter**, but is also the general term used to refer to the cleaned liquid leaving any **separation** process.
[*See also Centrate, Permeate*]

Filtrate collector, receiver. In simple terms this is the tank following a **filter** (or **centrifuge**) into which the **filtrate** (**centrate**) drains through the filtrate pipe system, prior to subsequent treatment or safe disposal. However, in many cases the filtrate leaves the **separation** equipment carrying considerable quantities of entrained air (especially from a centrifuge), in which case the receiver is designed to de-aerate the liquid.

Filtrate pipe. The pipe or pipe system connecting the active filtering units, such as the **cells** of a **rotary filter**, with the **filtrate collector**.

"Filtratest". A proprietary item of test equipment, made by Bokela, which is an elaborately instrumented **nutsche** filter, capable of identifying and monitoring all the stages in a **cake filtration** process.

Filtration. *See* pages 114–16

Filtration efficiency. A general term describing the efficiency of a **filtration** process, which should state the efficiency (in fractional or percentage terms) of removal of a fraction of the solids contained in the **feed suspension** by means of that process. Because most real suspensions and most real **filter media** have a range of sizes (of particle or pore diameters, respectively), then the only useful efficiency figures are those which are closely related to particle size.

Filtration pressure. In any **filtration** process, the driving force necessary to cause the fluid to pass through the **filter medium**, against the resistance of the medium (and any accumulated solid **cake**). The force can be created either by applied pressure above the suspension, by the action of gravity on the suspension above the filter medium, or by suction from below the filter medium. Overpressure can come from applied pressure in the fluid from a pump, or applied by a pressurised gas, or from the squeezing action of an impermeable membrane with a fluid pressure behind it; gravity force comes from the hydrostatic

Filtration. The process whereby solids suspended in a fluid (gas or liquid) are separated in part or totally from that fluid, by passage of the fluid through a permeable barrier, whose **porosity** is chosen to achieve the required level of separation. Together with **sedimentation**, filtration makes up the two basic mechanisms of mechanical **separation** of solid particles from fluids.

It involves a very wide range of processes, from the old saris in a drum used by the ladies of a Bangladeshi village to clean their river water, to the exceedingly complex **membrane** process used for the same basic purpose, the provision of drinking water.

Filtration involves the physical **separation** of one or more components from a **suspension** or **solution** in a fluid, by passage of that fluid through or across a barrier that is **permeable** only to some of these components. Because it involves passage *through* a barrier, filtration is almost entirely a characteristic of the size of the particle, droplet or molecule being separated, whereas relative particle density may be a more important feature in sedimentation.

The separation is achieved by physical or mechanical means, but without the involvement of a change in phase (such as the melting of a solid, or the evaporation of a liquid). As is often the case in attempting a perfect definition or classification, the boundaries involved in defining filtration in this way are less than absolutely precise. Thus: some membrane separations do involve a change in phase; some filtration processes involve electrical as well as mechanical forces; and some processes involve chemical forces as well as physical ones. Most significantly, the process of dry screening (**sieving** or **sifting**) involves the passage of part of a mixture of granular solids through a **screen**, with little or no passage of a fluid at all, yet this is clearly a filtration process.

Rotary drum filter

That said, the bulk of filtration processes involve the removal of particles, droplets or molecules, from a fluid, by means of a physical barrier, the **filter medium**, through which they will not pass by virtue of their size. A particular form of filter may be able to use a wide variety of filter media, to achieve the same, or different, separations.

There are two main purposes in filtration, either

- to remove impurities from a fluid (gas or liquid), or
- to recover valuable materials from suspension in a fluid (usually a liquid).

It is usually the case that a particular filtration operation is aimed at one or the other of these two purposes, although sometimes both are operative.

The first of these, **clarification**, normally employs finely porous filter media, and aims to remove as much of the impurity as possible, and preferably all of it. The second, **harvesting** or **recovery**, also aims for as complete a recovery as possible of the wanted material, but uses coarser media, mainly because the **cake** of recovered solid does most of the filtering, and the process is less concerned with the clarity of the **filtrate**.

Here again the division is not exhaustively precise: some harvesting processes remove waste solids for subsequent treatment, while some clarification removes only some of the suspended solids (ahead, for example, of a finer filtration process). Filtration may also be used to **classify** a suspended solid into two separate size **fractions**.

In terms of the way in which a particle is trapped by a filter medium, and so removed from suspension in the fluid, four basic mechanisms can be distinguished:

- *surface straining*, in which a particle that is larger in size than the pores of the filter medium deposits on the surface of the medium, and stays there until it is removed, while particles that are smaller in size than the pores pass through the medium;
- *depth straining*, which applies in filter media that are relatively thick by comparison with their pore diameters, when particles will travel along the pore until they reach a point where the pore narrows down to a size too small for the particle to go any further, so that it becomes trapped;
- *depth filtration*, in which a particle is trapped in the depth of a thick medium, even though it is smaller in diameter than the pore at that point,

by one or more of a complex mixture of capture mechanisms; and
- *cake filtration*, in which a layer or cake of particles accumulates on the surface of the medium, and then acts as the filter medium for subsequent filtration (cake filtration may follow an initial period of surface straining while the larger pores are blocked, but cake filtration can occur even when the particles are all smaller than the pores, by the bridging of the particles across the entrance to a pore).

Any real filtration process will probably involve a combination of two or more of the above mechanisms. The two straining processes will quickly blind the medium, as the pores progressively block, requiring some kind of cleaning process to be implemented. Depth straining and depth filtration are usually classed together, under the name of depth filtration, since both usually occur together.

The mechanisms just described are all variants of one major group of filtration processes, in which all of the fluid flows through the medium, leaving any separated material within or upstream of it. This is known as **through-flow** filtration (also as dead-end filtration), and is the traditional way in which filtration processes have been operated, probably from before the start of recorded time. An alternative process now exists, as a significant part of the filtration business, in which the main fluid flow is directed across the surface of the medium, with only a portion of the fluid passing, at right-angles to the main flow, through the medium. Material deposited on the upstream surface of the medium is then largely swept away by the fluid flow, which often runs in a recycle loop. This technique is known as **cross-flow filtration** (also as tangential or parallel filtration).

Unlike sedimentation, which will proceed naturally under the force of gravity, filtration will not usually occur until some **driving force** is applied to the system. The driving forces used are either the result of a hydrostatic head of fluid, in the form of a height of feed slurry above the filter medium, with ambient pressure below it, or the effect of a differential fluid pressure applied across a filter medium and any accumulated cake upon it. (These are, of course, both essentially the application of pressure, but it is convenient to divide them in this way.) The hydrostatic head is of little value when it is only influenced by gravity, but when amplified by **centrifugal force** it becomes a very potent means of achieving separation and the **centrifugal filters** are an important group of separating tools, often characterised by ease of washing and the achievement of high levels of dryness in the discharged solids.

Candle filter

The equipment items driven by differential pressure are traditionally the essence of filtration, with some of the best known types of filter in their ranks. The driving forces here can be divided into two classes; vacuum and over-pressure. **Vacuum filters** are limited in differential pressure to figures approaching 1 bar, but have the advantage of relative simplicity for the more productive types, with any filter cake available at ambient pressure for discharge. **Pressure filters** range from the simplest in-line filters to the very complex rotary drum or disc designs housed in a large pressurised shell. The driving force is higher (it can be in excess of 10 bar) and the filtration thus more efficient, but at the price of higher capital and operating costs for the biggest machines.

The nature of the suspending fluid obviously has an important effect on filtration processes. Where gases are concerned, the solids or droplet content is usually low, and clarification, probably by depth filtration, is the required process. (Where solids are at a high concentration in a gas, this would normally be dealt with by a **cyclone**.) Liquid suspensions are divided among those needing clarification, and those requiring recovery of the suspended solid, when **cake filtration** is the norm.

In the range of liquid filtration processes, the size of the suspended solids certainly affects the choice of filter. For the bulk of the history of filtration, it has been concerned with separating relatively coarse particles, to not much less than a millimetre in size. Much finer filtration is now demanded by almost every end user, and the old world of **macrofiltration** now goes down to 10 μm, while the newer world of **microfiltration** has appeared, and is currently being pushed into the sub-micrometre region, or even below 0.1 μm. From the other end of the spectrum, the exotic world of **reverse osmosis**, a membrane system developed to remove ions from water, has expended upwards into the **ultrafiltration** region, which now overlaps the bottom end of microfiltration, while the new region of **nanofiltration** has appeared between RO and UF.

To return, finally, to the function or purpose of filtration, the end user can identify two main functions for separation equipment. One is to enable the main production line to operate, where the filter or centrifuge is essential to it (as is a basket centrifuge to the manufacture of sugar) – these are the **process** uses. The other broad class sits in the services to the manufacturing process, where continuity of operation may not be essential – the **utility** applications. Such uses might lie in the cleaning of hydraulic or pneumatic circuits, or the clarification of steam condensate. The difference between these lies in the specificity of the equipment, which may be unique to one particular process, while a utility filter will be much the same whichever end-use industry might be using it. A philosophical problem here is that what might be a utility filter in one sector could be a process filter in another.

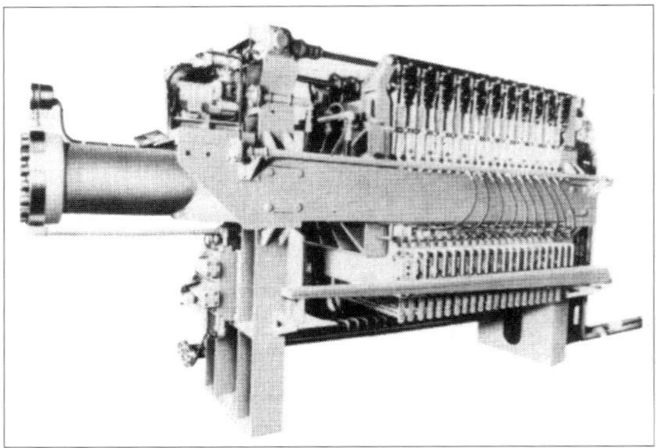

Filter press

head of the suspension above the filter (or from the equivalent but much larger centrifugal forces); and suction comes from a vacuum applied below the filter medium. This pressure needs to increase as a **filter cake** builds up on the medium – if it remains constant then the filtrate flow will decrease.

Filtration purpose. There are two main purposes for the use of **filtration** (and the associated technologies): to **clarify** a fluid, free of a relatively small quantity of contamination, and to remove (and probably recover for subsequent use) a high concentration of solids from **suspension**. These two closely match a different division by purpose: **utility** vs **process**, where process separation is usually very specific to the industrial process within which it is employed, whereas utility applications are much the same wherever they occur (hydraulic oil cleaning, boiler condensate treatment, etc.). Almost all utility applications are for clarification, and a high proportion of process applications involve solids recovery.
[*See also Filtration*]

Filtration rate. A rather imprecise term meaning the fluid flow rate through a **filter**, measured as total flow volume (or weight) or as flow per unit area of **filter medium**.

Filtration ratio. [*See Beta factor*]

Filtration resistance. A fairly general term, which acknowledges that there is resistance to the flow of **filtrate** through a **filter medium** and any collected **cake**. It should be given as the sum of the resistances of the medium and the filter cake.
[*See also Cake resistance, Filter medium resistance*]

Filtration time. A term used of a **filtration** process, without any special level of precision, referring to the time taken for the process to be completed. It relates to a **batch filtration**, and can refer to any spread across the time for a complete filtration cycle, including the total time for complete passage of **filtrate**, or the time taken to reach a certain pressure drop across cake plus medium.

Fine capillary system. In a draining **filter cake** (or any other porous mass), two broad groups of channels are recognised. The **coarse capillary system** is formed by the spaces between the particles of the cake, which are relatively easy to drain, whilst the fine capillaries are the **pores** within each particle, from which it is usually much more difficult to remove the **residual moisture**.

Fine filter. A general term referring to any **filter** capable of separating particles from a fluid down to a low **cut-size**. Its usage dates from a period before the availability of the modern **filter media**, which are capable of very fine filtration, and the specific use of **microfiltration** as a term for fine filtration.

A fine filter would normally be a **depth filter**, engaged in the **clarification** of a fluid down to particle sizes in the range of 1 μm.

Fine grain yield. A term sometimes used instead of **recovery index**, when applied to **classification** rather than **screening**.

Fineness of separation. A very imprecise term, referring generally to the **particle size** at which the **separation** process is taking place. It should rather refer to **cut-size**.

Fines. In any **classification** process, in which a mass of solid particles is to be divided into two (or more) **fractions**, measured by average **particle size**, then the fines are the fraction with the smallest average size. The particle size at the boundary between coarse and fine particles is the **cut-size**.

Quite often, fine particles are an unwanted fraction, in a process manufacturing a crystalline product for example, and they need to be removed, usually for recycle to the main production process (as seed particles for growth of larger crystals).

Fingering. [*See Channeling*]

Finishing process. The final stage of any manufacturing process, in which the product is prepared for its end-use. In **filtration** terms, this might refer to the provision of a **polishing filter**, to provide the sparkle in a liquid product.

It also has a very different meaning – that of the treatment of a material intended for use as a **filter medium**, following the main manufacturing process such as **weaving**, **needling** or **bonding**.
[*See also Calendering, Coating, Singeing*]

Fipoca filter. A form of constructed **cartridge**, made by Rellumit, in which a stack of annular discs, each with a set of grooves cut at an angle to the radius of the disc, forms the **filter medium**, with the pores formed by these grooves. Filtration down to 5 μm is possible.

Fish eye. A term used for lumps of **flocculant**, floating in a **suspension**, which have not gone into solution. Describes their appearance – they being an obviously unwanted waste of material.

Fish hook effect. The shape of a **grade efficiency curve**, named from its appearance, due to **dead flux**. It is characterised by a high concentration of fine particles in the **underflow**.

Fish tail (distributor). A device for distributing **feed slurry** or **wash liquid**, onto a **filter** surface below it, in which liquid flows down a sectioned sloping plate in the shape of a fishtail or fan.

The term fishtail is also used of the intake grid of small end-of-duct ventilation systems, for the removal of dusts and fumes from working machinery.

Fixed bed. Refers to a bed of granular particles, such as might be used for a **deep-bed filter**. Also termed a packed bed, it is a mass that remains in its container, as distinct from a **moving bed**, in which the particles are continuously fed in to the container, and correspondingly removed from the other end of the bed.

Fixing wire. A wire wrapped spirally around the outside of the **drum** of a **drum filter** to hold the **filter medium** in place, especially where a reverse flow of compressed air is used to detach the accumulated **cake**.

Flaky. A term descriptive of **particle shape** in a mass of particles, suggesting a plate-like structure to the particles.

Flameproofing. Where items of **separation** equipment, especially **centrifuges**, are working with flammable liquids, then measures must be taken to avoid ignition. This includes rendering all motors safe, and may involve the blanketing of the **filtration** zone with an inert gas, such as nitrogen.

Flash spun. One of the more recent forms of **non-woven** materials, which is formed by the **extrusion** of a thick solution of a **thermoplastic polymer** from a **spinneret**. When the fluid pressure is sharply

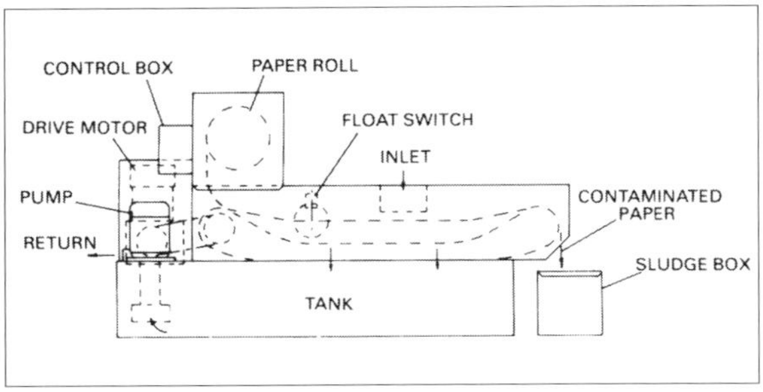

Flat bed filter

reduced as the solution leaves the nozzle, the solvent flashes off, leaving the polymer in the form of a very fine **filament**.

Flat bed filter. An early form of **filter**, employed in the clarifying of machine tool coolants and lubricants, in which the **filter medium** lays flat at the bottom of a trough into which the slurry flows. The medium, originally a **paper**, is wound across the base of the trough from a feed roll to a collector roll, from which the separated **swarf** falls into a drum, for disposal.

In the earliest forms of the equipment, the filter medium would be wound manually and intermittently, and the filtering of the fluid through the medium would have been under the influence of gravity. More modern versions use a **non-woven** polymeric material (such as a **spun bond**) as filter medium, have the medium moving slowly but continuously through the filter, and may even apply a slight **vacuum** below the flat medium.

Flat bottom hydrocyclone. A cylindrical **hydrocyclone** with a flat base, instead of the usual long conical section leading to the thickened slurry offtake. The induced liquid flow carries the separated solids to the centre of the flat bottom, whence they are discharged. The required **cut-size** in the **separation** can be controlled by the height of solids at the base of the hydrocyclone, and such units can separate at sizes up to 0.5 mm.

Flat sheet modules. A term used specifically of **cross-flow filters** using **membranes** as the **filter medium**, in which the membranes take the form of flat **sheets**. Flat sheet modules, also known as plate and frame modules, leaf modules or membrane stacks, are available in various configurations, but they all contain flat sheets of membrane, separated by spacers and built into a stack like a plate and frame filter. The membrane may rest on a grooved support plate (plate and frame module) or on a porous sheet (leaf module). In some designs, the **permeate** from each pair of membranes can be collected separately, so that damaged membranes can be identified and replaced relatively easily by dismantling the stack. The costs and fouling characteristics of flat sheet modules are intermediate between those of **tubular elements** and **hollow fibres**.

Fleece. A fleece literally is the layer of wool removed from a sheep's back at shearing time. Since the voyage of Jason and the Argonauts in search of the Golden Fleece, the world has been aware of this important early example of solid-liquid **separation**: the entrapment of alluvial gold particles in the hairs of a sheepskin pegged out on the bottom of a river.

To the modern walker in cool climates, a fleece is a soft wool-like garment made from **woven** or **non-woven** materials, worn as an insulating layer. Because of the word's connotations in English it is perhaps unfortunate that it is now used in the **filtration** industry to describe a range of **filter media**, mainly of modern **non-woven polymeric** materials, some of which can be as stiff as thick cardboard, and most unfleece-like in appearance. This usage has probably come from the direct translation of the German "Vlies" or "Vliesstoff", and the success of Freudenberg Vliesstoffe AG in making the non-woven materials. To others, the term is synonymous with **felt**.

Flexural rigidity. A measure of the stiffness of a material, defined by:

$$G = 0.00167ML^2$$

where G = flexural rigidity
M = mass of material per unit area
L = overhang length (= twice the **bending length**)

Flight. In a **decanter centrifuge**, this is the helical metal plate welded radially to the **conveyor** to

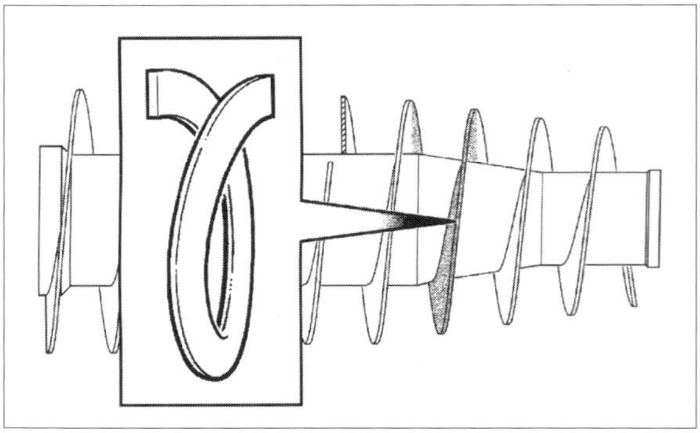

Decanter conveyor flight

provide the scrolling surface of the conveyor. It is one continuous plate wound end to end on the conveyor hub, and there may be more than one flight on the conveyor. In general terms, it is the "blade" of any screw conveyor.

Flight tip. The outer periphery of the **conveyor flight** in a **decanter centrifuge**, which creates most of the solid movement along the bowl, and therefore suffers the most wear. Accordingly, the tip will be made, at the very least, from a layer of hard wearing metal, or more likely from a layer of **tiles** fixed to the outer edge of the flight (which are more easily renewable than having to replace the whole conveyor).

Flint sand. A sand-like material made by the crushing of flint, and used as a **deep bed filter medium**.

Floater disc. A disc fitted at the **centrate** discharge **(front) end** of the **conveyor hub** of a **decanter centrifuge**, which dips into the **pond**, thereby holding back any solids floating on the pond surface, and so prevents their discharging with the centrate.

Floaters. Any particles that float on the surface of a liquid, either because they are less dense than the liquid, or because they are hydrophobic and attach to gas bubbles, in either case not separating by downwards **sedimentation**.

Floating conveyor. A **conveyor** in a **decanter centrifuge** whose overall density is less than that of the suspending liquor of the feed slurry, so that it floats in the pond. Used on some high speed decanters. (Also called a swimming conveyor.)

Floating scraper. A device on a **drum filter** for the removal of the **filter cake**. It is mounted in a balanced way, such that a minimal pressure is required for it to follow the contours of the drum.

Floc. In a **suspension** of solid particles in a liquid, a floc is a larger particle formed by the **aggregation** of otherwise dispersed particles. The particles in a floc join together either because of reduction in the strength of the electric field surrounding the particles, such that the attractive **van der Waal's** forces are able to act (**agglomeration** or **coagulation**), or because a chemical **flocculating agent** has been added to the suspension (**flocculation**).

By virtue of its larger size, a floc will sediment faster that the basic particles from which it is formed. It can include some of the finer particles of the suspension, thereby further improving the sedimentation performance. In the case of **cake filtration** a cake made of flocs will be much more permeable than that from the basic particles. A sediment or cake built up from flocs will also be much more compressible than if made from the basic particles.

Floc blanket clarifier. [*See Sludge blanket clarifier*]

Flocculant, Flocculating agent. A chemical added to a suspension to cause the **aggregation** of the basic particles in the suspension into **flocs**, and thereby make the suspension more easily **filterable**, or more rapidly **sedimented**. These chemicals are long-chain **polymers**, usually carrying electric charges along, and especially at the end of the chain, frequently called **polyelectrolytes**. They act either by forming a bridge between two particles, or by forming patches on the particle surface that are electrostatically attractive, one to the other.

The use of flocculants can greatly speed up a **separation** process, but they are quite expensive, and so their use is an interesting economic balance. They are also quite difficult to get into solution, and the polyelectrolyte dosing equipment will add considerably to the first cost of an item of separation equipment. Their use in the food and beverage industries may be restricted, but they are widely used in the treatment of sewage sludges, their arrival on the scene having made the **decanter centrifuge** a suitable tool for **sludge dewatering**.

Once they are added to a suspension, a finite time is necessary for them to take effect, and this flocculation time must be allowed for in the design of the separation systems in which they are used.

Flocculant chamber. In any item of **sedimentation** equipment, **flocculant** may have to be added to the feed slurry to increase the average particle size, and frequently this is done in a chamber ahead of the **feed zone**. In the case of the **decanter centrifuge** this is a compartment behind the feed zone in the **conveyor hub**, sometimes separated from it by a buffer chamber, used for receiving the flocculant before delivering it into the **pond** where flocculation occurs. In certain designs of decanter it can double as a rinse chamber. Flocculant (or rinse) is fed to the chamber along a floc tube, which is a concentric tube outside the feed tube, inside the hub.

Flocculation. The process whereby the particles in a **stable suspension** are caused to aggregate into **flocs**. Flocculation improves the separation behaviour of a stable suspension, whether by **sedimentation** (because of the greater sedimentation velocity of the larger particles) or by **filtration** (the larger particles are more easily filtered and form a more permeable cake).

Flocculation is caused by the addition of **flocculants** to the suspension, and should be differentiated from **coagulation** (which is a **destabilisation** process). The flocculant must first be made into a suitable solution (not easily done), and then this solution must be added to the suspension as quickly and as uniformly as possible with the minimum of **shear**. The mixture must then be allowed sufficient time, still under low shear conditions, for the flocs to form.

Flocculation time. [*See Flocculant, Flocculation*]

Floc density. The **bulk density** of a **floc**, which will normally be quite low, the primary separating parameter being its **effective diameter**. If the floc is compacted, then its mechanical strength (its resistance to shear stresses) is increased. This may be necessary prior to a **centrifugal separation**, where the shear forces are high, in which case the flocculated suspension is given a shearing pre-treatment, for example by mechanical stirring, to increase its floc density.

Floc factor. The **Richardson & Zaki equation** is used to describe the **sedimentation** of particles from a **suspension**. Where the suspension has been **flocculated**, it is necessary to use a floc factor to determine the volume fraction of **flocs** in it. This factor, k, is the ratio of the volumetric floc concentration, c_f, to the total solids content by volume, c:

$$c_f = kc$$

Floc strength. The ability of an aggregated **floc** to hold together once formed. This depends upon the nature of the **flocculating agent**. One method of conducting a floc strength test is to submit a sample of flocculated sludge to a standard stirring process, and then make a series of **capillary suction tests** at regular intervals of shearing time. A strong floc will show little change in capillary suction time.

Florentine filter. A continuous **vacuum belt filter** used for the dry fractionation by **filtration** of crystallised oils and fats. Operation may be at temperatures from 2 C to 45 C, with high concentrations of solids. A stainless steel belt is used as the **filter medium**.
[*See also Dewaxing*]

Flotation. A major **separation** process, classifiable as a **sedimentation** process, but working effectively in the opposite direction, against the force of gravity, by causing solid particles to attach to a gas bubble, and so float upwards through the **suspension**. [This is the main present meaning of the term, although it originated in the mineral dressing industry to mean literally the floating of particles on a dense liquid medium, with the gas bubble version known as froth flotation.]

The separated particles collect on the surface of the liquid and are skimmed off from this surface layer. It is obviously essential that the particles, once brought to the surface, do not break away from their supporting bubbles, i.e. the froth at the surface must be stable. Frothing agents may be added to the suspension to ensure such stability.

There are several types of flotation process, dependent mainly on the way in which the gas bubble flow is generated, such as **dispersed air flotation**, **dissolved air flotation**, and **electroflotation**. Although originating in the mineral processing sector, flotation is now widely used in **waste water treatment**.
[*See also Heavy media separation*]

Flotation column. A simple form of **flotation separator**, suitable for fine particles, and combining concentration and cleaning in a single unit. The column is about 10 m high and feed enters about 1 m from the top. Air is introduced near the bottom by an internal sparger or external bubble generator. As the bubbles rise they contact the descending solids in a collection zone. After attachment of the **hydrophobic** particles to the bubbles, these rise above the feed point to a cleaning zone where they meet a downward flow of wash liquid. This washes any **hydrophilic** particles off the bubbles, and prevents bubble coalescence. Hydrophilic particles exit at the bottom and hydrophobic particles are carried into a launder at the top.

Flow. [*See Flow regime*]

Flowing film separator. One of a number of simple forms of **gravity thickener** in which concentration is achieved by causing the feed to flow over a surface. These include **pinched sluices**, **Reichart cones**, **spirals**, **shaking tables** and **vanners**.

Flow moisture point. A **suspension** may be **thickened** through progressively decreasing moisture contents, initially staying fluid, then becoming pasty, and finally quite rigid. The choice of final moisture content for any **dewatering** process is determined by what is next to happen to the dewatered material. If it is to be transported any distance (as, for instance, might be the case of a sewage sludge being taken to a landfill site), then its behaviour under transportation conditions are of significant interest.

A pile of dewatered solids may appear stable at a particular moisture content, but may become more liquid if vibrated, as might happen on the bed of a truck. The flow moisture point is an indicator of such behaviour, and is determined by placing a hemispherical mass of the moist solid on a table and subjecting the table to a set of defined oscillations. At some specific value of the moisture content, the mass will become more liquid and will flow across the table. This value is the flow moisture point.

Flow rate. The rate at which material passes through a vessel or a system. Usually expressed in volumetric terms, as volume passing per unit time, but may also be expressed as a mass flow rate.

Flow regime. Any fluid will flow under the conditions of an applied pressure, and the manner of that flow will depend markedly upon its resultant velocity and the shape of the space through which it is flowing. In a smoothly walled channel, and at low velocities, the fluid flow will be **laminar**, with successive layers of fluid passing over one another in a smooth linear fashion. The layer of fluid in contact with the wall will be almost stationary, and layers will move progressively faster the further from the wall they are, until they reach the bulk velocity of the fluid. This is termed the laminar flow regime, within which the relationship between pressure and flow rate is determined by viscous forces.

If the channel is roughly surfaced, with many obstacles in it, and the applied pressure is high, causing flow at high velocity, then the flow is fully **turbulent**, with many eddies, and no significant region of linear flow lines. This is the turbulent flow regime, which results in high pressure drop between one end of the flow channel and the other, the energy loss being dissipated by the eddies. The pressure/flow rate relationship is now controlled by inertia forces.

As flow rates increase from the laminar flow values, some turbulence is seen, and this is the intermediate (or transitional) flow regime. The different regimes are characterised by their respective **Reynolds Numbers**, which will be low for laminar flow (below 2000) and high for fully turbulent flow (above 4000).

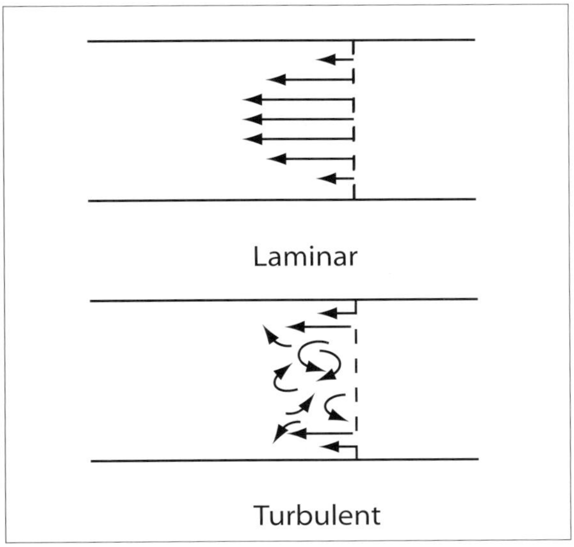

Flow regimes

Flow resistance. The resistance to fluid flow offered by a **filter medium** and/or **filter cake**. Measured in terms of the **pressure differential** required to give unit volume flow through unit volume of material of a fluid of unit viscosity.

Flow-time relationship. A representation of the behaviour of a **filter medium** as it is used and slowly blocks. Such a relationship can be developed under constant flow (i.e. increasing pressure drop) or constant pressure (i.e. decreasing flow) conditions.

Fluid. A term inclusive of two of the three states of matter: liquid and gas. In correct usage the word "fluid" should refer to either liquid or gas or both. Thus a fluid/solid separation process would be one in which solid particles are being separated from a liquid or a gas (or both).

Fluid split. A term sometimes used for the ratio between the **overflow** flowrate and the **feed** rate, based on liquid flows only. The corresponding term for slurry flows is volume split.

"Flush-flo" filter. Made by Auto-Klean, the Flush-flo is an automatically operated in-line **strainer**, employing a **wire wound medium**, which is rotated past a slotted tube. This tube scrapes off collected solids, and also enables a back flush of **filtrate** through the slot to wash off remaining solids.

Flush plate press. A term sometimes used for a (plate and frame) **filter press**. The term flush plate is used for the plate in such a press.

Flush water. The water used to wash out a piece of **separation** equipment before it is shut down, especially in the case of **centrifugal sedimenters**.

Flux. The term used historically for the rate of **permeate** flow in **membrane filtration**, typically measured in l/m^2h.

Flux decay, Flux decline. The reduction in **flux** through a **cross-flow membrane** in operation, caused by **fouling**.

Flux decline model. A model of the operating state of a **membrane cross-flow filter** when the **flux** starts to decline.

Flux density. [*See Magnetic flux density*]

Fly ash. The fine ash produced by the combustion of pulverised coal. Mostly a waste material, but can be used in cement formulation, and occasionally as a **filter aid**.

Foam. A dispersion of gas bubbles throughout a liquid, in such a way that the gas occupies almost all of the volume, surrounded by the liquid as the thin walls of the bubbles.

If the liquid were to set solid, a very light material is produced, but one of little use to **filtration**, because the **pores** do not interconnect. If, however, the foam is **reticulated**, by a chemical or thermal process that breaks down the bubble walls, then a useful **filter medium** can be created.

If the liquid is very pure, then a foam cannot form – the existence of a foam implies that the liquid contains one or more stabilising agents (such as **surfactants**).

Foam fractionation. A **flotation** process with a deep **foam** at the liquid surface.

Fog. A stable **suspension** of liquid droplets in a gas, whose droplet size is greater than that of a **mist**, to the extent that a fog significantly impairs visibility. Natural fogs contain droplets up to 50 μm in diameter.

Follower plate. [*See Queen plate*]

Forced discharge. Refers to the discharge of separated solids from a piece of **separation** equipment, using mechanical force or fluid pressure to move the solids in a controlled and guaranteed fashion out of the equipment, for disposal or subsequent processing.

Force fields. A term usually used to describe the application of **magnetic** or **electric fields** to enhance a **separation** process.

Formazin Turbidity Unit (FTU). A method of measuring **turbidity**, i.e. the concentration of a suspension. There are two standards; the European (EBC) and American (ASBC).
[*See also Jackson Turbidity Unit, Turbidity*]

Form drag. When a fluid flows over a surface, there is a drag on the surface, made up of two components: skin friction (viscous drag) and form drag, due to the formation of eddies. The latter only occurs during turbulent flow, and is of little interest to flow through porous masses, which mainly occurs well into the streamline regime.

Form filtrate. A term sometimes used for that **filtrate** collected during the **cake formation** stage of a **filtration** process

Forward edge. A term relevant to the **cells** of a **rotary filter**, marking that part of the cell that first enters the **suspension**, and which as a result has the shortest cake formation time. The cake will therefore be at its thinnest at this edge, and this will have an impact upon the **dewatering** process. Even more so is the effect with a **rotary disc filter**, where the region of the forward edge nearest to the **cell foot** is even thinner, and can pass the greater part of any air flow being used to **dewater** the cake.

Fouling. Can refer to the blocking of any **filter medium**, but has taken on major connotations in the case of **membrane** filters, operating under **cross-flow filtration** conditions. In their earliest days these were used almost entirely for the desalination of water, and the filtration of blood. These applications were then bedevilled by the formation of a slimy layer on the upstream membrane surface, the result of biological growth. As membranes have become used for an ever wider range of **separation** applications, then the composition of the fouling layer has become more diverse, but it remains the main problem in the implementation of membrane separation systems.

The main consequence of fouling is blockage of the membrane pores, and hence a reduction in **permeate** flow and/or a necessary increase in **trans-membrane pressure**. Methods used to combat fouling include the choice of a membrane material as little susceptible to fouling as possible, the dosing of the surface of the membrane with substances than inhibit fouling, and the increase of shear close to the membrane surface, either by increasing suspension flow rates, or by moving the membrane in relation to this flow (by rotation or vibration), or by mounting moving surfaces near to the membrane.

Fouling index. A method of characterising **fouling** of a **cross-flow membrane filter**.

Fraction. A definable component of a mixture, based upon some physical property, such as particle size, or density (or, in the case of distillation, by boiling point). This property is usually the one upon which a **separation** process is based, to separate that fraction from the mixture.

Fractional filtering factor curve. A term sometimes used for a **grade efficiency curve** when applied to **filters** rather than **sedimenters**.

Fractional grade efficiency. A precise measure of the efficiency of a **filter**, also known as a Tromp curve. It describes the separation characteristic T(x) of an item of **separation** equipment at the particle size x, and represents the quantity of particles of this size or above, $M_G(x)$, as a fraction of the total amount of particles, $M_A(x)$, both discharged in the coarse fraction from the equipment:

$$T(x) = M_G(x)/M_A(x)$$

Fractional mass recovery. A term sometimes used as equivalent to **total efficiency**.

Fractionation. The process by which a mixture of a number of components is separated into those components. This might be by means of a **classification** process, in say an array of **sieves**, to produce the **fractions** based on **average particle size**. The term is most commonly used in the separation of liquid mixtures by distillation (where it is also called fractional distillation).

Fractionation (efficiency) index. A term sometimes used to define the performance of a **classification** operation.

Frame. In a (plate and frame) **filter press**, the frame creates the space in which the **cake** forms.

A very different meaning is that of the cast or fabricated structure upon which any item of **separation** equipment is mounted, although the term is mostly used of the supporting structure for **centrifuges**.

Frame filter press. [*See Filter press*]

Frazier Permeameter. An instrument used in testing materials to determine their **porosity** as an air **permeability** figure, in the clean state. A typical determination would produce the volumetric flow of air through 0.1 m^2 at a differential pressure of 1.5 cm wg. Fabrics yielding a flow rate of 0.0005 to 0.005 m^3/s would be considered to be very tight materials, while flow rates of 0.2 to 0.24 m^3/s would be considered to be extremely porous. The test is applicable to both **woven** and **non-woven** materials.

Freeboard. The clear space between the top of any liquid contained in a vessel and the rim of that vessel. The leaving of sufficient freeboard may be necessary to allow for the formation of a **froth** layer during a **separation** process.

Free flow electrophoresis. An **electrophoretic** process that does not involve **filtration**.

Free settling. Sedimentation of a dilute suspension such that particles settle without interference from other particles.

Free settling classifier. A **hydraulic classifier** in which the **separation** occurs at a relatively low concentration, i.e. under **free settling** conditions.

Freeze-thaw process. A process specifically applied to **waste water treatment sludges**, in which the suspension is rapidly frozen and then allowed to thaw again. The result of the process is to convert gelatinous, stable suspensions into suspensions of rapidly settling, aggregated particles in a clear liquid, that are easy to **filter**. The cycle may need to be repeated twice or more, to achieve the required level of separation.

Friction. The force resisting the movement of one body across the surface of another. If the force perpendicular to the sliding surfaces (which will often be the weight of the upper body) is N, then the friction force (or the force needed just to overcome the effect of friction):

$$F = \mu N$$

where μ is the coefficient of (dynamic) friction. This

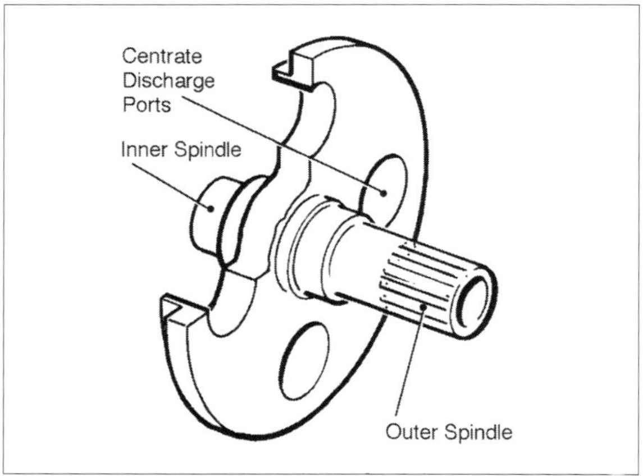

Decanter centrifuge front hub

coefficient is independent of the area of contact between the bodies.

Friction factor. [*See Fanning friction factor*]

Front, Front hub. The front of a **decanter centrifuge** is conventionally considered to be the end of the **bowl** from which the **centrate** is discharged. The front hub is therefore the flange, or end plate, which closes the bowl at the centrate end, and includes the spindles that support the front main and front conveyor bearings. It also supports the centrate **weir plates**.

Froth. A concentrated **dispersion** of gas bubbles in a liquid. Except in combination, the word is effectively interchangeable with "foam", although perhaps has an element of shorter life in its meaning.

Froth flotation. A flotation process in which a thin froth is formed at the surface.
[*See also Foam fractionation*]

Frothing agents. Froth stabilising agents, usually **surfactants**, which are added to the solids suspension in order to stabilise the **froth** or **foam** formed during the **flotation** process.

Frozen image centrifuge. An item of laboratory equipment used for designing **gravity thickeners**.

FTU. [*See Formazin Turbidity Unit*]

Full axial flow. A condition obtaining in some specialised designs of **decanter centrifuge**, in which the liquid flow along the **bowl** is axial (rather than the normal helical flow between the **conveyor flights**), and cannot deviate from it. This flow is achieved between radial vanes or **Sigma enhancement** vanes.

Fullers earth. A non-plastic clay, mainly **montmorillonite**, which has a high affinity for oils and fats, and which is therefore used as an **adsorbent** for such material. As its name implies, originally used in the processing of cloth, especially woollen items.

Full flow filter. A **filter** which accepts the total flow in a system, as distinct from a **bypass filter**.

Fume. A fine smoke of solid particles formed in air by the desublimation of vapours or by chemical reactions. Mainly concerns the metals processing sector. The suspended particles are mainly submicrometre in size, but **agglomerate** quite well.

Funda filter. A type of horizontal **plate pressure filter**, featuring disc-shaped plates with **cake** forming on the upper surfaces, from which **centrifugal** discharge removes the cake after it is **dewatered**. The set of plates is rotated on a shaft driven from above the filter (cf the **Schenk filter**, which is underdriven).

Fundabac filter. A **candle filter**, made by the DrM Dr Müller company, in which candles are bundled together, each bundle being covered in a bag of filter medium, which is cleaned by a reverse pulse of **filtrate**, the bundles being flushed clean in turn.

Fungi. A group of living organisms, including yeasts, rusts, smuts, mildews, moulds, and mushrooms. They are now considered to represent a separate kingdom from plants and animals, and are among the most widely distributed organisms on earth. Many fungi are free-living in soil or water; others form parasitic or symbiotic relationships with plants or animals, respectively. Their relevance to **filtration** lies in their ability to interfere with processes, and so create the need for exhaustive cleaning.

Funicular moisture. A term sometimes used to describe one of the forms of liquid held in a porous mass: funicular liquid is that in the voids, and so removable by mechanical compression. (The other two forms are **capillary** and **pendular**.) The funicular state generally corresponds with saturation levels between 30 and 60%.

G

GAF bags. An early, but well used, form of filter **bag**, fitting inside a **woven wire cage**, with flow from inside out. It was used singly, or in arrays up to eight in number.

Galileo number. [*See Archimedes number*]

Gangue. The term used in the mineral industry for the unwanted material naturally associated with a mineral, which needs to be removed from the valuable ore as a waste. This removal process is one of the most important uses of **separation** equipment of all types.

Garnet. Crushed garnet is a material increasingly being used in **deep bed filters**, because it has a density significantly greater than silica or **flint sand**, and so will return to its original position after the bed is disturbed during the **washing** process.

Gas blowback. A method of discharging a cake from a filter by blowing gas from behind the cloth. To be distinguished from **backflushing** wih **filtrate**.

Gas blowing. Dewatering a cake by applying pressure to force gas into the cake and thereby displace the liquid. The term is used obviously for **pressure filters**, but also for **vacuum filters**, where the gas is actually sucked into the cake. The method only works if gas flow is even through the cake – any **channelling** of the gas will render the process unsuccessful.

Gas bubble size. [*See Bubble size*]

Gas bypass. When a filter **cake** is being **dewatered** by an applied pressure of gas, the largest through **pores** will clear first. Once they are clear, gas is free to flow straight through them, and additional pressure will be required from the gas compressor to clear the reminder of the pores. This bypassing of the gas can add considerably to the operating cost of the filter.

Gas cleaning. The general term used for the removal of a small concentration of dust from any gas stream – from the air for a breathing mask to the exhaust gas from a large furnace. It specifically implies the removal of solid contaminants – the removal of gaseous impurities is the task of **gas purification**.

The main paradox with industrial gas cleaning, and the treatment of engine exhausts, is that most such exhausts are hot. Accordingly the good system designer looks to cool the gas to recover the heat content of the gas. But most such exhausts are sufficiently dusty as to make almost any heat exchanger inoperable in a very short time because of deposited solid particles. So the gas stream needs to be cleaned before it can be cooled. But most **filter media** are not capable of operating at the temperatures of most exhausts. So the gas needs to be cooled before it can be filtered, and so on.

The way to break this vicious cycle, of course, is to develop filter media that are capable of operating efficiently and for a long time at high temperatures – hence the strong current interest in **ceramic** filter media.

Gas filter. Any filter intended for **gas cleaning**. It will normally be operating as a **depth filter**, although the increasing demand for very low cut sizes is causing **membrane** media to be used, which operate by **surface filtration**.

Gas permeation. A **membrane separation** process that uses non-porous membranes, through which gases pass selectively by **diffusion** through the material of the membrane. The most important uses are in the recovery of hydrogen from waste gases, and in the separation of oxygen and nitrogen from air.

Gas scrubber. [*See Scrubber*]

Gas separation. A mixture of gases cannot easily be separated by the processes of **filtration** or other mechanical **separations**. The techniques available for this purpose are **absorption**, **adsorption** and **condensation**. However, some **membrane** separation processes are now able to effect separation, at least of **vapour** mixtures.

Gaudfrin (Universal) filter. A vacuum or pressure **leaf filter** in which the leaves are lifted and moved between sectors of the filter for different stages of the **filtration** process.

Gaussian distribution. [*See Normal distribution*]

Gearbox. Any device for adjusting the rotational speed of machinery, including **centrifuges**, but especially important for the **decanter centrifuge**, where it is the means whereby the **scroll** is caused to rotate at a slightly different speed from the surrounding **bowl**, and hence convey the solids along the bowl. It is usually mounted at the drive end of the rotating assembly.

Gearbox chatter. [*See Chatter*]

Gelatin. A natural organic material, derivable from bones, which has some use as a **coagulant**.

Gel filtration. A batch-wise **separation** process, akin to **chromatography**, for materials dissolved in a solvent, which separates high molecular weight solutes from those with lower molecular weights. The solution is passed over a bed of porous gel granules, and the smaller molecules diffuse into the gel, while the larger ones pass on with the solvent. After the batch has been processed, the bed is washed with a fresh quantity of solvent, and the retained solute is eluted from the gel.

Gel layer, Gel polarisation. The reduction in flux in a **cross-flow membrane filter** caused by a build up of fine particles (often macromolecules) on the membrane surface. It is thus a specific form of **fouling**.

G-factor, G-force. Alternative terms used, like **C-value**, to describe the much higher driving forces in **centrifugal separation**, as ratios to the force of **gravity**. G-force is probably the most common term used to describe the extra impetus given to a **sedimentation** or **filtration** process by carrying it out under centrifugal force. It is dependent upon the distance from the axis of rotation, so that, for a given angular velocity, a larger centrifuge will exert a greater g-force than a smaller one. G-force is dimensionless, and is actually defined as the ratio of (the square of the angular velocity times radius) to the gravitational acceleration, g.

Gill separator. A rotary fairly-**high gradient magnetic separator**, used for separating magnetic from non-magnetic material. It consists of a laminated grooved rotor at the centre of a set of electromagnetic poles. The feed flows along the grooves between a pole and the rotor. The non magnetics flow straight through, while the magnetic materials are carried forward to be collected separately.

Glass. The basic material of quite a wide range of **filter media**, which almost all start with the formation of glass **fibres** by some kind of extrusion process from the molten state. The production of glass as fibre had two early non-filtration uses – as pads of thermal insulation for buildings, and for inclusion in plastic sheets as a reinforcement. The ability to make glass into fibres led to their being **wet-laid** as an equivalent to **cellulose** fibres in paper, and **dry-laid** as pads for air conditioning filters. A prime advantage of glass as a filter medium is its relative chemical inertness.

Glass beads. [*See Ballotini*]

Glass bead test. A standard **challenge test** for **filter media**, in which glass beads, having a spherical shape, are suspended in a fluid and passed through the medium. The sample analysis involves actual counting of the beads, which can be a very time-consuming process, especially as the average bead size gets towards 1 micrometre.

Glass fibre. Molten glass can be drawn or blown into fibres by a number of different techniques, which are now capable of producing very fine fibres, measured in nanometres rather than micrometres. The finest fibres are **wet-laid** into papers for use in high efficiency filters. General purpose **HVAC** filters use coarser fibres as pads or **felts**.

Glass filters. [*See Sintered glass*]

Glass paper. Filter papers made with glass fibre are becoming increasingly common both in laboratory use, and for domestic and industrial air cleaning. They have taken a considerable share of the market from the more traditional **cellulose** papers.

Glaze, Glazing. The blockage of a filtering surface, making it impermeable, either by the pressure on a

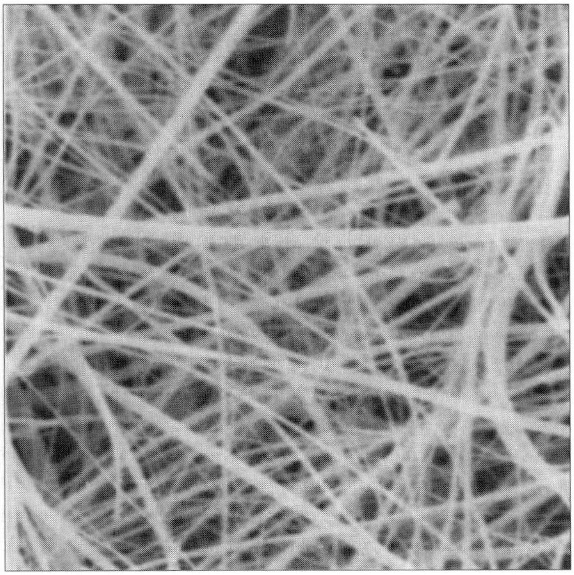

Glass fibre paper

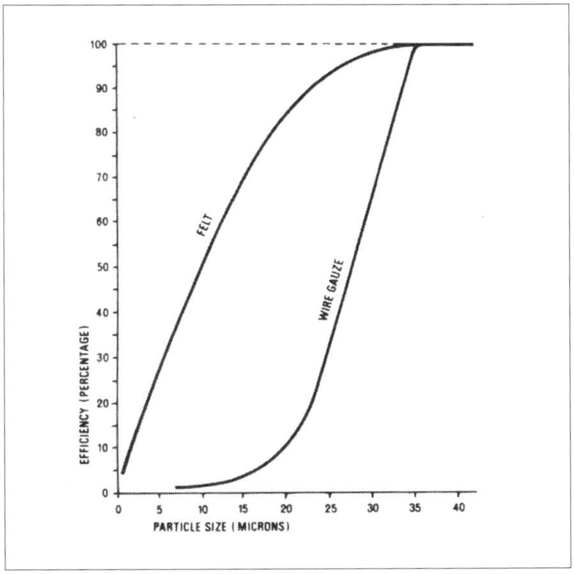

Media grade efficiency curves

cake heel of the discharge knife, which forces fine particles into the pores between larger particles, or if cooling or chemical reaction in the **filtrate** cause a component from the liquid to solidify or precipitate. If the glazing is caused by crystallisation, then it may be possible to render the medium-heel combination permeable again by washing, otherwise the medium must be removed and cleaned or replaced. A problem especially for **knife-discharge filters** and **centrifuges**, but also generally for the treatment of organic materials.

Glove box. [*See Cabinet*]

Gore-Tex. A material made by W L Gore and Associates, based on **PTFE**, which has become a very popular **filter medium**.

Grace medium resistance. A parameter in early work on the **filtration** of solids under **centrifugal force**.

Graded media. A feature of **deep bed filters**, in which the bed is made from several layers of materials, differing either in density or size (or both), such that the **filtration** becomes more efficient as the **suspension** flows through the bed, and also that the whole bed will re-assemble itself naturally after it has been thoroughly backwashed.

Also, but less often used, of laminated **filter media**, with successive sheets of decreasing porosity.

Grade efficiency curve. A method of showing the efficiency of a **separation** in terms of behaviour at different particle size levels. It is a plot of particle size against the fraction of the solids in the feed at that size that appears in the **underflow**.

Grain. A term used occasionally as synonymous with **particle**.

Grammage. A term used in the paper industry to describe the thickness of paper in terms of its weight per unit area, usually measured in g/m^2, and so used in the form "90 gram paper".

Granger filter. A development of the **filter press**, in which the **filter medium** is used first on one side and then the other, for **clarification** duties. Accumulated solids are washed off by a reversal of flow, the first **filtrate** after reversal being rejected.

Granular, Granular media. Materials, especially **filter media**, are described as granular when the particles of which they are constituted are approximately equidimensional, even though irregular, in shape.

Granular cake. A filter **cake** consisting of large particles, which therefore breaks easily into individual particles when removed from the filter.

Granular medium filter. A general term for a **deep bed filter**, used for the clarification of water, in which the **filter medium** consists of loosely packed granular material such as **sand**. The bed may be fixed or moving. These filters usually operate under the influence of gravity (i.e. the hydrostatic head of liquid), but some versions are pressurised to a moderate degree.

Grashof number. A **dimensionless number**, which represents the ratio of gravitational forces to viscous forces in a convective flow, which is useful in studying flow in an **inclined plate settler**.
[*See also Nakamura-Kuroda equation*]

Grate magnet. A set of magnets in the form of an array of tubes set across the flow of liquid, to remove a small amount of magnetic contamination.

Gravitational force fields. The controlling influence in any **sedimentation** process undertaken in the absence of **centrifugal force**. These fields are always present, but may be completely uninfluential compared with centrifugal or pressure forces.

Gravity. The force of attraction of any terrestrial body towards the earth. It is this attraction that gives weight to any mass, and it is the driving force for any simple **filtration** that employs just hydrostatic head to cause the **filtrate** to flow through the **filter medium** – such as in the domestic coffee filter, or the **flat bed filter** of many engineering workshops. Many **deep-bed filters** are gravity driven – as are all dry **screening** operations, and a wide range of **sedimentation** systems.

Gravity clarifier. A clarifier in which a low concentration of particles is separated from the liquid by gravity **sedimentation** alone.

Gravity concentration. [*See Gravity thickener*]

Gravity drainage. The emptying of liquid from the spaces within a porous mass, such as a **filter cake**, under the influence of gravity alone.

Gravity filtration. The **filtration** of a **suspension** solely under the effect of gravity (the hydrostatic head) on a layer of the suspension above the **filter medium**. Particular examples of gravity filtration range from the simple domestic coffee maker using a wire mesh, or perforated sheet, or piece of filter paper as filter medium, through **flat-bed filters**, to the large **deep-bed filters** used for water purification.

Gravity separation, separator. Any gas/solid or liquid/solid separation process that takes place under the influence of **gravity** alone. It is quite important in **filtration**, but much more so in **sedimentation** processes.
[*See also Gravity filtration, Gravity settler, Gravity thickener*]

Gravity settler. Any device, allowing the sedimentation of suspended solids under the influence of gravity. The purpose of a **gravity thickener** is to thicken, i.e. increase the solids concentration in the suspension, whilst a **clarifier** aims to produce an **overflow** that is as free of suspended solids as possible. Some settlers, of course, aim to achieve both objectives.

Gravity tables. Data tables relating API and Baumé gravity figures to liquid specific gravities.

Gravity thickener, thickening. Any **separation** system that, acting by **sedimentation** under the influence of **gravity**, causes a liquid **suspension** to be thickened. The process is usually carried out in large tanks, either rectangular in shape or (more often) circular. At the bottom of the thickener is usually a mechanical **rake** that sweeps the thickened sludge to the **underflow** exit. This rake will move from end to end in the case of rectangular thickeners, and radially for round ones, sweeping to a central discharge.

The dilute feed suspension, normally after **flocculation**, is fed continuously at one end of a rectangular thickener, or in the centre of a circular unit, and settles under the influence of gravity on the suspended solids. A clear zone forms in the upper region of the thickener, separated from the feed zone, and overflows into a collection trough at the discharge end, or around the outside of a circular thickener.

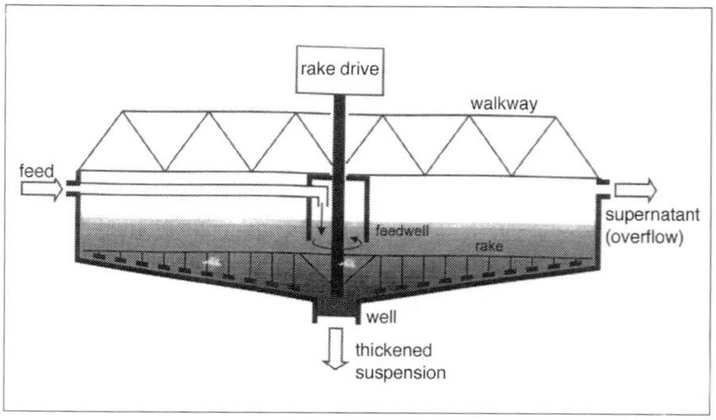

Circular gravity thickener

A sharp division can be seen between the supernatant clear liquid and the sedimenting solid. Below this interface is the **swarm sedimentation zone**, which changes to the **compression zone** in the lower part of the thickener. In the compression zone, the particles are close enough to affect the rate at which they all settle, and the degree of thickening achieved will depend very much on the thickness of the compression zone, and the time for compression.

High degrees of thickening can be achieved in **deep-cone thickeners**.

Grease filter. The separation of globules of grease from suspension in a liquid is achieved either in a **coalescing filter** or by **adsorption** on **activated charcoal**.

Grit. A coarse granular solid, usually of mineral origin, usually classified as greater than 75 μm in average particle size.

Grit box, Grit trap. A simple **sedimentation** vessel, usually a rectangular tank, in which very coarse solids can be separated from a liquid **suspension**. Its most common use is as the first stage of **separation** in waste water treatment works, when heavy rain falls have added a significant flow through the sewerage system, bringing with it road grit. The trap also collects any other large objects suspended in the sewage flow, and will usually be fitted with a mechanical rake to remove the settled solids.

The term grit box is also used to refer to the chamber fitted to the bottom of a **hydrocyclone**, which can be opened intermittently. (The term grit pot is also used here.)

Grizzly. A very coarse **sieve** in the form of heavy bars, mounted parallel to the flow of solids or slurry and used for removing large **tramp material** from water, or for the removal of large objects from a solid mixture. The bars (or sometimes chains) are inclined at 20°–50° to the horizontal and may be shaken or vibrated to assist with **classification** and to minimise blinding problems. The apertures are thus parallel to the flow and may taper along their length. Some designs have a comb-like **rake** to clean the slots.

Grooves, Grooved bowl. In a **decanter centrifuge**, the separated solid is carried to the exit of the **bowl** by a rotating **scroll**. Normally the solids are conveyed in a direction parallel to the bowl axis, but in some systems, the solids are carried with the rotation of the scroll. To prevent this from happening, axial slots are machined in the bowl wall or beach to encourage keying of the solids to the wall, and so to assist scrolling efficiency.

Guard magnet. A simple magnetic separator set upstream of another separator to protect it from ferrous tramp material.

Guar gum. A natural, water soluble polysaccharide that is used occasionally as a **flocculant**. Its solutions are neutral and non-ionic, so its flocculant action is little affected by variations in pH.

Gupte equation. An equation describing the permeation behaviour of a porous mass, similar in structure to the **Kozeny-Carman relation**. It is applicable for low values of **Reynolds Number** (Re<1):

$$(\Delta p/\rho v^2)(d/h) = 5.6\varepsilon^{5.5}Re^{-1}$$

where Δp = pressure drop across the porous mass
ρ = fluid density
v = average flow velocity through the mass
d = a characteristic length (usually effective particle diameter)
h = thickness of the mass in the direction of fluid flow (cake height)
ε = porosity of the mass.

Gyratory screen. A particular form of oscillating **screen**, used for dry and wet **separations**, in which the motion of the screen is controlled by two out-of-balance weights attached to the central drive shaft, such that the screen has a partly circular vibration as well as a linear one. The gyrations control the residence time on the screen, so as to achieve optimum performance in separating the particle size mixture fed to the screen.

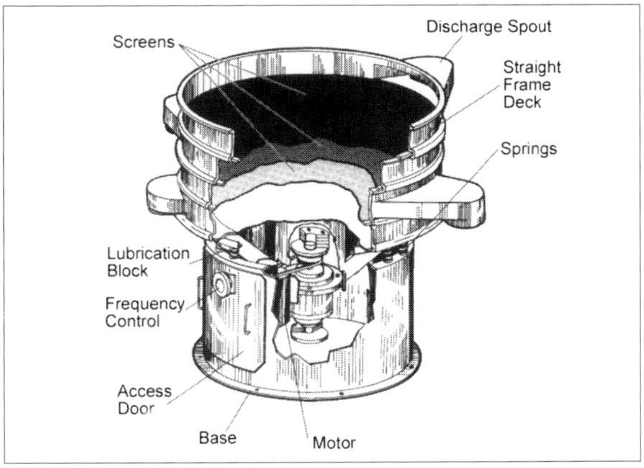

Gyratory screen

H

Hagen-Poiseuille equation. A relationship used in describing the flow of a liquid through a **membrane** in **cross-flow filtration**. It describes the **permeate** flux:

$$J = \varepsilon d^2 \Delta P / 32 \eta X \tau$$

where ε = **porosity** of membrane
d = mean pore diameter
ΔP = **transmembrane pressure**
η = fluid viscosity
X = membrane thickness
τ = **tortuosity** factor

The equation does not give a completely accurate prediction of membrane behaviour, but it is reasonably accurate for **ceramic** membranes where the pores are uniformly distributed and approximately the same diameter, the fluid flow is laminar, there is no **fouling** and negligible **concentration polarisation**. It is not valid when the transmembrane pressure has increased to the point where concentration polarisation is the controlling factor.

Hairy-random staple fibre. A description of natural **staple fibre**, as used in models of flow through **filter media** made from multifilament or multifibre yarn, where flow through the yarn may be as important as flow round it.

Half flow pore size. A term sometimes used for **mean flow pore size**.

Half plate. A plate at the ends of a **filter press** with drainage surfaces on one side only.

Hamaker constant. The property of a solid that determines the magnitude of **van der Waal's forces** of attraction between it and another solid.

Hansen filter. An early example of the use of **electrostatically** enhanced **filter media** was in the resin/wool mixture used by Hansen in a military gas mask, patented in 1931.

Happel cell model. A model of microscopic flow behaviour, of use in the study of **permeabilities**. A cell model views a mass of particles not as a set of tortuous flow paths, but as an array of interacting even if individual spheres, with the flow field around a typical sphere being described in detail.

Hardinge classifier. A gravity **sedimenting classifier** in the form of an inclined trough, up through which moves a set of transverse scrapers attached to a continuous belt. Each scraper carries with it a quantity of solids, which is stirred as it moves up the slope, releasing fine solids, which float up through holes in the belt. The stirred coarse solids are discharged at the top of the slope, while the fines exit with the liquid overflow from the other end.

Harvesting. A term used of **separation** processes in which the prime object is the recovery of a considerable amount of suspended solid from a suspension, by means of **filtration** or **sedimentation**. Harvesting is the prime purpose of a great deal of **cake filtration**.

Because recovery of solids is rarely a requirement in gas/solid separation, harvesting is mainly a feature of separation from liquid suspension. The main application for harvesting of solid/gas suspensions is at the terminus of a pneumatic conveying system, when the separation is normally done by **cyclones**.
[*See Clarification*]

Haze. A thin **suspension** of fine solid particles in a gas, sufficient to have some effect on visibility. The word is also used for a fine suspension in liquids, especially in beverages such as beer or wine.

Hazen effective size. A term used specifically in the characterisation of sand particles for use in **deep bed**

filters in water treatment. It defines the lower size of the bulk of the sand and is defined as that sieve size in mm that retains 90% by weight of the sand, and allows 10% to pass through it.
[*See Uniformity coefficient*]

Hazen uniformity coefficient.
[*See Uniformity coefficient*]

Head loss. A term descriptive of the changes in pressure in a fluid as it moves around a flow circuit. It is given by the difference between the hydrostatic heads at any two points in the circuit. It is equivalent to pressure drop.

Head plate. [*See King plate*]

Heap dewatering. The dewatering of porous solids by allowing water to drain, under the influence of gravity, through a pile of the solids, into drainage channels below the pile.

Heap leaching. The **extraction** of a soluble component from a mass of solid particles by pouring a solvent (usually water) over a pile of the solids, so that it percolates through the mass into a drainage channel.
[*See Percolation leaching*]

Heat treatment. The application of moderate to high temperatures (and sometimes increased pressure) to a material or a mixture in order to change its properties. The change may be chemical, as in the heat treatment of sewage **sludge** to render it innocuous, or semi-chemical as in the **coagulation** of colloidal particles by heating the **suspension**, or physical, such as the heat setting of woven materials to stabilise them.

Heavy media separation. A method of separating particulate solids into two **fractions** by means of their different **densities**. It is achieved by adding the particulate mixture to a liquid medium that has a density intermediate between those of the two fractions. The slurry is allowed to settle, and the **overflow** then carries the lighter fraction, and the **underflow** has the heavier. This process is mostly used in **mineral dressing**, to separate concentrated metal ores from the **gangue** in which they are found, after the mined material has been excavated and crushed. (Also known as dense medium separation.)

Heel. The residual layer of a **filter cake** left on the **filter medium**, after the bulk of the cake has been removed by a knife-like **blade** or scraper. The cutting edge of the blade must be kept away from contact with the filter medium, for fear of abrading the medium material, and this separating distance causes the heel to form. Over a number of filtration cycles, the heel will become clogged with fine particles, and the flow resistance will increase to an unacceptable value. On the other hand, a heel can act as a **precoat** layer, and prevent bypassing of fine solids at the start of each filtration cycle.

The need for a heel is particularly apparent with **centrifugal filters**, because of the high relative speed of drum and knife, and a special stage may have to be built into the operating cycle to condition the heel material, especially its surface, to increase its **permeability**. This is done usually with a liquid spray over its surface (especially effective if the solids are soluble in water), or by **backflushing**.

Also refers to the thin layer of separated solid of necessity left between the **bowl** and the **conveyor** of a **decanter centrifuge**, because of the need for a mechanical clearance.
[*See Glazing, Slurry heel*]

Heel torque. In a **decanter centrifuge**, the **conveyor torque** developed by the heel after feeding is finished and the bowl has emptied, apart from the heel.

Helium separation. [*See Gas permeation*]

Helmholtz vortex. [*See Potential vortex*]

Henry coolant filter. A horizontal, **flat bed**, **gravity filter** for machine tool **coolants**, in which the basic **filter medium** is a fixed **screen** made of **wedge wire** bars. A mat of swarf rapidly forms on this screen and is the actual separating medium. This mat is moved across the screen intermittently by two or more drag chain conveyors, to be discharged over the end of the screen furthest from the slurry feed.

Henry's law. A basic law of the chemistry of solutions, which says that the mass of gas that is dissolved by a given volume of liquid at constant temperature is directly proportional to the gas pressure (and it assumes that there is no chemical reaction between gas and liquid). It is used in the design of **dissolved air flotation** processes.

HEPA. [*See High Efficiency Particulate Air filter*]

Hermans-Bredée filtration laws. Relationships developed 70 years ago in an attempt to rationalise the study of **clarification** processes, i.e. the **filtration**

of dilute, non-cake forming suspensions. The most important mechanisms were identified as direct **sieving** action by the upstream facing pores and the direct **interception** of particles by the **filter medium**.

Heterocoagulation. The mutual coagulation of different species, some of which have negative charges while others have positive charges.

Heterodisperse. A term used of **aerosols**, to imply that the particles comprising the aerosol have a wide range of particle sizes.

Hexhlet collector. A dust sampling device, consisting of a horizontal **elutriator**, a critical orifice control, a dust filter and compressed air ejector.

Heywood tables. In the study of dilute **sedimentation** processes, relationships can be developed involving the **drag coefficient** for a settling particle and the **Reynolds number**. These are complex equations, which can be simplified in terms of two factors dependent only on system properties. Heywood developed two large tables of data relating these factors for spherical particles.

HGMS. [*See High Gradient Magnetic Separator*]

HGMS canister. A simple batch-operated **magnetic separator** consisting of a canister containing the magnetisable matrix and surrounded by an electromagnet. The matrix can be ribbon or wire wool. Gradients as high as 10^5 T/m can be used to trap submicrometre weakly magnetic particles. During operation, flow is usually upwards. The magnet is switched off to clean the system, and the separated magnetic materials can be recovered by removing the canister and replacing it with another one, or by flushing with wash liquid.

High-capacity thickener. [*See High-rate thickener*]

High Efficiency Particulate Air filter (HEPA). A particular reference to a range of fine air filters. In format, these are usually **panel filters**, in a rigid rectangular frame, intended to be installed in a matching housing to provide an inlet air filter to critical working environments. The **filter medium** is usually **pleated** to give a high **filtration area** to frame size ratio.

The filters will be specified against a **DOP** or other **challenge test**, and will have a specified maximum clean filter pressure differential for a rated air flow. [*See CEN/Eurovent*]

High Gradient Magnetic Separation. A magnetic separation process, for either **clarification** or **classification**, using a strong magnetic field with a fine matrix of magnetic material. Also known as high intensity magnetic separation (HIMS) and wet high intensity magnetic separation (WHIMS).

High Intensity Magnetic Separator. [*See High Gradient Magnetic Separator*]

High intensity press. A design of continuous **belt press**, working on **compressible filter cakes** that have already undergone some **dewatering**. Made by Andritz, the equipment features zones of compression using water cushions.

Highly compressible materials. An important class of solids that are separated by **filters** or **centrifuges**, where the accumulated solids can be **dewatered** to a considerable degree by mechanical **compression**. Most sewage sludges are of this kind, and it is common that such solids will have been **flocculated** before separation. Equipment used for their recovery includes **belt presses**, **filter presses** and **decanter centrifuges**.

High-rate thickener. A **gravity thickener** designed to give high throughput rates by careful attention to the **flocculation** stage. The high-rate thickener is similar to the conventional design but is always used with flocculants, and throughputs of 2 to 10 times higher than those in conventional thickeners are achieved by maximising the efficiency of the flocculation within the equipment. This may be done by design of the **feedwell**, which may have mixers to disperse the flocculant into the feed slurry. A variety

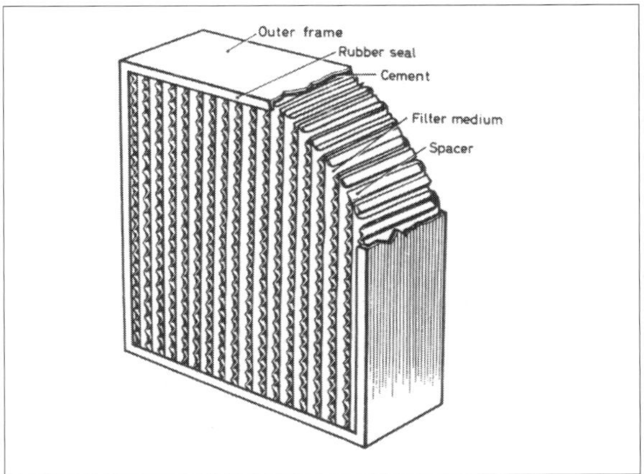

Typical HEPA filter

of designs are available with static or rotating mixers. Also, the feed may be introduced below the surface of the thickened slurry. High rate thickeners need good automatic control. (High-rate **clarifiers** are also known as **reactor clarifiers**.)

High-shear cross-flow filtration. A means of operating **cross-flow filtration** so as to create high rates of shear close to the **filter medium** surface, mainly to reduce **fouling**, but also to minimise **concentration polarisation**. It is achieved either by increasing the flow rate of the **suspension** across the medium's surface (at the cost of higher pumping pressures), or by mechanical agitation: rotating or vibrating surfaces close to the filter medium, or rotating or vibrating the medium itself.

HIMS. [*See High Intensity Magnetic Separator*]

Hindered sedimentation, Hindered settling. As a very dilute **suspension** settles under the influence of gravity, the particles initially are free to settle unaffected by the presence of nearby particles. At a particular concentration, however, the settling rate changes, as the influence of nearby particles starts to take effect, and the particles start to settle as a mass, rather than as a collection of individuals. Hindered settling can begin at quite low concentrations, say a few volume percent.

The rate of settling of a particle under hindered settling conditions is lower than that predicted by **Stokes' Law**, due to the presence of adjacent particles. Equations such as the **Richardson & Zaki** equation can then be used.
[*See also Swarm sedimentation*]

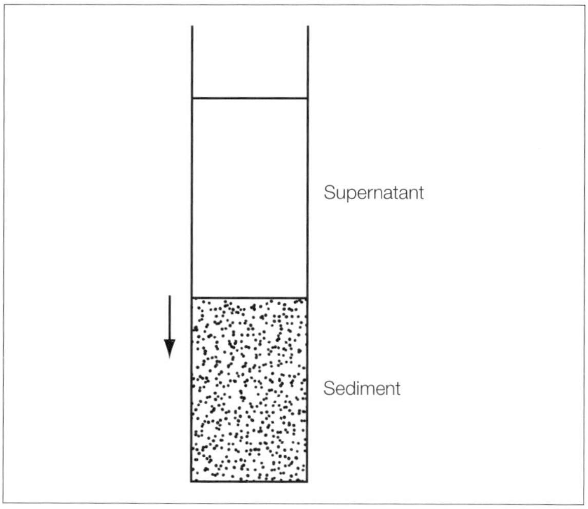

Hindered settling

Hindered settling classifier. A **hydraulic classifier** in which the separation of the different solids occurs in a fluidised bed.

Hindered systems. In batch **sedimentation**, hindered systems are the main alternative class to dilute systems, the difference being the proximity of one particle to another.
[*See Hindered sedimentation*]

Hoffmann filter. A vertical **leaf pressure filter**, in which the **filter elements** are filter leaves mounted radially about a central **filtrate** collection pipe. Cake collects on the outside of the leaves and is scraped off by a set of scraper bars moved manually at regular intervals. The filter found its main use in machine tool coolant filtration.

Holding capacity. [*See Dirt holding capacity*]

Hold up. Generally refers to the liquid held in a **filter** between the **filter medium** and the **filtrate** outlet. Hold up in small scale test equipment can make the data difficult to analyze.

Hollander cloth, weave. [*See Dutch weave*]

Hollow fibre. A form of **membrane filter medium**, in which the polymeric material is extruded as a fine hollow tube, after which it is treated to render it porous. In this form the membrane is able to stand up to very high **trans-membrane pressures**, and so it is the form in which the earliest applications of **reverse osmosis** were embodied, with fibre diameters of less than 1 mm.

Hollow fibre module. A single **hollow fibre** is of little use as a separating medium, because of its low

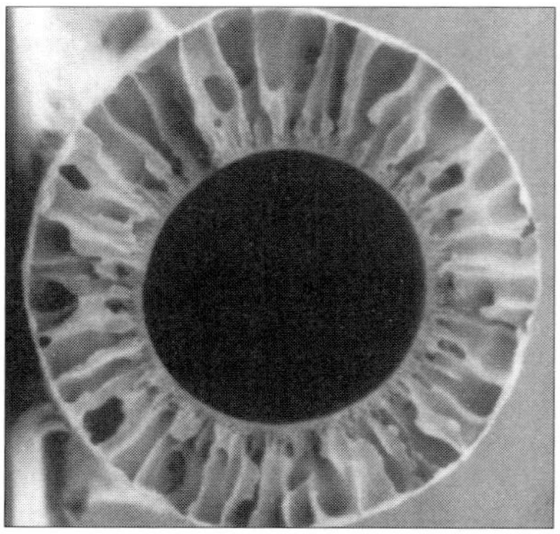

Magnified hollow fibre

Hollow fibre module

surface area. Accordingly, a great many fibres are bundled together in parallel, with their ends sealed into stout plates. The resultant **module** is the basic form in which hollow fibres are used for **membrane separations**.

The feed suspension enters the outer housing for the module at one end, and flows into and along the central channel of each fibre. The permeate leaves through the wall of the fibre, into the surrounding housing and out of a **permeate** discharge pipe, while the **retentate** flows through each fibre and out beyond the far support plate. The support plates must be sealed into the housing tightly enough to prevent leakage under the high operating pressures. When exhausted (i.e. when the **trans-membrane pressure** becomes uneconomically high), the whole module is removed for cleaning (or disposal). The need for the feed slurry to be able to flow through the central channel of the fibre explains the necessity for some kind of a **pre-filter** ahead of the module.

Homogeneous, Homogeneity. Of constant composition throughout, without any variation in concentration. The term is applied to mean similarity at a scale smaller than that which concerns the process of interest, but does not necessarily mean identity of mixture down to the molecular scale. Any homogeneous mixture would thus behave the same however small the sample, down to some limit below the practical requirement of the use of that mixture. Specifically refers to the uniformity of distribution of a solid in a filter **cake**.
[*See Isotropic*]

Homogeneous aerosol. An **aerosol** in which the concentration of the **dispersed phase**, and the size distribution of the particles that constitute this phase, are not dependent upon the position within the system. This is not to be confused with a monodisperse aerosol.

Homogeneous depth filter. A **filter** whose **medium** is a relatively thick porous mass, whose composition is the same in all directions. This describes most **depth filtration** media, in fact most filter media in general, the main exception being those media expressly made to be **asymmetric**, with a thin surface layer intended to act as a **surface filter**, supported on a much coarser layer of the same material.

Homogenisation, Homogeniser. The creation of a homogeneous mixture, especially where the resultant mixture is not intended to separate under the influence of gravity, as in the case of an **emulsion** for use or sale as such (such as a machine tool coolant or a mayonnaise). Homogenisers use a strong mixing action to achieve such a mixture, often combined with an extrusion through very fine holes.

Homopolymer. A **polymer** made up entirely from one **monomer**. This is the most common type of polymer
[*See Copolymer*]

Honeycomb cartridge. Another name for a **yarn-wound cartridge**, in which the successive spiral turns of the yarn are spaced well apart, to give the appearance of a honeycomb (even though with diamond-shaped spaces, rather than hexagonal ones).

Honeycomb weave. A **fabric** whose pattern resembles the hexagonal comb of the honey bee. Also called waffle weave, such fabrics are used for **drainage cloths** below the **filter medium**.

Hood. In working environments, hoods are employed over machines or process plant that generate dusts, fumes or noxious vapours, to carry them away and so avoid pollution of the immediate environment. A captor hood has an exhaust fan to draw in sufficient air to control the pollutant emission. A receptor hood collects the emissions under pressure from the emitting process, and needs a fan only to carry away the pollutants as they are emitted.

Unlike an enclosure, a hood is not sealed around the machine generating the polluting emissions. [*See Steam hood*]

Hopper clarifier. A **sludge blanket clarifier** consisting of a cylindrical or rectangular tank with a conical hopper bottom. The feed slurry enters at the base of the hopper section, and passes up through a fluidised mass of **flocculated sludge**, which retains the suspended material, allowing clear water to **overflow** from the top of the vertical section.

Hopper dewaterer. A top-fed **rotary drum filter**, operating under gravity or vacuum, in which the sides of the compartments are extended outwards with walls, so that each compartment forms a shallow hopper. The feed **slurry** and **cake** are thus held on the filter for a longer time than would normally be the case.

Hop-sack weave. An old name for **plain weave**, as used for making sacks for hops.

Horizontal belt filter. Any belt, or double **belt filter**, in which the orientation of the belts is horizontal rather than vertical. This is by far the most common form of belt filter.

Horizontal filter. A general term, sometimes used for any kind of **top-fed filter**.

Horizontal leaf, plate filter. A **pressure leaf filter** in which the leaves are arranged horizontally, one above the other. Most commonly the elements are plates, with **filter media** on the top side only. The **filter cake** can then be left in place between successive batches. Leaves with filter medium on both sides are used less often.

Horizontal tubular filter. Effectively, this is a **candle filter** on its side: a cylindrical pressure vessel, with its axis horizontal, contains a number of tubes mounted parallel to the vessel axis. Each tube is of a porous material, usually in the form of a **cage** with a **filter bag** inside it. Flow is from inside outwards, and the filter operates batchwise, with flow stopped and the vessel opened to allow the bags to be removed and emptied. Such a filter is best used for **clarification** purposes.

Hot filter press. A modification to a standard **diaphragm filter press** in which heating plates alternate with filtering plates. The heating plates generate steam which then presses the diaphragm to squeeze the accumulated filter **cake**. It then heats the cake to dry it.

Hot gas cleaning. The cleaning of gases, usually exhaust gases from industrial processes or engines, which are too hot to be filtered through standard **filter media** such as **woven** or **non-woven** fabrics. The media required for such purposes must be able to resist temperatures above 150-200C on a continuous basis, which means that they must be made of glass, ceramics or metals, although some polymeric materials can accept the lower range of temperatures quite well.

Housing. [*See Filter housing*]

Hub. Confusingly used to describe two distinct parts of the **decanter centrifuge**. The first usage, preferable referred to as the **conveyor hub**, is for the central part of the conveyor, to which the conveyor **flights** are attached, and through which passes the **feed tube**. The conveyor hub contains the **feed zone** and any other compartments needed to treat the separated solids. The other useage, referred to as the **end hub**(s), covers the end flanges of the bowl, each having a central spindle for supporting main bearings on the outside and conveyor bearings on the inside.

Hubless conveyor. In a **decanter centrifuge**, a **conveyor** that has no central hub, or just a very small diameter one. The flights are held on radial vanes joined onto a small diameter pipe support.

Hukki screen. A type of stationary **screen classifier**, consisting of a cylindrical **wedge wire** screen, mounted with its axis vertical, with a solid cone below it. Feed slurry enters centrally at the top and is distributed by a rotating mechanism. Fines pass through the screen with the **filtrate** and coarse material is discharged through the apex of the cone. **Wash liquid** can be introduced into the cone section.

Humidity. Most **filter media** are relatively unaffected by the humidity of the ambient air, but it becomes an important parameter when electrostatic charges are involved in the **separation**: such charges are much more likely to leak away at high humidity levels.

Humidity control. A feature of the production of compressed air, the act of compression can condense water vapour from the air, in the form of droplets of water in suspension. These need to be removed from the air (by a **coalescing filter**, for example), before any **adsorption** drying is attempted.

Hummer screen. A horizontal **screen** that is vibrated by electromagnets above the screen surface acting on armatures close to the surface.

HVAC. An abbreviation for "heating, ventilating and air conditioning" which covers all the processes, including **filtration**, which are used to clean and condition the air in a living or working space, intended to maximise the health and good feeling of the occupants of those spaces.

Hydraulic classifier. A **classifier** in which the mixture of solid particles, suspended in a liquid, are separated by **elutriation**, usually with the addition of extra liquid.
[*See also Mechanical classifier*]

Hydraulic conductivity. [*See Permeability coefficient*]

Hydraulic diameter, radius. A calculated mean size for a real **pore** in a porous mass. For a single flow channel that is not circular in cross-section, the hydraulic diameter is four times the ratio of the cross-sectional area to the wetted perimeter (i.e. the periphery of the pore, assumed to be in contact with the fluid).

Can also be used of a calculation of the average pore size of the whole mass:

$$d_h = 4\varepsilon/(1-\varepsilon)S_v$$

where d_h = hydraulic diameter
ε = porosity
S_v = specific surface of the porous mass.
[*See Sauter diameter*]

Hydraulic filter. A general term for any **filter** used in a **hydraulic system**, to remove insoluble **contaminants** for the hydraulic fluid.

Hydraulic filter press. A **filter press** in which the compression of the accumulated **cakes** is effected by a hydraulic ram. This can achieve high compression pressures, such as may be needed to express cocoa butter from ground cocoa nibs.

Hydraulic jig. A **jig** in which fluidisation of the mass of feed solids is achieved by pulsing the liquid through the solids.

Hydraulic oil filter, Hydraulic system filter.
[*See Hydraulic filter*]

Hydraulic press. A **separating** device removing liquid from a sludge by **expression**, in which the pressure is supplied by a hydraulic ram.

Hydraulic systems. Many operating systems employ a pressurised fluid to transfer power over a distance, and thus to operate machinery and control working systems. If that fluid is a liquid, then the system is called a hydraulic one, and a major requirement of such a system is that the fluid be kept free of suspended solid particles, which might damage some of the delicate components within the system. The removal of such particles is a demanding task for **filters**, especially of the **cartridge** type.

Hydrocyclone. A form of **inertial separator** used for the separation of solids from **suspensions** in a liquid, by means of **centrifugal sedimentation**. The **centrifugal force** is imposed upon the suspension by its swirling rotational motion in a stationary cylindrical tube, consequent upon the injection of the suspension at one end of the tube, through a narrow zone at the periphery of the tube and at a tangent to it.

The liquid develops a **vortex** flow and moves in this spiral pattern along the length of the cylindrical portion of the hydrocyclone. The suspended solids are sedimented into the liquid layer at the wall, and also move along the wall away from the feed point. The degree of sedimentation is determined by the **angular velocity** of the liquid, and a certain **cut-size** will be achieved depending upon that velocity and the dimensions of the hydrocyclone.

At the end of the cylinder furthest from the feed point, the shape changes to a long cone. At some position between the end of the cylinder and the apex of the cone, the main liquid flow reverses to return inside the primary or separating vortex to the feed end, where it leaves the hydrocyclone through a **vortex finder**, a central axial tube at that end, carrying the **fines** with it.

The thickened slurry continues along the wall of the cone, to discharge through the **underflow** tube at its end. The hydrocyclone thus acts as a **classifier**, producing a thickened slurry of the coarse fraction. It is a simple piece of equipment, and whilst any one unit at a given flow rate will have a fixed cut-size for the suspension being processed, a battery of hydrocyclones in series can produce a number of different size fractions – as is needed, for example, in the manufacture of starch or china clay. Hydrocyclones

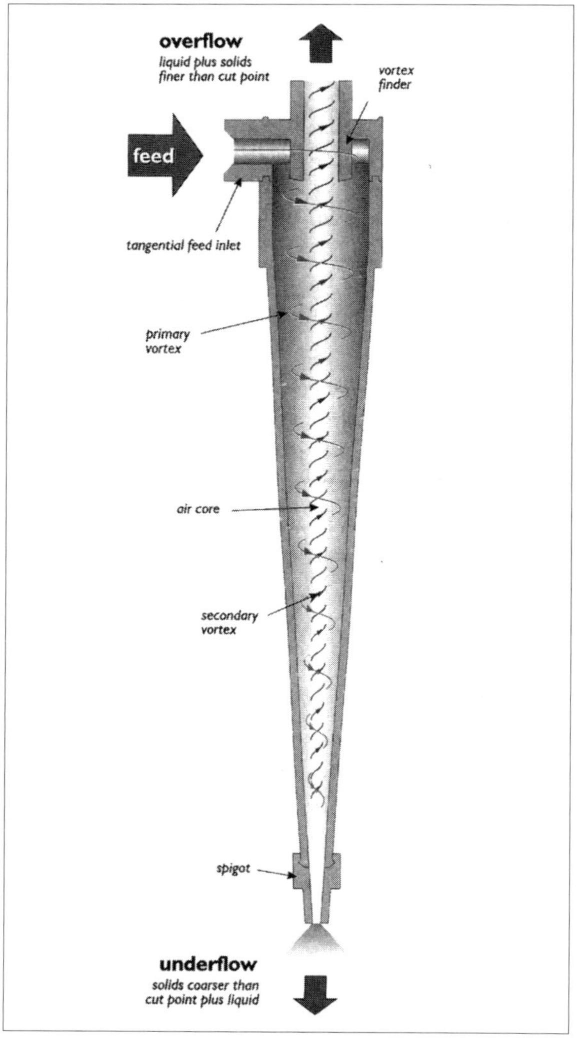

Hydrocyclone

must be used in parallel to achieve higher fluid throughputs, for the same cut-size.
[*See Cyclone, Flat-bottomed hydrocyclone*]

Hydrodynamic interaction. One of the capture mechanisms involved in **clarifying filtration**. In it the capturing particle is seen to influence the path of the particle to be captured, and results in a form of diffusion towards the capturing surface.

Hydrodynamic permeability. A flow characteristic of a porous mass, equal to permeability of the mass divided by the fluid viscosity.

Hydroentanglement. Also known as as spun lacing. The consolidation of a **felt** by the passage through it of fine jets of water at high speed (which acts, like **needle punching**, to entangle the fibres in the immediate vicinity of the water jet).

Hydrogen bond, bridge. The hydrogen molecule is both small and highly polar. As a result, it can attract other polar entities to itself, and bind them together, the hydrogen then forming a bridge between them. The forces involved in the formation of such a bridge are significantly greater than the **van der Waal's** forces of non-polar molecular attraction. Hydrogen bonds are the reason for the nature of water, including its solvation properties.

Hydrogen separation system. [*See Gas permeation*]

Hydromation flat bed filter. A pressurised form of **flat bed filter**, in which the filter operates and the accumulated solids are discharged cyclically.

Hydrophilic. A substance that is easily wetted by water is said to be hydrophilic. The **contact angle** will be less than 90°, as is the case with, say, water and glass, where the meniscus at the top of a column of water in an open tube is concave towards the air (i.e. the edges are higher than the centre). The more general term lyophilic is sometimes used, which also applies to non-aqueous liquids.

Hydrophobic. A substance that is not easily wetted by water, repelling it rather than attracting it, is said to be hydrophobic. The **contact angle** will be equal to or more than 90°, as is the case with, say, mercury and glass, where the meniscus at the top of a barometer column is concave towards the mercury (i.e. the edges are lower than the centre). Another common hydrophobic system is water and PTFE. The more general term lyophobic is sometimes used, which also applies to non-aqueous liquids.

Hydrostatic cake formation. In any **filter** in which the **filter medium** dips into a container full of a **suspension**, the **hydrostatic head** of the suspension around the medium may be sufficient to cause the start of filtration. The result may be an uneven laying down of **filter cake** over the surface of the filter during each cycle of operation. This is especially noticeable in **rotary disc filters** with large diameters, where the hydrostatic head is working, at progressively increasing depths of immersion, before the vacuum is applied inside the disc.

Hydroseparator. A term used for a **gravity settling** device, similar to a **bowl classifier**, but with a cylindrical section as well as a conical one, and with a **rake** designed to handle coarse solids as the underflow. Can be used as a **clarifier** but more commonly used as a **classifier**, although classification performance tends to be poor. The concentration of the **underflow** needs to be kept fairly low to prevent blockage. The rakes are deeply submerged and any change in speed cannot be relied on to alter the efficiency, so hydroseparators do not handle changes in particle size well.

Hydrosizer. A gravity **classifier**, operating in the **hindered settling** regime, with a grid at the base, through which the **elutriation** flow enters and fluidises the settling solids.

Hydrostatic flotation. A **dissolved air flotation** system in which the air is dissolved under a hydrostatic head and is released as it passes through the flotation column.

Hydrostatic head. The pressure exerted at the base of a column of liquid by virtue of the weight of the liquid in the column. In simple **filters**, this pressure may be enough to cause the suspension to be filtered by a **filter medium** at the base – a **gravity filter**.

Hydrostatic head flow test. A method of measuring the **permeability** of clean **filter media** by allowing a head of liquid to fall by a given amount.

Hydrostatic precipitator. A specialised form of **gas cleaning** equipment, in which the suspended dust is washed out of the gas by a combination of a water spray and centrifugal separation, the **centrifugal force** being developed by rapid changes in fluid flow.

Hygroscopic. Hygroscopic materials have a high affinity for water, and will attract it to themselves out of the air, and so will become damp.

Hyperbaric filter. Literally any **filter** working under pressures greater than atmospheric, but specifically refers to a class of continuously operating **rotary drum** or **disc filters** mounted inside a pressurised shell, which contains all parts of the working filter. Such units can be very large, with filtration areas of more than 100 m^2 in one vessel, and can work at pressures of up to 8 bar. The main problem with such filters is the removal of the separated **cake** from the pressure vessel on a continuous basis, which requires some kind of sealed valve or lock hopper

Hyperfiltration. [*See Reverse osmosis*]

I

Idle time. That period in any cyclic process during which the main activity of the process is not being undertaken. Thus, in a **batch separation** process, the idle time covers the period during which the machine is being filled with **suspension**, and the time taken to discharge the collected solid after it has been **dewatered**, and any time required to prepare the equipment to receive the next batch of slurry. The total cycle time is then the sum of separation and idle times

Imhoff cone test. A laboratory test for determining the amount of air in a **flotation** cell.

Imhoff tank. A simple, small-scale waste water treatment device, in the form of a tank with two compartments, one above the other. In the upper tank, which is in the form of a series of channels, suspended solids sediment out from the liquid. The separated solids fall into the lower compartment in which **digestion** takes place.

Immedium filter. The first design of **deep bed**, sand filter, devised in Holland in 1945, to overcome the earlier problem of deep bed filters with the finest sand at the top, in which the normal down flow of raw water quickly blocked this fine layer. In the first version of the Immedium filter, the **filtrate** offtake manifold was buried below the surface, so that about 20% of the raw water flow was downwards, and the rest upwards from the base, through the coarser layers of the bed. In this way, the bed did not expand when it became blocked – a situation prevented in later versions by the fitting of a stationary grid of metal bars in the upper surface of the sand.

Impactor. Any device in which suspended **particles** impact upon a collecting surface by virtue of their **inertia**.
[*See also Cascade impactor*]

Impact ring centrifuge. A continuous **filtration centrifuge**, rotating about a vertical axis, with a conical **basket** opening upwards, from a **feed zone** at the bottom. The **separation** zone of the basket is divided into a number of annular segments by a series of rings, each standing proud of the separating surface. A concentrated slurry is fed at the narrow end of the cone, and the solids are accelerated across its surface as far as the first ring. Here the impact of the solids on the ring causes their **dewatering**, as they overflow the ring and move on to the next. The separated liquor leaves through slots or holes in the basket, and the dewatered solids overflow from its outer periphery.

Impeller. In principle any device that imparts motion to something else, such as in a pump. Specifically, in the case of a **decanter centrifuge**, it is a small blade fixed to the outside of the **beach** at the **cake discharge**, to impel the cake away from the discharge area, or around and out of a **collector**.

Imperfection. A measure of the sharpness of cut in any **classifier**, given by $(d_{75} - d_{25}) / 2d_{50}$, where d is the sieve aperture diameter allowing passage of 25, 50 and 75% of the solids respectively. (Sometimes called the imperfection index.)

Imperforate basket centrifuge. A **basket centrifuge** with its axis of rotation vertical, open at the top, used for separating immiscible liquids, which can flow out of the basket once separated. Any solid suspended in the liquids will be deposited on the **bowl** wall, such that the centrifuge will have to be cleaned at intervals.

The term is sometimes used to include the range of **solid bowl centrifuges**, such as the **tubular bowl**, **decanter** and **disc stack** variants.

Impermeable. A material is said to be impermeable if no fluid, nor any particle, nor molecular nor ionic

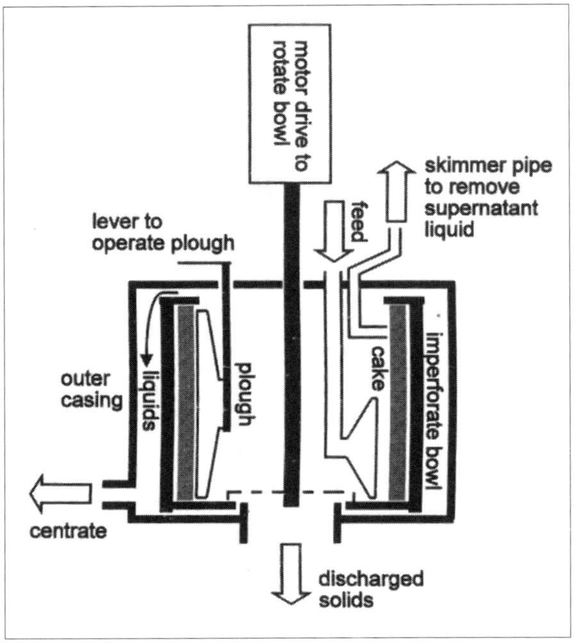

Imperforate basket centrifuge

species, can pass through it, either by bulk flow through physical pores, or by diffusion.
[*See also Permeable, Semipermeable*]

Impingement. One of the mechanisms of particle or droplet **capture** in a **depth filter**. It involves the actual striking of a fibre or other solid collecting surface by the moving particle, caused by the **inertia** of the particle, as the suspending fluid changes direction at the collecting surface, leading to the capture of the particle by means of whatever adhesion forces are operative. The key features of this process are that the particle is relatively small in terms of the size of the channel though which it is passing, and that it crosses the flow lines of the fluid immediately before it is captured.
[*See also Interception, Sieving*]

Impingement separator. Any separation device involving a deep **medium**, in which the **capture** of particles is largely by **impingement**. The term usually refers to the separation of particles or liquid droplets from a gas, the comparable process in liquid suspensions being termed **straining**, although impingement processes do occur in **depth filtration** of liquids.

The collecting surface in such equipment may be a simple set of plates, or a much more complex system, such as **knitted wire mesh**. A major feature of impingement separation equipment is that its design should be such as to avoid re-entrainment of the separated solids.

This kind of equipment can be classified within a larger group of **inertial separators**.
[*See also Demister*]

Inclination angle. The angle subtended to the horizontal by the plates of a **lamella separator** (normally around 35°), which is a major parameter in determining the performance of the separator.

Inclined axis centrifuge. A **peeler centrifuge** with an inclined axis such that the discharge end of the **screen** is at the lower point, to allow easy **cake discharge** – which just falls out of the open end of the **basket** when cut away from the **screen**. This is not a very common design of centrifuge.

Inclined plate clarifier, Inclined surfaces.
[*See Lamella separator*]

Inclined screen. Any **screen** inclined to the horizontal to allow flow of slurry or mixed solids across it, usually in the direction of the slope, although if the screen is vibrated then solids flow may be upwards.

Inclined wall centrifuge. A term sometimes used for any of the range of **cone screen centrifuges**.

Incompressibility, Incompressible cake. Some layers of solid **particles**, especially those composed of coarse particles with diameters of more than 100 μm, cannot be compressed by the imposition of a mechanical pressure upon them (unlike layers of very small particles or **flocs**, which show a distinct **compressibility** and so can be mechanically squeezed). Incompressible cakes cannot therefore be processed in **belt presses**, for example, but have to be **dewatered** by a superimposed gas pressure, or by **centrifugal** action.

Incremental analysis. A method of studying the process of **filter cake formation** by splitting it into a series of small increments accruing with each element of formation time.

Indexing. In any mechanical process in which a set of operations are performed in series upon some material, in a different place for each operation (rather than one after another in the same place), then the material has to be moved from one station to the next. This process of successive precise relocation is called indexing.

Indexing belt filter. Term sometimes used for **intermittent belt filter**, in which the belt indexes from station to station.

Indifferent electrolyte. An electrolyte that, when added to a **suspension**, only alters the composition and extent of the **electrical double layer** around a particle and does not cause **adsorption** of **ions** at the particle surface.

Induced magnetisation. The magnetisation of a material in a magnetic field, which is one of the bases of **magnetic separations**. It is a function of the field intensity and the magnetic permeability.

Inert gas, Inert gas blanketing. In strict chemical terms, an inert gas is one of the noble gases: helium, neon, argon, krypton, xenon and radon, which are chemically unreactive. In **separation** terms, an inert gas is one that does not react with, nor have any other impact upon, a mixture being separated. This generally refers to the use of a nitrogen (or, less often, helium) atmosphere to protect part or all of the slurry from contact with oxygen. This protects the system against oxidation, and perhaps even explosion.

Inertia. The property of a rotating body that opposes any change of rotational speed. A function used to calculate run up times of rotating bodies, such as **centrifuges**.

Inertial lift. A factor in one of the **capture mechanisms** working on a particle in **clarifying filtration**, whereby a particle is seen to wander between different flow **streamlines**, and so become captured by a **filter medium** particle or fibre which straddles several streamlines.

Inertial impaction. [*See Impingement*]

Inertial separator. Any **separation** device in which the solid **particles** or liquid **droplets** are made to leave the **suspension** as a result of their inertia caused by their movement within their suspending fluid. The two broad classes of inertial separator are those relying on impingement on collection surfaces, and those, such as **cyclones** and **hydrocyclones**, which convert the inertia into **centrifugal force**.

Inflatable dam. An elastomeric device fitted inside the **front hub** of a **decanter centrifuge**, to achieve an overflow **weir** by hydraulic inflation, and so to control **pond** depth.

Inflatable seal. A component of a **decanter centrifuge**, similar to an **inflatable dam**, but used to control wash-out by acting on valve ports at a larger radius than the **pond** surface.

Influent. The stream of fluid (usually a suspension) at the inlet of any **separation** device. Also termed **affluent**, and the opposite of **effluent**.

Inhibitor. A chemical agent added to a liquid in order to limit or eliminate corrosion, oxidation or bacterial action in any system containing that liquid. This beneficial action must be balanced against unwanted action in **separation** processes, such as coalescers.

Inhibitors are used in **flotation** processes selectively to suppress the flotation of a specific mineral.

Inhomogeneous depth filter (medium). Any **filter medium** whose **porosity** varies with distance through the thickness of the medium. Although this could apply to any material, it is mainly seen in **asymmetric membranes**, which have a finely porous upper layer with coarser pores beyond this surface layer.

Inlet cone. A feed distribution device, frequently used in **centrifugal filters**, to allow the even distribution of a feed slurry around the entry zone of the centrifuge **basket**. The inlet cone may be stationary, or, more usually, it will rotate with the basket, in which case it also acts as an **accelerator** to bring the slurry up to the rotational speed of the basket.

In-line conditioning, dosing, mixing. The addition of one component to another within the pipework system through which one is flowing, and their mixing together, at the system flow rate, without therefore needing to have separate dosing and mixing vessels. The development of improved additives, and especially of efficient in-line mixers, has made this process feasible.

In-line filter, In-line strainer. A simple **strainer** or **cartridge filter** installed in a **housing** connected directly to the pipeline carrying the fluid to be treated. It treats the full fluid flow, and is intended to interfere with that flow as little as possible.
[*See also Air-line filter*]

Inner liquid. A term sometimes used of the liquid held in the finest capillaries, or inside isolated voids, in

the particles of a **filter cake** or other layer of particles. In the particular case of biological cells, the inner liquid is isolated from the cell's surroundings by the cell membrane, and can represent a large volume fraction of the particle layer. Such liquid is inaccessible for mechanical or centrifugal removal unless the cell wall is broken.
[*See also Interstitial liquid*]

Inorganic coagulants. Most inorganic **coagulants** (salts of aluminium, calcium or iron, for example) give rise to multivalent metallic ions. These improve **agglomeration** of particles both by reducing the **electrical double layer** and by specific adsorption of ions at the particle surface. They have a long history of use in water and waste water treatment, and on the bases of relative cost and performance can still hold their place in competition with polymeric coagulants.

Inorganic membranes. The first **membranes** used in **filtration** were made from organic **polymers**, which still supply the bulk of such material. There is now, however, an increasingly wide range of **filter media** in the membrane separation fields that are made from inorganic materials: **ceramics**, **glass** and metals.

Instantaneous wash curve. The plot of concentration of **solute** in the **filtrate** (usually in dimensionless form) against **wash ratio** in a **cake washing** process.

Integrity (of filter). The condition of a **filter** or **filter element** that ensures its operation according to its specification. This primarily means that the element has no mechanical fault in it that will allow unfiltered solids to pass through it. This is, of course, a generally important characteristic of any filter element, but may be of critical importance, for example, in pharmaceutical or biochemical applications.

Integrity test. Any non-destructive test carried out to establish the **integrity** of a **filter** or a sheet of **filter medium**. This could be done by a **bubble point test**.

Interception. One of the main mechanisms of capture of a **particle** or liquid **droplet** flowing in a **suspension** in the interstices of a **depth filter**. It involves the passage of the particle in the flow lines of the fluid to within a centre-to-surface distance of less than half the effective radius of the particle, so that the particle touches the collecting surface and is held by it, by means of whatever adhesion forces are operative. The key features of this process are that the particle is relatively small in terms of the size of the channel though which it is passing, and that it does not cross the flow lines of the fluid before it is captured.
[*See also Impingement, Sieving*]

Intercrimp. It is normal in the manufacture of woven **wire cloth**, especially from larger gauge wires, to impress a **crimp** in each wire at the point where **warp** and **weft** wires cross. This gives stability to the resultant mesh. In some wire cloths, especially of large aperture and/or light wires, it is advisable to impress an extra crimp, between the wire contact points, to add stability to the finished cloth.

Interface. The surface between two immiscible phases in contact.

Interface profile. In the ideal case of **displacement washing** of a **filter cake**, the **wash liquor** steadily pushes the **mother liquor** ahead of it. The interface profile is then the boundary between wash and mother liquors.

Interfacial tension. The force operating in the boundary between two immiscible phases, measured by the work necessary to increase the boundary area by a given amount. The molecules in the body of the denser phase are attracted equally in all directions by other molecules, except at the surface, where there is a net attraction inwards. This net attraction causes the surface to contract to its minimum possible value – which is why small liquid droplets are spherical. Interfacial tension is measured in Newtons per metre, and is reduced by the presence in the liquid phase of surface-active agents, and by increasing temperature.
[*See also Surface tension*]

Intermediate blocking filtration. A **filtration** process that occurs between **straining** and **cake formation**. It is included as one of the **Hermans-Bredée filtration laws**, and assumes that a particle may block a pore and subsequent particles may build on that blocked pore, so that there is a probability that any particle may block a pore.

Intermediate suspensions. A term used in the settling of suspensions under the influence of gravity. It is the region within the range of **hindered settling** where instabilities in the settling process can result in **channelling**.

Intermittent belt filter. A **vacuum belt filter** in which there are separate positions along the filter (for example, a set of **trays** under the belt) and the belt moves from one station to the next and pauses until the task is complete. Thus there might be a station for feed **thickening**, **filtration**, **washing** and **dewatering**.

Internal discharge. A **cake** discharge method for a **pressure filter** in which the filtering **element** remains inside the filter **housing** during discharge.

Internal drum, Internally fed drum filter, Interior drum filter. A vacuum **drum filter** that has the vacuum compartments formed on its outside, and the **filter medium** mounted on the inside of the drum, with a deep flange on one or both sides of the drum to retain the suspension being filtered. The feed slurry is introduced from one side of the drum, and forms a pool at its base. **Dewatered cake** is dropped from the cell at the top of the rotation, and is removed by a conveyor. A relatively uncommon form of filter, it is used for strongly sedimenting suspensions.
[*See also Dorrco filter*]

Internal moisture. A term sometimes used to describe the liquid held in the internal **pores** in a **particle**.

Internal structure (of filter cake). The internal structure of a filter cake – its **porosity**, the shape and **tortuosity** of its pores – affect both the flow of fluid through the cake and its ability to be **dewatered** by **compression**.

Interpleat. The formation of a pleated **filter medium** from a layered material, such that the layers are pleated together.

Intersection cut size. In a **classification** process, this is that particle size at which the cumulative percentage **oversize** in the **underflow** is equal to the cumulative percentage **undersize** in the **overflow**.

Interstices. The various spaces and openings within a porous mass – otherwise referred to as **pores** or **voids**. The actual interstices may constitute a set of continuous channels through the mass, or they may finish in a dead end. The surfaces of all connected interstices, dead end or not, are available for **adsorption**, but only the pores that run completely through the mass are of interest to **filtration**.

Interstitial liquid. In a **cake** of solids that has been **dewatered** by **drainage**, by gas pressure, or by squeezing, some liquid will remain at the contact points between solid particles, held there by **capillary forces**. The amount of interstitial liquid will depend upon the particle shapes, and upon the **wettability** of the solids, but can be as much as 20% of the volume of a fully saturated cake. The interstitial liquid can be reduced by **centrifugal** action at high speeds. The removal of interstitial liquid is characterised in the fourth region of the **Bond diagram**.

Interstitial velocity. The velocity with which a fluid flows through the **interstices** of a porous mass. It is calculated from the empty vessel flow rate divided by the area of the pores (i.e. superficial velocity divided by **porosity**).

Inverting basket centrifuge. A batch-operated **centrifugal filter**, introduced by Heinkel, in which the removal of the collected solids is effected by turning the **basket** inside out. A flexible cylinder of **filter medium** is attached all around the outer edge of a perforated **drum**, and at its other end to a movable piston. When the **separation** and **washing** stages are complete, the piston moves forward along the axis of the drum, until the filter cloth is inverted, and the **cake** has been discharged by the **centrifugal force**. This method of discharge leaves no **heel** on the cloth (as would be the case with a **peeler centrifuge**). The basket is in an enclosed **housing**, so that

Inverting basket centrifuge

the **filtration** can be aided by a super-ambient gas pressure. This machine is used in the separation of fine chemicals and pharmaceuticals.

Ion. Any atom or molecule, or part of a molecule, that has acquired an electric charge by the gain or loss of one or more electrons.
[*See also Anion, Cation, Co ion, Counter ion*]

Ion concentration. A parameter in the determination and control of the **electrical double layer** thickness and **zeta potential**. Usually, a higher ion concentration will lead to a thinner double layer, and easier **agglomeration** of the particles.

Ion exchange. An exchange process between ions in solution and ions on the surface of a solid that enables a change in the composition of a solution. The solids used are zeolites (natural and synthetic) and synthetic polymers with basic or acidic side groups. By using mixed bed exchangers, all electrolytes can be removed from water, yielding exceedingly pure water.

Ion flotation. A **flotation** process in which **ions** can be removed by first precipitating them with a **surfactant**.

Ionicity. [*See Charge density*]

Ionic strength. A measure of the effectiveness of inorganic **flocculants** in the **agglomeration** of particles. Generally, the higher the strength (i.e. the higher the valency of the ion) the more effective the flocculation.

Irradiation. The use of high energy radiation (such as X-rays) is believed to improve **filtration** behaviour as well as to sanitise the **suspension**, but cost usually tells against this means of filtration enhancement.

Irreducible saturation. The final level of **saturation** in a **filter cake** achieved after **gas blowing** the cake for a long time at high pressure.

Irrigated precipitator. An **electrostatic precipitator** in which the collector plates are washed with an appropriate liquid to remove the collected particles, which might otherwise interfere with the efficient operation of the equipment.

ISO, International Standards Organisation. The primary standards body, producing manufacturing and testing standards for all kinds of materials and equipment, into whose standards all national standards are slowly being subsumed.

Isoelectric point. At the isoelectric point of a **suspension** the **zeta potential** is zero, **electrophoresis** is not possible, but the ability to **agglomerate** separate particles in the suspension is maximised.

It is indicated by the pH value of the solution when dipolar ions are at a maximum.

Isokinetic sampling. The withdrawal of a fluid sample from a flowing fluid in such a way that the sampling device causes minimal interference with the flow of the fluid. The device must be designed in such a way that the flow velocity into the device is identical to that of the fluid flowing around it.

The sample probe will normally be an upstream facing, sharp edged tube. Isokinetic conditions are vital for sampling a gas/solid **suspension**, where the suspended particles are greater than 5 μm.

Isotonic. Two separate solutions are isotonic if their **osmotic pressures** are identical.

Isotropic. An isotropic material is uniform (or uniformly mixed) so that its properties are the same in any direction. An isotropic mass of particles would be one in which the centre of each particle has the same probability to be located at any random point in the mass at any random point in time.

Iwasaki equation. An equation developed in the study of uniform (homogeneous) **deep bed filters**. For a constant **feed suspension**, the **removal efficiency** is the fraction of particles in the suspension removed per unit bed depth:

$$dC/dy = -\lambda C$$

where C = volume fraction concentration at any level in the bed
y = bed depth
λ = filtration constant

J

Jackson Turbidity Unit. A means of measuring the cloudiness of a fluid. The oldest of the methods for turbidity measurement, based upon the viewing of a lighted candle through a calibrated glass tube.
[*See also Formazin Turbidity Unit*]

Jar test. Laboratory test to measure **sedimentation** characteristics. Jar tests include bulk settling tests, detention time tests, **long tube tests**, **short tube tests**, and ultimate pulp density tests. The **flotation** jar test is used for flotation.

In the case of sedimentation, the method employed is the monitoring of the rate of fall of the interface between the clear supernatant liquid and the top of the thickening **suspension**. Jar tests are particularly useful in assessing the effectiveness of **flocculants**.

Jenike shear cell. Laboratory test equipment for measuring the shearing properties of a porous mass, either instantaneously or after a period of consolidation, and hence determining the flow characteristics of a **filter cake**.

Jet collector. A term occasionally used for a **fabric filter**, where the **cake** of solids accumulated on the outside of the **bags** is blown free of the bags by a jet of air inside the bags (**reverse jet**).

Jet counter. A dust sampling device in which the sample is drawn by a hand pump at high velocity through a slit, before impinging on a glass plate. The tube leading to the slit is lined with damp absorbent paper, in order to maintain high humidity in the collector, which aids the retention of **aggregates**.
[*See also Jet impactor*]

Jet impactor. Any of a range of sampling devices for solid/gas **suspensions** in which the airborne dust or droplets are blown or sucked through a jet onto a plate, where the suspended matter is collected by inertial impaction. They all suffer from low efficiencies towards very fine particles, and from a tendency to break up any **aggregates** present in the suspension.

Jet filter. A fairly coarse **filter**, typified by the Doucet water filter, acting as an **in-line strainer** mostly for industrial cooling water flows. A cylindrical shell contains a **wedge wire element**, with the bars running parallel to the axis of the shell. The water to be cleaned enters at the axis of the element and flows outwards between the bars, leaving solids deposited on the inside of the element. The solids are removed by a short burst of backwashing filtrate.

J factor. An empirical parameter introduced (by Tiller) to account for **compressibility** effects which had been neglected in earlier theories of **cake filtration**.
[*See also Tiller-Shirato model*]

Jig. Equipment used in the mineral industry for **separating** solids by size, shape and density, by intermittently fluidising a mass of ground solids. It involves a fixed **screen**, mounted at a slight angle to the horizontal, through which water flows in a pulsating manner. A bed made up of granules of different mineral materials passes over the screen, and the pulsations in the water flow cause the granules to stratify according to their physical properties. At the lower end of the jig, the different strata leave through different exits.

Jones separator. A form of **high gradient magnetic separator**, in which a set of collection boxes rotates within an array of iron core magnets. Used for ore **separation** or **clarification**.

JTU. [*See Jackson Turbidity Unit*]

K

Kappa factor. A factor in **cake filtration** that relates **cake thickness**, h_c, **filtrate** volume, V, and the **filter area**, A, and is also a concentration factor, relating solids volume concentration, c_v, to cake **porosity**, ε:

$$\kappa = h_c A/V = c_v/(1 - \varepsilon - c_v)$$

Kelly filter. An old-fashioned design of batch-operated **leaf filter**, usually operated as a **pressure filter**, dating from the start of the 20th century, in which a rack of vertical filter **leaves** is mounted on sliding rails inside a horizontal pressure vessel. The leaves are manifolded to a common **filtrate** off take through the openable end of the tank, through which the rack of leaves is withdrawn at the end of the **separation** and **washing** parts of the cycle, to enable the collected solids to be washed or blown off the leaves into a collection hopper beneath them. The Kelly filter is often used with **filter aids** for slurries that are difficult to filter.

Kieselguhr. [*See Diatomite*]

Kelvin Voigt component. The viscous component of deformation during the **compression** of a **filter cake**.
[*See also Voigt element*]

Kinematic viscosity. The (coefficient of) kinematic viscosity, ν, is used in modifying the equations of motion of a perfect fluid to allow for the properties of a real fluid. It is the ratio of the (coefficient of dynamic) viscosity, η, to the fluid density, ρ:

$$\nu = \eta/\rho$$

Kinematic viscosity is measured in m^2s^{-1} (or in Stokes, where 1 St = 1 cm^2s^{-1}).

King plate. The plate at the fixed end of a **filter press**.

Knife, Knife discharge. A device for the removal of a (washed and) **dewatered filter cake** from the **filter medium** in a batch or continuously operating **filter** or **centrifuge**. It can actually have a cutting action, as is the case for the knife in a **peeler centrifuge**, or act as a deflector plate (in some **drum filters**). Although an efficient means for removal, it does leave a **heel** of solids on the filter medium, to prevent damage of the medium by the knife.

In the case of a filter, the knife will be fixed close to the medium surface, to slice off the accumulated cake, which will fall away under its own weight. Some cakes are less easy to remove, and an air blast from underneath the medium is then needed to supplement the knife action. The knife can then be fixed, in which case the filter medium must also be restrained. Alternatively, the knife can be floating over the cake and lightly loaded by springs, so that it is free to follow the movement of the medium, which is then allowed to billow out away from the drum surface under the pressure of the air flow.

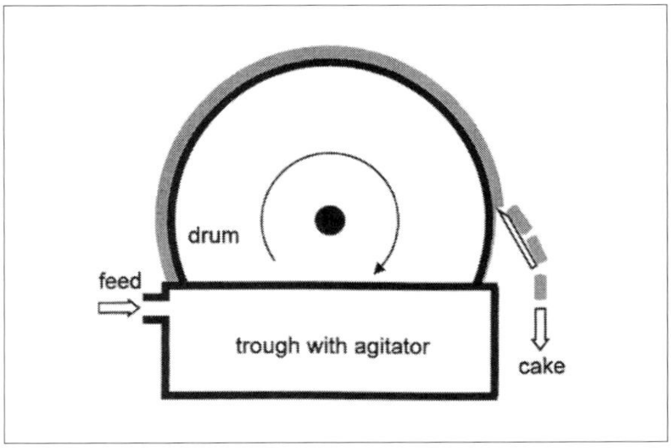

Drum filter with knife discharge

In the case of a **basket centrifuge**, the knife will normally be held well away from the cake whilst **filtration** and **dewatering** are under way, and then lowered gradually into the cake, to peel it away in layers (hence the name: "peeler" centrifuge). The knife has a minimum distance of approach to the filter medium, to avoid damaging it, hence it leaves a heel.

Knitted, knitting. A knitted fabric is produced by the interlocking of a series of loops made from one or more yarns, with each row of loops caught into the preceding row. Loops running lengthwise are called wales, those running crosswise courses.

Knitted fabrics are not much used in filtration, but knitted **wire mesh** is a very common material used to provide the bulky medium necessary for the **capture** and **coalescence** of mists.

Knudson flow. The flow regime operating at very low pressures (high vacuum), when the mean free path of gas molecules is larger than the dimensions of the confining space. Any resistance to motion is now due to molecular collisions with the passage walls, and not because of fluid viscosity.

Knudson number. A **dimensionless number** relating to low pressure flow. The Knudson number:

$$Kn = \lambda / r$$

where λ = mean free path length for gas molecules
r = a characteristic dimension, either particle mean diameter, or fibre diameter

Konfiltro. A **belt press** designed by Bayer and BHS that has a heated permeable pressing device placed on the top of the **filter cake**.

Konimeter. A particular kind of **jet impactor**, in which **aerosols** are sucked through an orifice, and impacted upon a collection plate, which is usually coated with an adhesive. Particularly useful for the comparison of different aerosols where the type of dust remains constant, but the concentration varies.

Kozeny-Carman equation. A relationship, analogous to **Darcy's law**, used to quantify the permeation velocity, v, through a porous mass of particles. It relates the superficial velocity to the pressure drop per unit thickness of the **cake** (in the direction of flow), by means of a constant factor that includes the specific surface area of the particles in the cake (i.e. surface area per unit volume), the fluid viscosity and the **porosity** of the cake:

$$v = 1/k(\varepsilon)\ \{\ \varepsilon^3\ \Delta p/(1-\varepsilon)^2\ S^2\ \eta\ h_c\}$$

where ε = porosity
Δp = pressure drop across cake
S = surface area of cake per unit volume
η = fluid viscosity
h_c = thickness of cake in direction of flow

The proportionality constant $k(\varepsilon)$ has to be determined experimentally, but has the value of about 4 for values of porosity between 0.3 and 0.65. The equation loses its accuracy at sizes below 10 μm.

This relationship can be used to determine the specific surface of a powder from measurements of **permeability** in packed beds of the powder.

Kozeny diameter. The average size of a particle determined from a measurement of **cake permeability**. The basic relationship derives from the **Kozeny-Carman equation**.

Krämer filter. A mechanically complex **filter**, which takes the form of a horizontal pressure tank, with an axial rotor on which are mounted a set of disc-shaped **membrane** filter units. In the **thickening** stage, the discs rotate to prevent the formation of a **cake**, but when rotation is stopped, a series of press **diaphragms** squeeze the cake to dewater it, before the accumulated solids are discharged. The filter is used for the separation of high-value pharmaceuticals.

Kynch theory. A theory of **batch settling** that leads to a design method for **gravity thickeners** (and which is similar to that given by **Talmage and Fitch**). The theory starts from the assumption that the settling rate of a **suspension** is solely a function of the solids concentration.

L

Laboratory tests. Small scale tests done at the laboratory bench to determine the **filtration** or **separation** behaviour of a **suspension**. These may be as simple as a **jar test**, or may involve special equipment for the purpose. The object is, as far as possible, to discover what degree of separation can be expected, and to set the dimensions of the full scale equipment accordingly. It may, however, be too large a jump from laboratory to full scale, and an intermediate pilot scale test may be needed.

Labyrinth. A **settling** tank fitted with **baffles**, designed both to increase the residence time in the tank, and also to promote **plug flow**, so as to reduce the residence time distribution.

Lace weave. A version of **twill** (or **linen**) **weave** in which the **warp** and **weft** threads, which cross one another at right angles, are of different diameters.

Lagoon. A large open pond, used for storing large flowrates of waste slurries, containing low or moderate concentrations of solids, and thereby **clarifying** them by **sedimentation**. The settled solids collect on the bottom, from which they must eventually be scooped up and removed. The solids can be recovered (e.g. by a drag line), but they are more usually a waste product, which is sent to landfill. The overflowing liquid has then been more or less clarified, but may need subsequent treatment before it can be released to a water course. The base of the lagoon should be lined to prevent the liquid leaching into the ground. If the settled particles are very fine they may need to be consolidated before landfilling, so that they will form a load bearing material. (In the mineral dressing industry, also called a tailings dam.)

Laid fabric. A material made only from **yarns** running in one direction, held together by rubber latex or some other adhesive, without using **weft** yarns.

Lakos separator. A cylindrical **classifying separator**, acting in the same way as a **hydrocyclone**. The feed enters an upper chamber and from here enters the main cylinder through tangential slots. The **underflow** can be discharged continuously but, because Lakos separators are normally used with dilute feeds, the solids are usually discharged intermittently from a **grit box** at the base of the cylinder. A plate near the bottom deflects the liquid upwards towards the **vortex finding overflow**.

Lambda. In a **decanter centrifuge**, lambda is the ratio of the **bowl length** (from feed point to **front hub)** to bowl diameter. It is thus a measure of the relative **clarification** ability.

Lambda value. A term used in the empirical modelling of the **dewatering** of filter cakes under the influence of **centrifugal force**. It is used in a relation between **residual moisture content** and elapsed **dewatering** time, and is defined by:

$$\lambda = \eta h_c / \rho C g\, r_h^2 t$$

where λ = lambda value
η = viscosity of the liquid phase
h_c = thickness of cake
ρ = liquid density
C = **C-value** of centrifuge
g = gravitational acceleration
r_h = mean **hydraulic radius** of the pores in the cake
t = dewatering time.

Lamella. Literally a thin layer of material, the word is used in **separation** technology to refer to a thin separating surface mounted as one of a parallel set, used in oil-water separation, for example.

Lamella clarifier. [*See Lamella separator*]

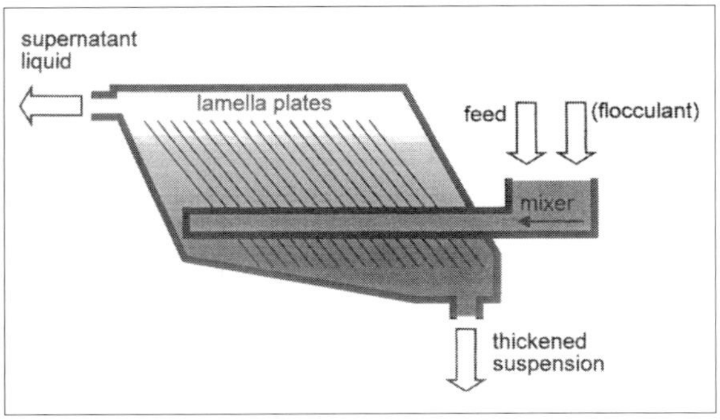

Lamella separator

Lamella separator. A **gravitational sedimentation** device, fitted with a set of **plates** (lamellae), joined parallel to one another, and with the whole set tilted at an angle to the horizontal. The presence of the plates in the **suspension** flow stream gives a very short settling distance (for a suspended particle or liquid droplet) before it hits a solid surface, and so leaves the suspension. The effect of the plates is thus to provide a very large area for sedimentation within a relatively small ground area.

The vertical distance between the plates varies from a few millimetres to a few centimetres, depending upon the size of the particles to be separated, and upon the required flow conditions. Separated solids can slide back down the plates and leave at their lower ends (which risks their re-entrainment at the entry points) or they can be carried on and discharge over the top edge. The clarified liquid is discharged through special rising channels.

Lamella separators can be used as **liquid/liquid separators** (as in oil-water separation), or as **clarifiers**, for the removal of small amounts of suspended solids – an application of increasing use in water treatment. Liquid/liquid separation and solid removal can also be achieved in the one separator. The device can, in addition, be used as a **thickener**, in which case there will be a sludge collection space below the plate assembly, which must be deep enough to allow the formation of a sludge compression zone.

Laminar flow. At low rates of flow, a fluid will move smoothly over a surface, through a channel or around an obstacle in its path, in such a way that small parcels of the fluid will move past one another without intermixing of the fluid. As flow rates increase this laminar (or streamline) flow will be disturbed by the formation of local eddies with resultant mixing of the fluid. In the laminar flow regime, flow characteristics are governed mainly by the viscosity of the fluid.

The extent of the laminar flow regime is indicated by the value of the **Reynolds number** for the flow system. The same sense of smooth flow is applied to the settlement of a particle in a fluid, for which system again a Reynolds number can be determined:

$$Re = vd\rho/\eta$$

where v = settling velocity
d = particle diameter
ρ = fluid density
η = fluid viscosity.

If the Reynolds number, measured in this way, is less than 0.2, then the particle movement is laminar, and settling is said to be in the **Stokes'** region.

Laminate, Laminated, Lamination. If two or more layers of material are laid one above the other, the resultant mass is called a laminate. The layers may simply adhere naturally, as is the case with **webs** of natural fibre making a **felt**, or they may be **bonded** together with an adhesive, or they may be fused together. Many types of **filter media** are made by lamination, and it is the standard method for making sturdy **membrane** media.
[*See also Composites*]

Langmuir equation. [*See Peclet number*]

Lap. A term used to refer to a relatively thin layer of **felt** or other **non-woven medium**, held between sheets of **woven fabric** to provide strength, and used in **air cleaning filtration** systems.

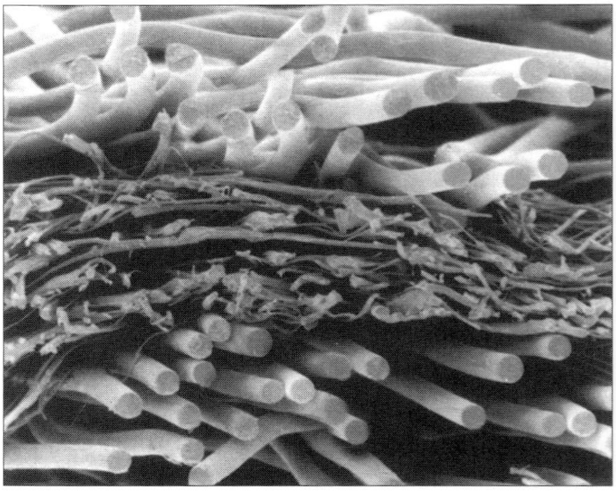

Three layer laminated medium

Laplace equation. An equation that states that the pressure required to dewater a **capillary** is equal to twice the surface tension times the cosine of the **contact angle**, divided by the pore radius. (Also known as the Young equation.)

In any capillary system where an interface exists between two fluids, one at least being a liquid, the shape of the meniscus at the interface is described by the general Laplace equation, which relates capillary pressure to the interfacial tension:

$$p_c = \gamma(R_1^{-1} + R_2^{-1})$$

where p_c = capillary pressure, or the pressure required to dewater the capillary
γ = interfacial (or surface) tension
R = main curvature radius of each of the two liquids.

In the particular case of a cylindrical capillary, holding a gas/liquid interface:

$$p_c = 2\gamma \cos\delta / r$$

where δ = wetting angle (contact angle) between liquid and solid surface
r = capillary radius

Large-scale filters. A term sometimes used for the production-scale **process filters**, to distinguish them from **laboratory** or **pilot-scale** filters.

Larox filter. A **diaphragm filter press**, with the **plates** horizontal, stacked one above the other. The **filter medium** is a single band of cloth, moving from section to section through the tower of plates.
[*See also Tower press*]

Latex particles. Spherical particles of latex can be generated in quite narrow ranges of particle size, and thus make excellent **challenge suspensions** for liquid **filters**.

Launder. A wide flat channel used for conveying liquids, often featuring as the delivery channel for a **slurry** to a **separator**.

Lauter tun. An exceedingly simple form of **filter** in the form of a large cylindrical vessel with a perforated base and a slow stirrer within the **suspension**, used for separating spent grains from the wort (sugar solution) prior to fermentation of the wort to beer.

Laval. Gustaf de Laval was a Swedish scientist and engineer, who, on the world stage, is most identified with the development of the high-speed steam turbine, but in terms of the **filtration** industry is best known as the inventor of what became the **disc-stack separator**. From the age of 27 he worked in a steel works, and built his first impulse steam turbine in 1882. A patented design of reaction turbine followed in 1883, and by the turn of the century, most steam turbine work was based on his inventions.

It was in 1878 that he invented the **centrifugal** cream **separator** and set up Separator AB to make and market it. The disc stack patent was actually awarded to von Bechtolsheim (in 1888), whom Laval persuaded to come to work for Separator – which, in the fullness of time became Alfa-Laval.

Layer formation. Refers specifically to the formation of a **gel layer** on the **membrane** surface in **cross-flow filtration**.

Layering. A method of operating **vacuum filters** handling **suspensions** whose **cakes** are prone to cracking. The cake, after **filtration** in the first stage of the filter, is moved to the next stage without cake removal. The slurry then fills the cracks with cake in the second stage, and the final cake can be **dewatered** efficiently, without bypassing of the dewatering air.

Layer thickness. A general term describing the thickness of a mass of porous material, which could either be a **filter cake**, or a **sediment**.

Lay up. A term sometimes used for the shutting down of an operating **filter** or **centrifuge**.

L/D ratio. In general terms, for any cylindrical structure, this is the ratio of the length of the cylinder to its diameter. If applied to a **filter cake**, formed in a cylindrical vessel, and if L is the height of the cake and D is the particle diameter, then this gives the effective number of particle layers in a cake, a term of use in predictions of **washing** performance.

Leaching. The **extraction** of a component from its mixture in a solid mass, either by simple solution with an appropriate solvent, or subsequent to a chemical process that renders the component soluble. The term is most often used for the recovery of valuable components from their ores, but has two more specific meanings in relation to **filtration**:

- the removal of a soluble component from a **filter cake** using **wash liquid**, or
- the dissolution of a soluble component from the **filter medium**, which may then contaminate the **filtrate**.

Lead. The term used to denote the number of **flights** wound onto any **conveyor hub** (e.g. single lead, double lead, etc), and especially for the **scroll** of a **decanter centrifuge**. The more leads, the more expensive the conveyor, but the lower the scrolling load on each flight: halved with two leads, and so on. (Sometimes called single start, double start etc.)

Leading edge. [*See Forward edge*]

Leaf. [*See Filter leaf*]

Leaf filter. A discontinuous **pressure filter** in which a set of **filter leaves** is mounted inside a pressurised shell. The **filtrate** flows into the centre of each leaf from its outside, allowing the **filter cake** to build up on the outer surfaces of the leaves. The filtrate passes from the interior of each leaf to the filtrate exit through a manifold. The dewatered and washed cake may be removed mechanically from the surfaces of the leaves, or by **backwashing**. Leaf filters are used for suspensions that are difficult to filter, and are often employed with **precoat** or **filter aids** to assist the separation process.
[*See also Moore filter, Niagara filter*]

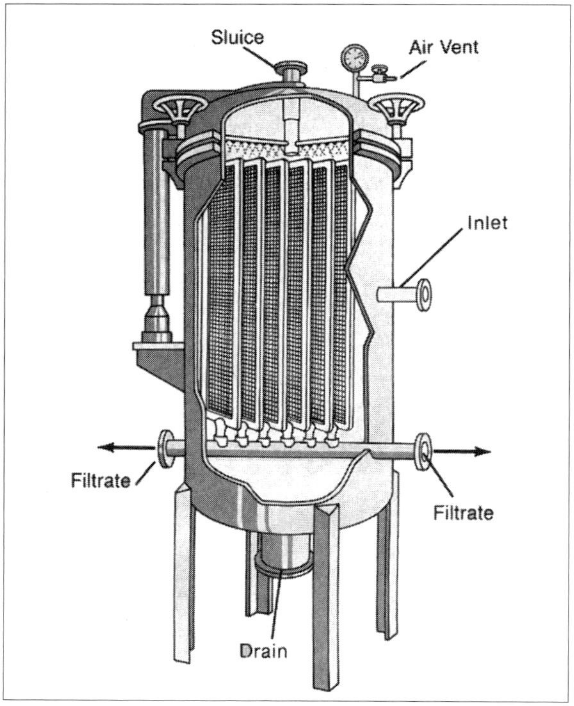

Vertical leaf filter

Leaf test. A laboratory test procedure, of great importance in the assessment of a **filtration** operation (provided that the sample being tested is truly representative of the actual feed material). It can give information on the nature of **cake formation**, the ease of pick-up by the **filter** of the agitated **slurry**, the phenomena surrounding **cake dewatering** and drying, and the effectiveness of **cake washing**.

The test involves setting up a piece of the appropriate **filter medium** to cover a circular or rectangular cell in the form of a shallow chamber fitted with a supporting screen, and clamped onto the chamber so as to give a defined **filtration area**. The medium can be fixed to one side, or both sides, of the chamber. The assembled leaf is then attached to a vacuum source, or placed in a pressure vessel, and immersed in the slurry to be tested.

Leakage air. [*See Secondary air*]

Leeds column. Originally designed as a column for **percolation leaching**, the Leeds column has since been developed for **flotation**. It is a vertical cylindrical vessel, divided into compartments (one above the other) by sets of rods. These rods can be depressed by the weight of the solids on them, so as to open a channel that allows solids to flow down and liquid to flow up.

Lees. The solid residues left in the bottom of a fermentation vessel after the fermentation is complete – and from which the product needs to be **filtered**.

Legionella. The active bacterium, *Legionella pneumophila*, responsible for **Legionnaires' disease**.

Legionnaires' disease. A form of pneumonia named after a group of American war veterans, at whose convention the disease was first identified in 1976. It is caused by the bacillus *Legionalla pneumophila*, which grows slowly in a moist environment, and in a narrow range of pH. It is most lethal originating from air conditioning systems involving air-cooling towers, whose circulating water systems require an extremely efficient **filtration** system to remove the bacilli.

Lenticular cell cartridges

Lenticular. An object shaped like a double convex lens is described as being lenticular in shape. The term can be used to describe the shape of particles, but it is more commonly used of particular forms of **filter element**, being a **capsule** that is made in this shape, and then used, either in this form, or stacked, one on top of the other, to form a type of **cartridge**.

Level controller. A device used to indicate and control the level of liquid in any vessel, but a very necessary accessory in **batch centrifuges** to ensure that they are not overfilled in the fill part of their operating cycle. Krauss-Maffei has a novel design that can distinguish between the surface of the suspension and that of the formed **cake**, thus showing when drainage is beginning.

Life expectancy, Lifetime. The operating period that can be expected of a **filter element** or piece of **filter medium** before it can no longer be adequately cleaned, and must therefore be replaced.

Limiting flux. The maximum flux achieved in a **cross-flow membrane** process, which is reached as the **feed** pressure is increased.

Linear screen. [*See Travelling band filter*]

Linear tipping pan filter. [*See Travelling pan filter*]

Line filter. [*See In-line filter*]

Liner. A removable bag for collecting solids, which can be removed manually from the separator for easy discharge. The bag may be permeable (as in a **filter**) or impermeable (as in a **sedimenting centrifuge**). The purpose of a liner is usually to provide easy discharge, but may also be to provide a finer **filtering medium** than the supporting screen.

The term also applies to removeable inserts (fitted, for example, inside the **bowl** or **beach** of a **decanter centrifuge**), intended to protect a surface from the effects of erosion, or to make that surface more suitable for the transport of accumulated solids across it.

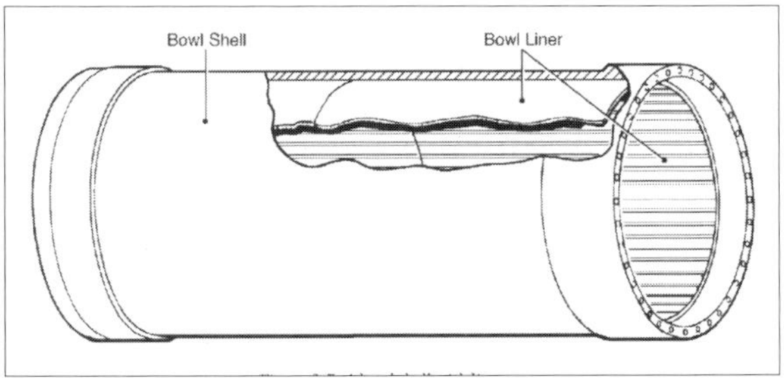

Decanter bowl with liner

Linen weave. A term sometimes used for **plain weave**.

Line settling. A characteristic of **zone settling** behaviour, where a visible interface can be seen between the top of the settled zone and the clear liquid above it, and that interface moves downwards with time. **Clarification** behaviour would see an interface moving upwards through the suspension. (Sometimes used as a synonym for zone settling.)

Liquid bridge. [*See Interstitial liquid*]

Liquidity index. A parameter used for defining the flow characteristics of a **filter cake**.

Liquid limit. One of the two **Atterberg limits**, used in measuring the flow characteristics of a **cake**.

Liquid/liquid separation. Most **separation** equipment, whether from liquids or gases, is designed to remove suspended solids. However, there are many industrial separations where the dispersed phase is a liquid, in the form of distinct droplets, and special types of equipment exist to effect this liquid/liquid separation. (Where the two liquids are fully miscible, then quite different processes, such as distillation, must be employed.)

These separations can be achieved mainly by **sedimentation**, although **filter coalescers** act in much the same way as **filters** for solids, with droplet entrapment on an open mass of **fibres** or **filaments**. Sedimentation can be effected under the influence of gravity or **centrifugal force**. The **lamella separator** and the **disc stack centrifuge** were developed primarily for liquid/liquid separation, and the simple **tubular centrifuge** is almost entirely used for this purpose. The mechanical engineering sector used a very wide range of simple devices for oil/water separations.

Liquid load. A term sometimes used for the proportion of **residual moisture** in a porous mass, especially a **filter cake** after **dewatering**. The load is defined as the ratio between mass of liquid and mass of dry solid.

Liquid viscosity control. The viscosity of a suspending liquid can have a very important influence on **separation** processes, and it may be necessary to submit a suspension to a **pre-treatment process** whereby the viscosity is changed in order to improve the separation efficiency. The two main methods for viscosity control are temperature (slight increases in temperature can make significant changes in **non-Newtonian liquids**), and dilution with a less viscous (but fully miscible) liquid.

Load, Loading. A term with several meanings in **separation** technology. The load stage in any operation is that part of a cycle when the **filter** or **centrifuge** is being filled with the **slurry** to be separated.

A filter that has reached the point in its cycle where it can take no more material for processing without requiring an excessive pressure drop is said to be fully loaded, while the loading of a filter at any stage in its operation is the weight of accumulated solids per unit **filtration area**.

Locus zero vertical velocity. In a **hydrocyclone**, there is a region of downward moving liquid close to the outer wall, and a region of upward moving liquid around the axis. The LZVV is the boundary between these two regions, being a three dimensional "surface" sitting roughly midway between wall and axis.

Log normal distribution. A method of representing a particular continuous distribution, for instance of **particle sizes**, in terms of ratios of deviations from a geometric mean value. A log normal distribution is often a close approximation to the actual distribution obtained by the crushing and grinding of a mineral. Its value is that it converts a curve into a linear representation.

Log reduction value. The ratio of reduction in concentration of a particular contaminant between the **feed** and the **filtrate** in a **filtration** process, expressed as a logarithm. Used especially of reduction in bacterial contamination.

London forces. [*See Van der Waals forces*]

Long arm centrifuge. A laboratory test **centrifuge** in which the tubes or beakers are mounted at a considerable distance from the axis of rotation.

Longitudinal baffle. A rectangular baffle plate, fitted between two consecutive **flights** of the **conveyor** of a **decanter centrifuge**, between the **feed zone** and the **beach**, to give a back pressure to the scrolling solids.

Longitudinal dispersion. A parameter involved in the **displacement washing** of **filter cakes**, especially as **wash ratios** increase. Rather than **plug flow** of the wash liquor, it becomes dispersed longitudinally through the cake.

Long tube test. A laboratory test used to size **gravity clarifiers**.

Looped wedge wire. One form of wire format used in **wedge wire screens**. The basic component is a thick round wire that is first looped at regular, well spaced intervals, and then pressed to give it the deep wedge section in the straight parts, and flat at the loops, which then act as spacers between adjacent wires, when a number are formed into a screen. The loops accept cross bars to support the **screen**, which now becomes an accurate **sieving** device.

Loose medium. Any unbound particulate **filter medium** material, such as granular solids and **filter aids**.

Looped wedge wire and screen

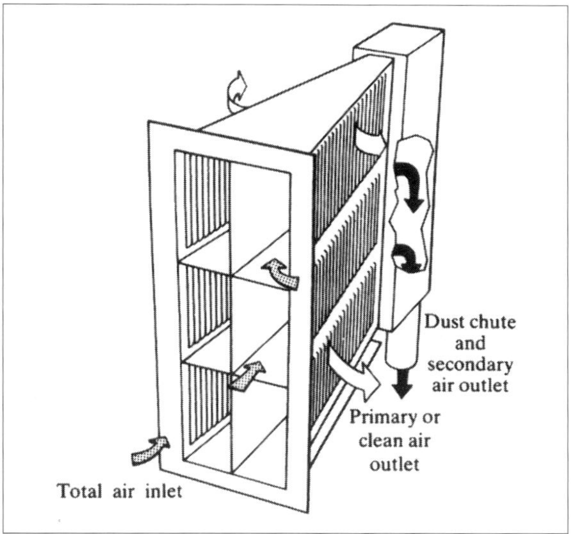

Louvred air cleaner

Louvred granular beds. Downward moving granular beds are of increasing interest in **gas cleaning**, since they can operate at high temperatures. The flow of hot gas is horizontal, across the flow of the solids, which are contained within two closed walls on opposite sides, and two walls through which the gas flows, made of horizontal louvres, angled inwards and downwards, so as to prevent flow of solids out through the spaces between the louvres.

Louvres. Devices used to separate dust from a gas flow as a result of **inertial forces**, as the gas is made to reverse its flow around sets of vertical vanes. At each reversal, the dust continues into a collection zone, from which it is removed by a fraction of the gas flow, the secondary flow.
[*See also Aerodynamic cleaner, Baffle collector*]

Lower casing. In the case of a **decanter centrifuge**, the lower half of the **casing**, which is fixed and attached to the **frame**.
[*See also Upper casing*]

Low vacuum belt filter. A general term for any **belt filter** used for **separating** and **dewatering** coarse particles, and therefore operating at a vacuum of less than 5kPa.

L-type filter. A **cartridge filter housing** that has its inlet and outlet ports at right angles, and the main axis of the **housing** parallel to one or the other.

Lug screen. A form of **well screen**, which is a **strainer** fitted at the bottom (i.e. the inlet end) of a well pipe. The lug screen has a **wire screen** mounted on vertical bars outside a perforated piece of pipe, allowing free entry of well fluid over the whole area of the screen.

Lundin filter. A type of **drum precoat filter**, developed for the beverage industry. Some versions have a very small feed trough at the side, while others have a pivoting feed trough that allows the submergence of the drum to be varied. There is also a spray bar for **wash liquid**, which allows the maximum quantity of beverage to be recovered. The Lundin filter exists in sizes up to 16 m^2.

Lyophilic. Colloidal substances that are **solvated** and therefore in a stable suspension are said to be lyophilic. Sometimes used as synonymous with **hydrophilic**.

Lyophobic. Colloidal substances that are not **solvated** and hence are easily coagulated are said to be lyophobic. Sometimes used as synonymous with **hydrophobic**.

Lysis. The breaking of a cell structure, in particular to release the product of a biochemical process. **Centrifugal filtration** is frequently the process step which follows lysis.

LZVV. [*See Locus zero vertical velocity*]

M

Machine direction. For **non-woven** roll goods, there is no equivalent to the **warp** and **weft** directions of a weaving loom. Nevertheless, there are some characteristics of the finished material that relate to the way in which it is made, whether these relate to the original width of the machine, or the direction in which the roll was moved off the end of the machine, known as the machine direction.
[*See also Cross-machine direction*]

Machine parameters. The key dimensions of the machine used for **separation** – **filter** or **centrifuge** – such as the **drum** width and **cake formation angle** for a **rotary filter**, or the diameter and length of the cylindrical part of a **decanter centrifuge bowl**.

Macrofiltration. A term that has been relatively recently introduced into the filtration vocabulary, to refer to the wide range of filtration applications at particle sizes greater than those now referred to as **microfiltration**, i.e. at sizes above say 5 μm. For a considerable period of time, this was the whole of **filtration**.

Macroflotation. A term sometimes used for normal **froth flotation** of suspended particles, to distinguish it from **microflotation**.

Magnesia. A white, amorphous, mainly insoluble powder (magnesium oxide) used sometimes as a challenge dust for testing **filter efficiency**.

Magnetic belt. A continuous **belt** fitted with magnets for the clarification of liquids containing ferrous contaminants, or for separating magnetic from non-magnetic solids. The magnetic materials will be carried on the belt, out of the **separator** trough to a collection zone, where they will be scraped off the belt.
[*See also Crockett belt separator*]

Magnetic cartridge. A magnetic separator that consists of a magnet inside a non magnetic, cylindrical chamber which in turn is surrounded by a ferrous cage. It is for the removal of ferrous contaminants from a **suspension**, and can be used in pipelines, in tanks or in channels.

Magnetic conveyor. A device used for separating magnetic materials from a **suspension** in a liquid. The suspension flows over a layer of non-ferromagnetic material (such as stainless steel), under which a series of powerful magnets move, in an endless chain. Mainly used for the removal of **swarf** from **cutting fluids**.

Magnetic disc filter. A rotating disc fitted with radial magnets, whose polarity alternates around the periphery, which dips into a trough through which the slurry flows, in order to remove ferrous contaminants from machine tool coolants. Collected particles are scraped off when the disc rotates out of the slurry. These are usually associated with a group of machine tools and have capacities up to 50 m^3/h of liquid.

Magnetic drum. A rotating **drum** carrying magnets. Magnetic drum **filters** are used for **clarifying** cutting oils and machine coolants. The magnetic rotating drum is usually covered with a disposable paper medium, so that the magnetic solids can be easily discharged. A compression roll may be used to recover coolant or oil from the solids. Drum diameters are up to 0.7 m.

Most low intensity magnetic **separators** are in the form of a rotating drum. There are 3–10 stationary magnets of alternating polarity arranged radially and covering about 180° of the drum. Electromagnets can be used but permanent magnets are more common. The drum is covered with non magnetic material and rotates at about 40 rpm. In the cocurrent design, the feed stream flows in the same direction as the rotation of the drum. The magnetic material passes through a gap where it is compressed and dewatered before being scraped off the drum. This design is used for producing a clean magnetic product from relatively coarse particles.

In the counter-rotation design, the feed flows in the opposite direction to the rotation of the drum. This design can handle high solids loadings and also surges in the feed. Losses of magnetic material are low, but the magnetic product may not be as clean as with the cocurrent design.

Another countercurrent design is also known as a Steffenson tank, in which the non-magnetics are forced to flow in the opposite direction to the rotation of the drum. This design can give high recovery of magnetic materials and a clean magnetics stream. Coarse particles tend to settle, so the design is used mainly for fine particles. Fluctuating flows can be handled.

Magnetic filter or separator. Any device for removing magnetic materials from their mixture with non-magnetic materials, or from suspension in a liquid. The magnetic force may come from permanent magnets, or from electromagnets (where the magnetic field is only active if the electric current is switched on).

The **separation** may be done by means of a dry separator, such as a conveyor belt with magnets close overhead (to separate magnetic ore, for example, from a crushed mineral, or tramp iron from foodstuffs), or in a magnetic filter, where the filter element is made of magnetic material (used for **clarification** of a recirculating fluid, such as in the removal of metallic particles from a hydraulic fluid circuit).

Magnetic flux density. A measure of the strength of a magnetic field, given by the number of lines of magnetic flux in a given area normal to that flux. It is equal to the permeability of free space times the sum of field intensity and induced magnetisation. Also given by permeability of the material times the field intensity. The tesla is the unit of flux density.

Magnetic hydrocyclone. A **hydrocyclone** can be placed in a magnetic field to separate magnetic solids. The magnetic field and **centrifugal forces** can be arranged to reinforce or oppose each other. If they reinforce, then the cut size will be lowered. If they oppose one another, the hydrocyclone can be used to separate magnetic from non-magnetic solids.

Magnetic susceptibility. The ratio of the intensity of magnetisation produced in a material to the field intensity causing that effect. This is a measure of the suitability of using magnetic methods of separation.

Main frame. A term sometimes used in the description of **centrifuges**, especially the **decanter centrifuge**, for the frame on which the centrifuge is mounted, when there is a **sub-frame** in use.

Mantle. A region surrounding the **vortex finder** of a **hydrocyclone**, as a hollow cylinder at the top of the **LZVV**, in which the **radial velocity** is zero.

Manual discharge. The manual removal of a cake of solids collected in a **filter** or a **centrifuge**. This may be done by means of a removable **bag**, or by scraping or flushing. Manual discharge is not an acceptable method for large volume separations, and tends to be found only in small-scale production processes.

Martin's diameter. A statistical determination of **particle size**, which works on the projection of the particle on the graticule of a microscope, and is the mean value of the lengths of the line across the particle that is parallel to the traverse of the microscope, and that divides the projected area into two equal parts.

Mashing. The process of creating a **suspension** of a dry solid in a liquid, particularly used in the fermented beverage industries for the preparation of the suspension of malted barley and hops, or crushed grapes, prior to the main fermentation process to produce beer or wine, respectively.

Mass balance. A design calculation process for any **separation** system, which is based upon the conservation of mass among feed, product and waste streams from the system.

Mass bubble point. [*See Boil point*]

Mass cake. A term sometimes used in **cake formation** calculations, to refer to the combination of **cake thickness**, solids density and cake **volume fraction**, and is thus the mass of dry cake deposited per unit area of the **filtration surface**.

Mass concentration. The **concentration** of dry solids in a **suspension**, as measured in mass terms, and quoted as mass of solids per unit volume of suspension. This can be found by weighing, either after thorough **filtration** and **dewatering** from the **suspension**, or by **evaporation** to dryness.

Because several suspension separation parameters are dependent on particle size, rather than mass, a **volume concentration** may be more useful.

Mass distribution. A feature of a **particle size distribution**, being the relative frequency of distribution by mass of particle. It is usually represented by a graph of **cumulative percentage undersize**.

Mass median diameter. In a **particle size distribution**, the mass median diameter is that diameter below which lies 50% of the total mass of that distribution.

Mass recovery. [*See Total efficiency*]

Mass settling. A term used in describing the various types of **gravity sedimentation** process, referring specifically to the settling of concentrated **flocculated suspensions**, as opposed to the **clarification** of dilute suspensions, or the **compression** of settled **sludges**.

Mass throughput. The quantity of material transported through or separated in an item of **separation** equipment, per unit time. Mostly used of the solid mass throughput, and, in the case of **filters**, usually expressed also on a unit filter area basis.

Mass transfer. A term used to refer to the movement of substances through a continuous phase, or between distinct phases, by convection or diffusion. It also is used to classify all the processes that operate in this way, such as **distillation**, **adsorption** or **extraction**.

Mat filter. [*See Pad filter*]

Material feeder. In some types of **separation** equipment, special devices may be needed to ensure that the **feed suspension** is evenly distributed across or over the separation zone. This is especially so with thick or strongly **flocculated** suspensions, being fed, say, to a **belt press**.

Mathematical models. Procedures intended to make the design of complicated processes more simple. They are often based on empirical observation, but can also give reasonable insight into the nature of an operating process. Many of the design procedures for **filters** and **centrifuges** are based upon such models.

Matrix. In a composite material, the matrix is the **continuous solid phase** into which the particles or crystals of the **dispersed phase** are embedded.

Matted cake. A **filter cake** containing interlocked **particles** or **fibres**.

Mattress. [*See Drainage grid*]

Maximum pore size. The diameter of the largest **pore** in a porous mass, assumed to be a **capillary** of circular cross section. It is this pore that is identified by the **bubble point test**.

Maxwell-Davidson filter. An early form of automated **filter press**, which included **cake compression** by **diaphragms**.

Mean filter rating. A term sometimes used for the performance rating of a **clarification filter**, based on a measurement of **pore size**.

Mean flow pore size. The mean size of the **pores** in a porous mass, as measured by a **bubble point** apparatus.

Mean flow rating. The particular **pore diameter** in any porous mass such that all the pores in the mass larger than that value carry a volume of fluid flow equal to the flow through all of the pores that are smaller than that diameter.

Mean particle passed test. A performance test for a **filter** or piece of **filter medium**, in which a test **slurry** of hard spherical solids (such as **ballotini**) is passed through the sample. The **filtrate** is passed through a very fine **membrane filter**, which collects any particles that pass through the sample. These collected solids are examined through a microscope, and the largest particle that can be seen is measured to give an **absolute rating** for the filter.

Mean particle size. A term used generally for the mean or average size of a mass of particles, as determined by a range of possible methods.

Mean pore flow test. A test procedure for **filter media**, especially for **membranes**, particularly specified by **ASTM** (ASTM F316). It is a gas flow test, in which the flow rate is determined for a series of applied pressures, both to the dry medium and the fully wetted medium. The plot of flow vs pressure is linear for the dry medium. This is drawn, and below it is then drawn the line for half of the flow rates. The point at which this half flow rate line crosses the curve for the wet medium defines the pressure equivalent to the mean flow pore size.

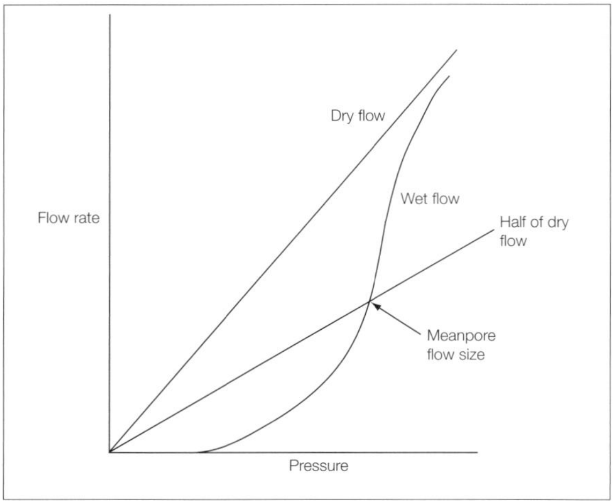

Mean pore flow test

Mean pore size. A general term referring to a mean or average pore size in a porous mass, as determined by any one of several possible experimental procedures.

Mechanical cell. A **flotation** cell with a mechanical agitator in it, to disperse the gas throughout the cell.

Mechanical classifier. Any **gravity sedimentation** vessel from which the solids are removed continuously by some kind of mechanical conveyor (drag line, screw, etc).
[*See also Hydraulic classifier*]

Mechanical dewatering. In **filtration** and **separation** terms, this refers to the removal of water or any other liquid that is held in the pores of a mass of solid material, such as a **filter cake**, by mechanical means, rather than by thermal methods (evaporation or drying). Mechanical methods include **drainage** under the influence of gravity or **centrifugal force**, squeezing by a piston or a flexible **diaphragm** (such as a **belt**) or a **screw press**, or by applied gas pressure.

Where the solid mass is to be completely **dewatered**, then the last liquid removal will need to be done thermally, because mechanical methods will always leave a little **interstitial moisture**. However, thermal energy is usually more expensive than mechanical, so it follows that the maximum of mechanical dewatering should be achieved before the solids are dried thermally.

Mechanical dewatering boundary. The mechanical **dewatering** of a porous mass reaches its usual limit when the **coarse capillary liquid** is all removed. This still leaves the **fine capillary liquid**, and that held in closed pores and at the particle touching points. Some of this fine and interstitial liquid can be removed under the higher **centrifugal** forces, or by the readjustment of particle positions under the force of a squeezing **belt**, but the boundary or limit of mechanical dewatering will always leave some liquid within the mass.

Mechanical jig. A **jig** in which the fluidisation of the **slurry** is achieved by vibrating a sieve plate under the solids.

Mechanical syneresis. A process within the group of **agglomeration** phenomena, in which a loose **floc** is densified by mechanical forces, which are applied locally and unevenly, fluctuating over the surface of the floc.

Mechanised filter press. [*See Automatic filter press*]

Media separation. A term sometimes used for **heavy media separation**. This is an important process in

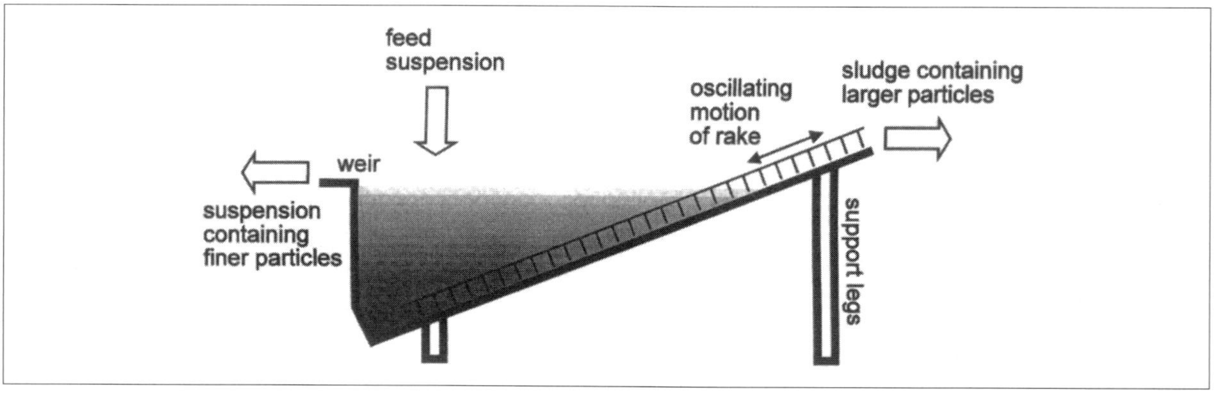

Mechanical classifier with rake discharge

mineral dressing, and the selection of the right medium or media for the separations required is a key stage in their success. Not only must the media offer the right density to allow floating and sinking of the different components, but they must also have no solvent action on any solid present, nor should their trace presence in the product affect its value.

Medium, Media. [*See Filter medium*]

Medium filter. A term sometimes used for those **filters** using a relatively coarse **medium** (**strainers** or **cake filters**) to distinguish them from **membrane filters**.

Medium migration. [*See Shedding*]

Medium resistance. [*See Filter medium resistance*]

Melt blown. Synthetic **thermoplastic** polymeric **filaments**, extruded continuously from a **spinneret**, are broken up by jets of air, and laid down on a moving belt as a mass of **fibres**. The product makes an extremely fine **filter medium**, which is now an important member of the **non-woven** materials spectrum. The fibres may also be laid down on a moving core as a thick **bonded cartridge**.

Membrane. *See* pages 164–6

Membrane bioreactor (MBR). A combination of a bioreactor (most commonly an activated sludge waste water treatment system) with a **membrane filter**, usually operating in **cross-flow.** This has become a major system in the treatment of waste water.
[*See also Submerged membrane bioreactor*]

Membrane cartridges. [*See Cartridge, Membrane*]

Membrane cleaning. All **filter media** must be cleaned at some stage (unless of the use-once-only type), but in no part of the filtration spectrum is cleaning more important than with membranes, which, because of their fine pores, are easily **fouled**, and only cleaned with difficulty. Several different processes are used, including *in situ* washing, and acid and alkali cleaning.

Membrane distillation. A **separation** process between two miscible liquids, which takes place by the vaporisation of one of the liquids at the upstream mouths of the fine pores of the membrane (which is not wetted by the liquid). The vapour is drawn through the pores by vapour pressure difference with the downstream side of the membrane, where the emerging vapour is condensed.
[*See also Pervaporation*]

Membrane filter press. [*See Diaphragm filter press*]

Membrane filters. Filters employing **membranes** are of two types, working in completely different ways. In the first, the membrane is completely impervious, and acts as a squeezing device to compress the **filter cake**, and so to **dewater** it. Examples of this usage are the **diaphragm filter press**, the **tube press**, and the compression belt in a **belt press**.

The other usage has the membrane permeable to at least one component of the feed suspension or solution. This is, of course, the basis of the large and rapidly growing field of membrane filtration.

Because of the major importance of the **semi-permeable** membrane as a **filter medium**, the alternative use of the word (widespread in the construction industry, for example) for an impervious material is discouraged in the **filtration** industry, and the word **diaphragm** is to be preferred.

Membrane filtration. [*See Membrane*]

Membrane filtration index. A method of characterising the **fouling** of **cross-flow membrane filters**, given by the slope of a **t/V vs V** plot.

Membrane fouling. [*See Fouling, Membrane*]

Membrane stack. A term sometimes used for **membranes** when used as flat **sheets**, in a **module** containing several sheets.

"Mercobowl" centrifuge. A range of **decanter centrifuges** developed by Dorr Oliver, now subsumed into Alfa Laval.

Merco strainer. A simple vertical cylindrical **screen strainer**, developed by Dorr Oliver, with **filtrate** flow outwards through the screen, allowing collection of dirt on the inside, from which it is continually scraped or brushed, to fall to an outlet at the base of the **housing**.

Mercury intrusion porosimetry. A laboratory method for the determination of the **pore size distribution** of a porous mass. Mercury is forced into the pores of the evacuated mass under a pressure that increases in a series of steps, and the amount of mercury taken up, at equilibrium, is noted for each pressure value. The pore size corresponding to each

Membrane. Originally implying a thin, **micro-porous** or **semi-permeable** plastic sheet, the term "membrane" is now applied to any **filter medium** that is capable of removing particles below 0.1µm. The membrane represents probably the fastest growing part of the filtration media market (especially if ceramic membranes for hot gas filtration are included).

Modern membrane technology began with the development of the first high performance membrane for the desalination of salt water by **reverse osmosis**, as reported in Loeb and Sourirajan's seminal paper of 1963. From these beginnings as a thin, flexible semi-permeable sheet of regenerated **cellulose** material, developed to separate species at the molecular and ionic level, the membrane has expanded enormously, to embrace solid inflexible ceramic and sintered metal, and an ever-increasing group of polymeric materials, and to applications that now extend well into the **microfiltration** range. The existence of the membrane as a very effective filtration medium led to the development of the whole field of **cross-flow filtration**, which also now extends well beyond its reverse osmosis origins.

In view of these rapid developments into such a wide range of formats and applications, any attempt to provide a succinct but comprehensive definition of a membrane is very difficult. To many people, a membrane remains a thin flexible material, but in filtration terms the word now covers any medium that can achieve separations at 0.1µm or below (down to molecular and ionic sizes), and which may be thick or thin, flexible or rigid, organic or inorganic. In addition, many membranes are now employed in microfiltration applications at cut sizes well above 0.1µm.

The membrane is essentially a **surface filtration** device, with little or no **depth filtration** involved in its use. In practice many membranes are of **asymmetric** structure and effectively comprise two layers. The active, surface layer is a very thin skin, the **permeability** of which is of critical importance. The lower, thicker layer is of more open structure, its role being to serve as a mechanical support for the active layer.

Large membrane plant

The fine surface structure of all membranes implies the need for significant pressure drops across the medium in order to achieve adequate fluid fluxes. As a result, membranes need to be contained in pressure tight **housings**, and considerable ingenuity is required to achieve sound and efficient operation.

The main processes in which membranes are used in industry are:

- the filtration of fine particles, down to less than 0.1µm in effective diameter, from suspension in liquids or gases (microfiltration);
- the removal of very large molecules and colloidal substances from liquids (**ultrafiltration**);
- the selective removal of some ionic species from solution (**nanofiltration**);
- the removal of effectively all dissolved and suspended matter from water and other solvents (reverse osmosis);
- the selective transport of ionic species only (**electrodialysis**);
- the separation of mixtures of miscible liquids (**pervaporation**); and
- the separation of gas mixtures, including mixtures of gases and vapours (**gas** and **vapour permeation**).

Most membrane processes operate by means of cross-flow filtration, in which only part of the fluid passes through the membrane as **filtrate** (or, more correctly, **permeate**, since some membrane processes operate by **permeation** rather than filtration); the retained part, or **retentate**, consequently becomes more concentrated in particulate or solute species. Whereas microfiltration utilises both **through-flow** and cross-flow filtration, cross-flow is the usual mode for the other membrane filtration processes, and has thereby grown to its present level of importance.

Depending on the properties of the material used, membranes may be produced in the following geometrical forms:

- **flat sheets**, which are self-supporting or backed by a supporting substrate;
- **spiral wound**, which is made by laying a series of membrane sheets and spacer sheets alternately, and rolling this array up into a cylinder;
- tubes – self-supporting or backed by a supporting substrate, typically12-24 mm in internal diameter;

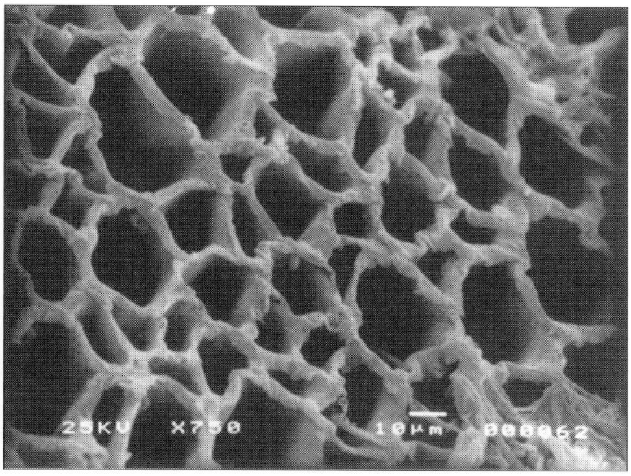

Membrane cross section

- perforated blocks, circular or hexagonal in cross section, perforated parallel to the axis of the block by a set of tubular spaces, on the surface of which the membrane is laid;
- **hollow fibres** – typically 40 μm internal diameter and 80 μm outside diameter, used in bundles sealed into plates at each end of the module.

Polymeric membranes are used in all these forms except the perforated block, which is very largely restricted to inorganic materials.

In all of the above formats the membrane medium is stationary, and the fluid flows across (or occasionally through) it. However, the problems of **fouling** and **concentration polarisation** can be at least partly reduced if an element of **shear** can be applied to the boundary layer at the membrane surface. This can be achieved by causing the membrane either to rotate or oscillate with relation to the fluid flow. Special equipment designs are being developed to achieve this.

Because of the very fine nature of the membrane media, it is normal practice to employ a filter, ahead of the membrane unit, that is intended to remove any particulate material that might interfere with the membrane process. This is especially necessary where the flow passages are very narrow, such as in hollow fibre membranes. In fact, some membranes themselves are used as **pre-filters** to membranes operating at a finer degree of separation. Thus there will normally be a microfilter ahead of an ultrafiltration or reverse osmosis membrane, but there may also be an ultrafiltration membrane ahead of a reverse osmosis step.

The build-up of a fouling layer on the surface of a membrane is one of the most serious problems in membrane processes. The term "fouling", rather than the more familiar "**filter cake**", arose from the origins of membrane processes in molecular separations, where macromolecular proteins would separate on to the membrane surface as a slimy, gel layer, which rapidly reduced the flux through the membrane. Fouling layers have to be removed periodically by cleaning, but much ingenuity is employed by membrane system designers to minimise fouling layer formation.

The extent to which cross-flow successfully prevents the surface of the membrane being fouled by deposited particles is dependent on a variety of factors, especially the cross-flow velocity. Chemical and/or mechanical procedures are usually required to clean (and sterilise) the membrane, which must be able to withstand the associated mechanical, chemical and thermal stresses.

Another operating problem, concentration polarisation, affects the membrane processes dealing with suspended or dissolved species. The molecules to be separated (i.e. kept in the retentate) diffuse through the liquid close to the membrane surface and become much more concentrated at the surface, creating a different kind of barrier to liquid flow, and so reducing flux. In the same way, the particulate matter accumulates in the liquid as it approaches the boundary layer, creating a similar resistance to liquid flow.

As well as local shear, there are two other types of method employed to reduce fouling and/or concentration polarisation, and so increase flux rates:

- changes in the surface characteristics of the membrane; and
- conditioning of the feed slurry/solution.

The surface of the membrane needs to be as smooth as possible, and the slurry or solution as free as possible of material that will foul the surface. Operational modifications are generally designed to create some kind of shearing or scouring of the fouling layer. Some of these are mechanical, while the use of a two-phase (gas/liquid) flow is growing in importance.

Factors that have a profound influence on the structure or morphology of a particular membrane are the nature of the process by which it is manufactured and the form of the raw material used. The main manufacturing processes are:

a. **sintering** of fine graded particles;
b. solvent casting or phase inversion, involving the stage-wise evaporation of a solution of polymer in a mixture of solvents;

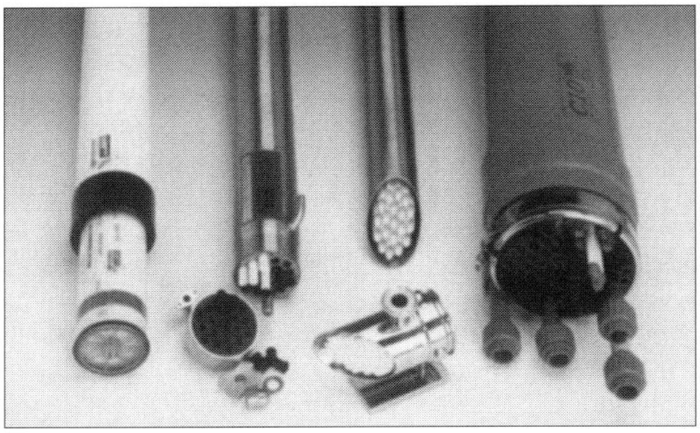

Membrane formats

c. **irradiation** and etching of an impervious film; and
d. **stretching** an impervious film to cause multiple ruptures.

To be effective for separation, membranes should exhibit appropriate characteristics, such as good chemical resistance (to both feed and cleaning fluids), mechanical stability, thermal stability, high permeability, high selectivity and general stability in operation. Originally, all membranes were based on natural materials or derivatives of natural cellulose. Whilst cellulosic media continue to play an important role in certain areas of application, the major source of membranes is now synthetic polymers. There is an immense variety of polymeric materials available as membrane media, including grades with specially developed properties: hydrophobic or hydrophilic, anionic or cationic, for specific filtration applications.

During the last twenty years or so, inorganic materials such as ceramics and metals have become of increasing significance as membrane materials. The introduction of these, despite their being nearly an order of magnitude more expensive than their organic counterparts, has occurred because of their much-improved operating lifetimes, their robustness, their greater tolerance to extreme conditions of operation, such as higher temperature and aggressive chemicals, and the subsequent overall saving in lifetime costs.

The irregularity of the pores of most membranes, and the often irregular shape of the particles being filtered, results in there not being a sharp cut off size during filtration. With **symmetric** membranes some degree of depth filtration could occur as smaller particles move through the tortuous flow path. To counteract this effect, asymmetric membranes, which have surface pore sizes much less than those in the bulk of the membrane material, are used to trap the particles almost exclusively at one surface (the membrane skin) whilst still offering low hydrodynamic resistance.

pressure level is then determined from the **Laplace equation**.

Mesh. A material woven in **plain weave** from a relatively stiff **monofilament yarn**, a mesh, in **filtration** terms, is characterised by well defined **apertures** between the yarns. Most commonly made from metal **wire**, but increasingly also from polymeric materials, the stiff filament enables the relative position of the yarns to be fixed (especially if the finished mesh is then **crimped** or **sintered**), and so provide an accurately sized aperture. A set of meshes of different sizes can then be used to undertake a particle size analysis on a mixture of solid particles.

Meshes are used for **screening** applications, and are the truest **surface filter media**. As such, of course, they can also be used for **cake filtration**.

Mesh size. The size of the **apertures** in a **wire** (or plastic) mesh, **screen** or **sieve**. Traditionally, mesh sizes are quoted as number of apertures per linear inch of the mesh, but the conversion from this number to the equivalent particle size obviously depends then on the diameter of the **filament**.

Metafilter. A **cartridge filter** in which the **filter element** is constructed from a stack of scalloped rings mounted on a central grooved rod.

Metal edge filter. An **edge filter** where the components are made of metal, a common usage in the automotive sector. The components for the **cartridge** are made of machined metal, such that quite precise apertures are left between them. These components may be rings or discs in stacks, or **wedge wire** or ribbon wound around a central former.

Metal fibre paper. A **filter medium** made, like **paper**, from the **wet-laying** of metal **fibres**.

Metal filters. A rather unhelpful term, which can either mean a **filter** made out of metal (as distinct from, say, plastic), or a filter for the removal of impurities from **molten metals**.

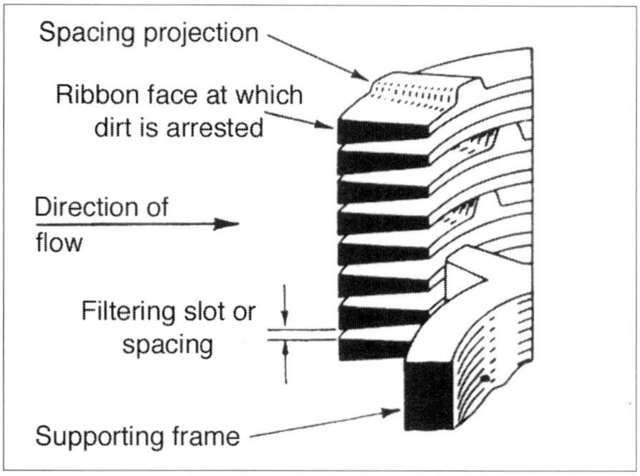

Metal edge ribbon filter

Metallic sheet media. Filter media made from sheets of metal, which usually means by the machining, or other processing, to produce precise apertures in the sheet. This can be done by drilling, or pressing, or by electrochemical erosion.
[*See also Expanded metal, Veco screens*]

Metal swarf. The small pieces of metal that are removed from a metal surface by any of the machining operations such as boring, milling or shaping. These pieces are carried away in the **cutting fluids** used to lubricate the operation, and they must be removed from that fluid before it can be recycled, it being too valuable a liquid, because of its special composition, to be discarded. The **filtration** of cutting fluids is a major application for **filters** and **centrifuges** of a wide range of types.

Methylene blue test. An early form of **filter efficiency test**, in which an **aerosol** of methylene blue droplets is passed through the test filter, and samples are taken on special paper upstream and downstream of it. The samples are steamed to intensify the colour, and the intensities of colour compared. (Now replaced by other methods, such as the sodium flame test.)

Micelle. A loosely bound **aggregation** of molecules, which may form one particle of a **colloid**. The term is used mostly of **surfactants**, which will distribute themselves over the surface of a liquid. At low surfactant concentration, the surface may not be completely covered. Then, as the surfactant concentration increases, the surface becomes completely covered, and any more added surfactant will form micelles on the surface. The concentration at which the surface is just completely covered is the critical micelle concentration (CMC). If the surface area is then increased, the micelles will disperse.

Microbe. Any minute living organism, but generally used as an inclusive term for **bacteria** and **viruses**.

Microbubble flotation. [*See Aphron flotation*]

Microfibre. A general term, which has come into use over the past decade, to mean very fine **fibres** with diameters in the region of a few micrometres or less. These have been developed for many different purposes, but have enabled the makers of **filter media** to produce the materials (especially **non-wovens**) necessary for the achievement of the much finer degrees of **filtration** currently being sought.

Microfiltration. That part of the **filtration** spectrum by size (from the very finest to the coarsest) which occupies the finer end of true filtration (i.e. the removal of suspended solids). At first, the term meant just the finest levels of filtration that could be achieved with available **filter media**. It has now become somewhat more precise, with its upper levels at **cut-points** of around 5–10 μm, extending downwards in size to cover the removal at least of bacteria (about 0.1 μm) and possibly viruses. It is bounded at its lower end by **ultrafiltration**, with which there is some overlap, and at its upper by **macrofiltration**.

Microflotation. The **separation** of **colloidal** particles by **flotation**.

Microflow lamella separator. A form of **lamella separator** in which the **plates** take the form of rectangular tubes, with the inclination in the opposite sense in adjoining packs.

Micrometre, Micron. A unit of length, for long the basis of sizing in much of **filtration** work – although the nanometre is of growing importance. Micron is giving way to micrometre as the preferred usage.

Micron rating. A somewhat outdated term, being a measure, expressed in micrometres, of the ability of a **clarification filter** to remove particles, with the rating expressing the filter's **cut point**. It may be an **absolute** rating or a **nominal** rating.

Microorganism. Any living body so small that it can only be seen under the microscope. A term largely replaced by **microbe**.

Microporous. Describes any mass of solids that has a mean pore size of 5 μm or less.

Microscope count. A technique for the determination of the mean **particle size**, within a certain size range, by the actual counting of the numbers of particles observed within the graticule of the microscope. The method requires a count of 500 or 600 particles to avoid serious error.

Microscreening. The separation of small amounts of **tramp material** in the inlet flow of water treatment and cooling systems, to remove such material down to finer **cut points** than was once the norm.
[*See also Microstrainer*]

Microstrainer. A gravity driven **rotary drum strainer** used for **clarification**, especially for intake water streams. Also called a microscreen, the microstrainer consists of a perforated drum, closed at one end, rotating about a horizontal axis. The open end fits into a wall which partitions the contaminated liquid from the **filtrate**. The feed suspension enters the inside of the drum, at the bottom, and the filtrate flows to the outside. Sprays at the top wash through the drum and discharge the solids into a trough running parallel to the axis of the drum, containing a screw conveyor. The drum is usually covered with a woven fabric with an aperture size of 25 μm although particles as fine as 3 μm can be collected because a cake is allowed to build up.

Middlings. A term used in the mineral industry for a fraction collected between the **concentrate** and the **tailings**.

Migration test. A test for any **filter**, but especially for filters intended for a **clarification** duty, using clean water for the feed, flowing through clean **filter medium**. Any contaminant found in the **filtrate** must have come from the medium.
[*See also Extractables, Shedding*]

Miller filter. An intermittent **vacuum belt filter** with a long cycle time. It has a short operating length of belt, which rests on a grid over a vacuum chamber. A lid is closed over the cloth to form an upper chamber, in which the cake is formed. The filter can be **precoated** and the **cake** can be gas blown or washed. At the end of the cycle, the lid is opened and the cloth **indexed** forward to discharge the cake, and present a fresh section of belt for filtration. The filter is enclosed, so it can be used for reasonably hazardous materials.

Mineral dressing, Mineral processing. The industrial sector concerned with the treatment of mined ore/gangue mixtures that enables the valuable ore to be separated from its associated waste materials. The processes used in mineral dressing include crushing, **screening**, **heavy media separation** and **flotation**.

Mineral wool. A finely fibrous material made from or spun out of a range of mineral materials, mainly for use in thermal insulation of buildings, but also usable as a **filter medium**. Asbestos would once have been included in this category, but is now no longer used because of its long term toxic effects.

Miscible. Two liquids are said to be miscible if they can be completely mixed together to form a homogeneous solution.

Mist. A suspension of liquid **droplets**, probably of **colloidal** dimensions, in a gas, such as to have some effect on the visual range. When this range is much reduced, the suspension is termed a **fog**.

Mist eliminator. A **filter** that removes the liquid **droplets** of a **mist** from the suspending gas. This is usually effected by means of a thick bed of **wire mesh** (frequently **knitted**) to provide a very large surface area on which the droplets can **coalesce** and fall into a collector.
[*See also Demister*]

Mixed bed media. A **filter medium** in a **deep bed filter** that is made of two or more different materials, classified either by size or density or chemical nature. The mixture of materials may make the **filtration** more efficient, or it may make **backwashing** and resettling of the media granules easier.

Mixed media filters. [*See Mixed bed media*]

Mixing cell model. A model used in the study of **displacement washing** of a **filter cake**, in which the cake is assumed to be made up of a series of mixing stages.

Mobility. In **filtration** terms, this refers to a parameter defining the **compressibility** of a **cake** in a **filter press**.

Moderate settling suspensions. It is convenient to classify **suspensions** according to their ease of **filterability**, which can be associated with the speed with which they settle in a jar test. A typical classification separates rapid settling, moderate settling and slow settling suspensions (which require very different types of **separation** equipment).

Module. A replaceable sub-assembly of **filter media**, that fits into a **filter housing**. The term is particularly used of **membranes**.
[*See also Element*]

Moisture content. A fairly general term used as both a description and a measure of the amount of liquid in a **cake** at any part of a **filtration cycle**. It is a function of **saturation** and **porosity**.

Moisture (residual). [*See Residual moisture content*]

Molecular diffusivity. In **filtration** terms, this usually refers to the **diffusion** coefficient of a **solute** in the **wash liquid** during a **cake washing** process.

Molecular sieves. Generally crystalline solids, the molecules of which are in a regular array, such as to leave channels and holes of molecular dimensions within the array. Relatively small molecules can then enter the array from outside the crystal, and become occluded in the holes. This provides a means of separating small molecules, especially water from gases (thereby drying them).

Molecular sieve materials are mostly natural and synthetic **zeolites**, which have been developed to be extremely efficient in separating organic molecules. It should be noted that this separation is based on particle size (like some **membrane** separations) and does not rely upon electric charge (like **ion exchange**).

Molecular Weight Cut-Off (MWCO). The measure used to determine the **efficiency** of separation in **ultrafiltration**. This is a process that is used to separate very large molecules from **suspension** or **solution**. The largest molecule not retained by the UF membrane determines its MWCO, which is usually measured in **Daltons** (or kD).

Molten metal filtration, Molten polymer filtration. A considerable range of materials and objects are made by the extrusion of molten metal or polymer through an appropriately shaped die. In many cases, the die is quite small (such as in the spinning of **microfibres**), and would thus be easily blocked by an unwanted particle in suspension. The necessary **filtration** to ensure freedom from such particles is a very severe duty for a **filter medium** – which tends to be a **ceramic foam** for metal filtration and porous metal or **sintered metal** mesh for polymer filtration.

Monodisperse aerosol. An **aerosol** in which all the particles are of the same size, or very close to the same value.

Monofilament. A single extruded endless **fibre**, which is used as the basis for a wide variety of **filter media**. Materials woven from monofilaments are used for **surface** and **cake filtration**, and have little application in the **microfiltration** range, mainly because of their low strength. The term is normally used of synthetic polymer extrusions, but can equally well refer to a metal wire. Both materials can be woven into **mesh**.

Monolayer filter. A **deep bed filter** with a single homogeneous layer of filter material, operating as a **rapid filter**, sometimes with a layer of **precoat** material. Its main use is in fresh water treatment, with the incoming water flowing downwards through the bed. It is regenerated by a short period of reverse flow, to backwash the layer.

Monoplate filter. A simple **batch filter**, operating under vacuum, gravity or pressure, with a horizontal perforated plate carrying a sheet of **filter medium** on its upper surface. Thick layers of **cake** can be formed, and the format is frequently the basis of laboratory test filters.
[*See also Büchner funnel, Nutsche filter*]

Montmorillonite. A clay mineral, comprising aluminosilicates of variable composition, making the main component of **bentonite**. One form is highly absorptive, for example, of water, to make a firm, mud-like structure.
[*See also Fullers' earth*]

Moon dam. A device in a **decanter centrifuge** that controls the rate of **centrate** discharge. It is a weir plate, in the **front hub**, which has a curved inner surface, parallel to and close to the **pond** surface profile.

Moore filter. A **vacuum leaf filter** (most leaf filters being pressure operated) with a long history. The **filter leaves** are connected to a vacuum source, and are carried successively into separate tanks for **filtration**, **washing** and **cake discharge**.

Most penetrating particle size (MPPS). A parameter involved in a method for determining the

separation efficiency of an **air filter**. A plot of particle penetration vs particle diameter gives a curve with a pronounced maximum – the MPPS. It is suggested that the efficiency at the MPPS would be a good measure of a filter's performance (since it is this size that is the key determinant of air cleanliness). **CEN** has a standard based on MPPS (EN1822).

Mother liquor. A term loosely used to refer to the liquid in which the solid particles are suspended and from which they are to be separated. It derives from (and is most correctly used in) **crystallisation**, in which the mother liquor is the highly saturated solution of crystallisable solids, from which the crystals are born in the crystalliser. Once crystal formation is sufficiently complete, the suspension of crystals and mother liquor (now a much weaker solution) is separated, usually by **filtration**, and any residual mother liquor left on the crystals is washed away from them (quickly, so as not to redissolve the solids).

Moulded. In the present context, refers to the formation of **filter media** into shapes by moulding – the resultant format may be held in shape by a bonding resin, or by thermal treatment. This largely refers to the production of moulded **cartridges**.

Moving bed filter. Any of a number of designs of **deep bed filter**, using **granular media**, in which the medium is recirculated continuously between a **filtration** zone, and a regeneration zone, in which the medium is **washed** free of collected solids.

Moving end plate. A term sometimes used for the **queen plate** of a **filter press**.

Moving layer. In a **decanter centrifuge** the liquid of the **pond** flows against the motion of the **scroll**. The upper turbulent liquor layer on top of the pond is called the moving layer, and the same term is applied to the theory that predicts its thickness.

MPPS. [*See Most penetrating particle size*]

Mud. A term generally equivalent in use to **sludge** or **slurry**, but specifically used to refer to the drilling muds employed in the lubrication of the drill bits used for oil and gas exploration.

Mud line. A term used occasionally for the interface between a settling slurry and the clear supernatant liquid in a **gravity thickener**.

Mud sump. Although used generally as an equivalent to **grit trap**, it is used specifically for the collection basin for settlable solids, ahead of a **coalescing filter**.

Mullen burst test. A measurement of the force necessary to burst a given area of **filter medium** (paper or cloth), under fluid flow condition. Usually expressed in cm of water pressure for a 5 cm diameter specimen.

Multi-chamber centrifuge. [*See Chamber bowl*]

Multifilament. Used to refer to both the **yarn**, and any material woven from it, the term relates to a yarn made from a number of single **filaments**, twisted together into the yarn. The resultant yarn has a greater tensile strength than that of the individual filaments, and cloth woven from it is correspondingly stronger. When used as a **filter medium**, **filtrate** can flow between the yarns and also through the yarns themselves. There is thus an element of **depth filtration** with multifilament yarns, and the consequent problem of blockage after an extended period of use.

Multilayer filter. Used occasionally to mean a **deep bed filter** employing layers of different **filter media**.
[*See also Mixed bed media*]

Multi-lead. Refers to any **screw conveyor** where the screw has more than one flight, but specifically to the **flights** on the **scroll** of a **decanter centrifuge**.

Multipass test. A laboratory test of the performance of a **filter** in which a batch of suspended solids is passed repeatedly through the filter without its being changed or cleaned. Particle measuring devices are fitted upstream and downstream of the filter to observe the change in **filtration efficiency** with time and degree of solid loading.
[*See also Single pass test*]

Multi-purpose filter. A filter that can be used for a number of unit operations, usually a **nutsche** filter with a number of accessories.
[*See also Filter dryer*]

Multi-stage pusher centrifuge. A **pusher centrifuge** fitted with multiple **screens**, each having its own pusher ring. One or more of the screens may move between stationary screens either side, to give an alternating push action as it oscillates.

MWCO. [*See Molecular Weight Cut-Off*]

N

Nakamura-Kuroda model. A design model for continuous **settling** in **lamella separators**, dating from 1937. The theory involves two **dimensionless numbers**, particular forms of a **Grashof number** and a **Reynolds number**.

Nanofiltration. A **membrane** separation regime that has recently been made up from the lower pressure end of **reverse osmosis** and the higher pressure end of **ultrafiltration**. Its main function is in fresh water treatment, and its main purpose is the selective removal of some ions from a solution, allowing the passage of others (e.g. retaining sulphate ions, whilst allowing passage to chloride ions).

Nap. Raised woolly layer of hairs or **fibrils** on the surface of a textile achieved by brushing or other means. Any nap on the surface of a **filter medium** may cause the accumulated **filter cake** to adhere more firmly to the surface of the medium. Filter media made from natural fibres are frequently surface treated to remove the nap, such as by **singeing**.

Napping. A finishing process for fabrics that raises short **fibrils** above the surface of the medium. This may produce a material that is more effective in trapping particles from a **suspension**, but the **naps** may make cake release more difficult.

Natural fibre, Natural filament. Materials derived from animal or vegetable sources: fibres from **cellulose**, **cotton** and **wool**, and filaments from **silk**, although flax/linen, jute and other fibres are used.

Navier-Stokes equation. A relationship describing fluid flow when the fluid is incompressible. **Stokes** used it as the basis for the derivation of **Stokes' law**.

Near size particle. A particle whose dimensions are close to those of a **screen aperture**, and so may pass through or may remain trapped on the surface.

Needled felt, Needlefelt. An ordinary **felt**, made of natural or synthetic **fibres**, which has been strengthened by **needling**. The felt is made in the normal horizontal format, with layers of fibre **web** built up to the required thickness. The resultant **batt** is then conveyed into the needling zone where it is vertically punched by the barbed needles, to entangle the fibres in the zone close to each needle. The felt may be punched from one side only, or, more usually, from both sides. There is a high density of needles, perhaps 250/cm^2, so that the resultant needlefelt is relatively uniform across its surface.

The needling gives considerably greater strength to the felt, but may also provide thin spots, where the pore size is greater than in the body of the felt. As with ordinary felt, needlefelts provide a good **depth filter medium**.

Cross-section of needled felt

If additional strength is required in the felt, than a **scrim** layer is included at a mid-point of the batt, with a coarse aperture, such that the threads of the scrim are unlikely to be broken by the needling.
[*See also Hydroentanglement*]

Needling. The punching of a layer of randomly arranged **fibres** to give greater stability to the regions around each needle, and to the material as a whole. Needling is done by passing the material horizontally under a closely packed array of needles (or between two similar sets, moving together). The needles may be as close as 250/cm^2, and they cause considerable disturbance, and resultant entanglement, of the fibres as they pass them.
[*See also Needlefelt*]

Negative dam, Negative pond. In some modes of operation of a **decanter centrifuge**, it is beneficial for the liquid level to be deeper than the level at which the solids discharge from the **bowl**. The negative dam is the **weir plate** in the **front hub** that produces that deeper **pond** (and there must, of course, be devices at the **beach** end of the bowl to prevent overflow through the solids ports).

Nellie filter. A particular type of **sheet** (or **pad**) **filter**, made by British Filter for many years, as a **clarifying** filter, especially in the brewing industry. It consists of a horizontal cylindrical vessel containing a pack of circular sheets held together so that there is no space for **cake** build up – the liquid flow is radially through the thickness of the sheets.

Nephelometer. An instrument for measuring **turbidity** by observing scattered light in a direction perpendicular to a light beam. As a consequence, turbidity of a liquid or gaseous **suspension** can be expressed in nephelometric turbidity units (NTU). The term is sometimes also used for instruments measuring transmitted light but here the term absorptimeter is better used.

Newtonian fluid. A Newtonian fluid (effectively only relevant to liquids) is one in which flow starts immediately when a force is applied, i.e. it has no yield point. Under conditions of **laminar flow**, layers of fluid flow across one another, generating shear between the layers. The shear stress is directly proportional to the fluid **viscosity**, and the velocity gradient over the layers of fluid:

$$\tau = \eta(dv/dy)$$

where τ = shear stress
η = fluid viscosity
dv/dy = velocity gradient across the layer of fluid

The viscosity of a Newtonian fluid is constant at all shear strains, unlike **non-Newtonian** fluids, which exhibit a viscosity dependent on the value of the shear strain.
[*See also Thixotropy*]

Newtonian suspension. A **suspension** that behaves in the same way as any **Newtonian fluid**, namely it has a viscosity independent of shear stress. Suspensions are much more likely than pure liquids to show anomalous behaviour, of **shear thinning** or even **shear thickening**.

Newton's law. This refers specifically to the law relating to motion in a viscous fluid, and gives rise to an equation for the rate of **settling** of very coarse particles. Newton's law in this case relates the **drag force** on a moving particle to the projected area of the particle, the fluid density and the square of the particle velocity, all multiplied by a **drag coefficient**, which is a function of the particle **Reynolds number**.

Niagara filter. A batch-operated, **pressure leaf filter**. Long established in a wide range of applications, the Niagara filter has rectangular **filter leaves**, suspended in a vertical pressure vessel, from a common manifold for the discharge of the **filtrate**. A **feed suspension** is pumped into the holding vessel, and **filter cake** builds up on the leaves as the filtrate passes through the **filter media** fitted on their outside. Operating pressures are in the region of 5-6 bar, and the system is well suited to the use of **precoat**. Finished **filter cake** can be removed as a **slurry**, by **backflushing**, or in the dry state.

Nikuradze chart, plot. A graphical representation of flow conditions in pipes, including artificially roughened pipe surfaces.
[*See also Reynolds number*]

Noils. The short **fibres** produced as waste from the combing of natural fibres, such as **wool** or **silk**. They can be made into a **filter medium** for relatively undemanding duties.

Nominal rating. An arbitrary size nominated by a manufacturer of a **clarification filter** to indicate the size of particles that might be trapped by it, generally such a size that will be retained at an efficiency of around 96%. Such a rating will be of use within the manufacturer's own range, but is unlikely to be very meaningful in comparisons with the products of other manufacturers.

Nonionic. A nonionic molecule is one that does not dissociate into **anion** and **cation** when it is dissolved in any solvent. It may still be possessed of charges at various parts of the molecule.

Non-Newtonian fluid, suspension. Only normally applicable to liquids, a non-Newtonian fluid is one whose viscosity does depend upon the rates of shear applied to the fluid (unlike a **Newtonian fluid**). It may, for example, be as stiff as a gel when stationary, but become fluid upon being stirred (**shear thinning**).

Nonwoven. Any material made from **fibres** or **filaments** that are randomly laid, rather than woven or knitted, is termed nonwoven. The **webs** of fibre or filament are formed either by **carding** processes, or by air laying – the blowing of individual fibres onto a collecting drum or a moving belt. The resultant webs may either be fully random in orientation, or have some imposed direction, and may cohere by natural features of the fibres, or be consolidated by **felting** or **bonding**.

The complete range of nonwoven materials now includes a wide range of very good **filter media**. **Felts** were probably the first materials ever to be used for **filtration**, and **paper** was soon used for the same purpose. Modern nonwovens such as **spun bonded** and **meltblown** materials are expanding rapidly in filtration usage.

Normal consolidation. The uni-dimensional **compression** of a **filter cake**, in the opposite direction to its formation, as occurs, for example, in a **diaphragm press**.

Normal distribution. A frequency distribution is a means of describing how a particular attribute is distributed among the members of a group, and in particular the distribution of a set of measured quantities about their mean value. The distribution can be expressed as a table, or as an equation or as a graph. If the deviations of the members of the set are the sum of a very large number of small independent deviations, then the distribution is said to be normal, and its graph is the familiar bell-shaped curve. This curve is symmetrical about the single mean value, and its mode and median are equal to its mean.

The mathematics for the curve were developed by the English mathematician De Moivre, but also associated with it were other great mathematicians, Gauss and Laplace, the distribution also being known as a Gaussian distribution. It frequently fits the distribution by particle size of a mass of particles produced by condensation or precipitation.
[*See also Log normal distribution*]

Nozzle-bowl centrifuge, separator. A special design of **disc stack centrifuge**, used for the separation of a small amount of suspended solid material from a liquid or a mixture of two liquids. The separated solids collect as a thick **suspension** at the widest part of the double-conical **bowl**, from which they are discharged into the outer **casing** through a set of permanently open nozzles. The nozzles are sized to match the solids concentration in the feed to the centrifuge, such that there will always be a layer of settling solid inside the nozzle, to prevent the flow of liquid out from the nozzle. The solid is discharged as a thick sludge. The use of nozzles requires that the feed liquid has at least a minimum value of solids concentration – below that value, then a **valved nozzle** or **opening bowl** type of machine will be needed.

NTU. [*See Nephelometer*]

Nuclepore membrane. A polymeric **membrane** made from an impermeable sheet of polymer (usually **polycarbonate**) by irradiating it with an appropriate high energy radiation of charged particles. Each particle passes relatively straight through the sheet, leaving a weakened path behind it, which can then be chemically **leached** to form a reasonably uniform and straight circular **pore**. The resultant membrane has quite a high initial **flux**.

Number distribution. The frequency distribution of a range of **particle sizes** plotting number of particles with a particular size against that size. The number median diameter is then that particle diameter below which in size are 50% of the number of particles.

Number of revolutions. Rotating equipment can be found in the **filtration** and **separation** industry in the form of a **rotary filter**, a **rotating screen** or a **centrifuge**. One of their main parameters is their speed of rotation, as generally expressed in number of revolutions per minute (RPM). While centrifuges can rotate at a few hundred up to several tens of thousand RPM, screens and filters are much slower, at 0.1 to 10 RPM.

Pressurised nutsche filter

Numerical index. A term sometimes used in **screening** processes as a measure of **separation** efficiency.

Nutsche filter. One of the simplest forms of filter, operating batch-wise but used at every processing scale, is the vertical cylindrical vessel with a perforated base, on which a piece of **filter medium** is mounted. It is closed at its top by a domed lid, carrying the feed pipe, and at the bottom by the **filtrate** container. The bulk of the vessel is filled with **suspension** which is then filtered, possibly under its own hydrostatic head, but more usually by applied suction beneath the filter medium, or pressure applied above the suspension. A **filter cake** will form on the filter medium at the base, and this can be discharged after each batch of slurry, or allowed to accumulate through several batches.

This format allows a range of additions to the simple filter: an agitator can be mounted on an axial shaft to provide stirring during the filtration or any other activity; **washing** of the collected cake is very easy, as is thermal drying. The system can be used for a chemical reaction or a crystallisation step before filtration. The whole unit can sometimes be rotated about a horizontal axis, in order to empty out the accumulated cake.
[*See also Filter dryer*]

Nylon. A generic name for a group of thermopolymers, which are basically polyamides. These are both engineering polymers, and the material for a range of textile **fibres**. They are condensation products of a diacid and a diamine, and are given type numbers derived from the carbon atom content of the two components. Thus Nylon 66 is formed from hexamethylene diamine and adipic acid.

O

Odour control. The removal of noxious odours from a gas stream is usually a problem of **adsorption** or **wet scrubbing**, and hence outside the general scope of this work. However, the most common form of odour control, namely the packed bed of **activated charcoal**, is increasingly being incorporated into a **combined filter**, whose **filter medium** does at least two jobs – filter out fine suspended particles and remove odours. For this purpose, a **composite** material is used, which includes a finely porous material supplemented by activated charcoal, perhaps as a layer of charcoal cloth, or as particles of charcoal embedded in a **non-woven** medium.

Off-line filter. Most **separation** equipment works on-line, i.e. it takes the full flow of the process or utility fluid and treats it as necessary at the full-flow capacity. However there are occasions when this is not the best way to achieve the required degree of separation, perhaps for a particularly difficultly filterable suspension, which may then be taken from the process line to a separate **filter** for this purpose.
[*See also On-line filter*]

Ogawa efficiency. An index introduced by Ogawa to characterise the performance of a **classification** operation.

Oil bath filter. By now a rather outdated **filter** for engine **air intakes**, it draws contaminated air in through a layer (the bath) of oil, which removes some of the contaminant. The air flow induces an oil mist flow with it into a **filter element**, which operates as much by viscous **impingement** as any other method to remove the remaining impurities. This has largely been replaced by better dry intake filters.

Oil dewaxing. [*See Dewaxing of oils*]

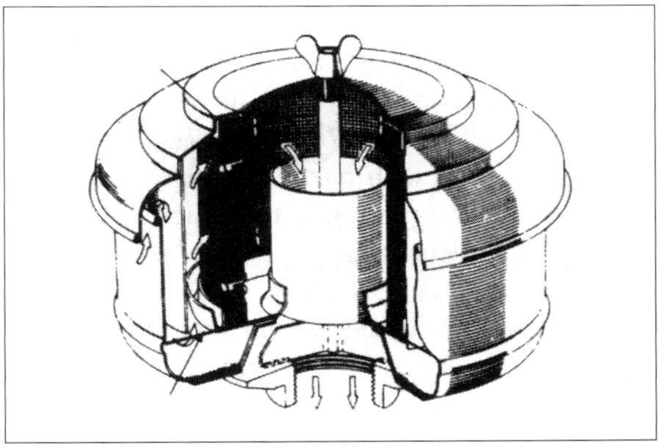

Oil bath air filter

Oil nozzle. A tube over whose edge the light phase of a **three-phase decanter** flows. It is used for controlling the surface level of the light phase, and it is adjusted by shims.

Oil remover. A device in the delivery side of an air compressor that removes any oil from the compressed air, either that was in the inlet air or that which has been entrained in the air from the compressor lubricant. Large droplets will be removed in an **impactor**, while small droplets will be separated in a **coalescer**.

Oil seeds. A major raw material for vegetable oils, which forms the feed (after appropriate crushing) to a wide range of mechanical **expression presses**.

Oil/water separation. The most common form of **liquid/liquid separation**, with a wide range of equipment suitable for mobile engine use as well as for process separations.

Oligomer. Low molecular weight **polymer**, made of only a few **monomer** units.

Oltmann construction. An empirical method of designing **gravity thickeners**. It is based on a plot of the **batch settling curve** (interface height against settling time), from which the compression point must be identified, leading to the **critical thickener flux**.

OMD filter. The OMD (Open Mud Discharge) filter was developed by Industrial Filter & Pump, in the USA, as a form of **variable volume** filter. It has a horizontal cylinder within which are a set of vertical circular **plates**, with **filter cake** forming on both sides. A pair of impermeable **diaphragms** are positioned midway between each pair of plates, and when the cake reaches a diaphragm, flow is stopped, and the diaphragms are hydraulically pressed onto the cakes, dewatering them by compression. Finally, the shell is withdrawn and the plates vibrated to release the cake.

Omega. The Greek letter omega is the normal symbol for **angular velocity**, so the word has come to be verbal shorthand for angular velocity in descriptions of **centrifuges**.

One pass clarifier. A term sometimes used for the traditional **gravity clarifier** with no recirculation of the solids.

On-line filter. Any **filter** that operates within a flow system such that the full system flow passes through it for treatment.
[*See also By-pass filter, In-line filter, Off-line filter*]

Open area. On a **screen**, the area of the openings as a percentage of the total area of the screen surface. On any other porous medium it is the **pore** area, usually expressed also as a percentage of the total **filtration** area.

Open bubble point. [*See Boil point*]

Opening bowl (disc stack) centrifuge. A **disc stack separator** made in two halves, which can be separated momentarily (under the control of the increasing solids level inside the **bowl**) for solids discharge.

Operating pressure. The normal pressure at which a system operates, usually implying the pressure upstream of a **filter**.

Optimum time cycle. In the operation of any **cake filter**, batch or continuous, the equipment should be designed to maximise throughput and/or output. This results in the search for the optimum time cycle, dependent on all of the components of the **filtration** cycle – feeding, **cake formation**, **draining**, **washing** and final **dewatering**, plus any **dead time**. The creation of a thick cake may not be better than operating with a thin one, if other stages are thereby made quicker.

Orbit theory. This (often termed equilibrium orbit theory) is a design method employed for **hydrocyclones**. It involves the balancing of the forces exerted on a particle by the **radial flow** of the liquid towards the axis and the **centrifugal force** outwards.
[*See also Locus zero vertical velocity*]

Orifice plate. A means of determining the flow rate of a fluid through a pipe or duct. A plate is fixed across the pipe, and at its centre is a carefully machined orifice of a precise diameter. Fluid pressure readings are taken immediately adjacent to the plate, both upstream and downstream from it, and the flow rate can be calculated from the pressure differential across the plate.

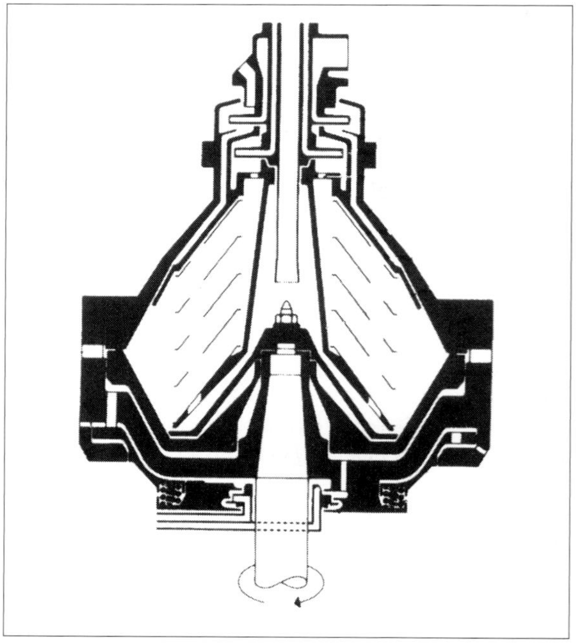

Opening bowl disk centrifuge

Orthokinetic collisions. Collisions among particles in a **suspension** caused by fluid movement. Without inter-particle collision, **aggregation** is much less likely.
[*See also Perikinetic flocculation*]

Orthokinetic coagulation, Orthokinetic flocculation. The **agglomeration** of particles held in **suspension** in a liquid, induced by the movement of the liquid causing local **shear**. It is much slower than **perikinetic flocculation**.

Oscillating centrifuge. [*See Tumbling centrifuge*]

Oscillating screen. [*See Gyratory screen*]

Oseen equation. A relationship developed for the system of **dilute sedimentation** (**free settling**), in which the drag coefficient on a particle is calculated without assuming that inertia forces are not negligible. For a sphere, it is a simple relationship between **drag coefficient** and **Reynolds number**, accurate up to Re =1.

Osmosis. If two vessels containing a solution in the same solvent of the same solute, but at different concentrations, are connected by a free passageway, then the solvent will diffuse from the more dilute solution into the more concentrated until the concentration is the same in both vessels.

If two vessels containing a solution in one and the pure solvent in the other are joined through a semi-permeable **membrane**, with the same hydrostatic head on either side of the membrane, then solvent will move through the membrane by osmosis, from the pure solvent side into the solution, in a similar attempt to equalise the concentration on both sides of the membrane.

As the solvent moves in this way, the level on the pure solvent side falls, and that on the solution side rises, thus creating an increasing hydrostatic head on the solution side. There comes a point where the driving force due to osmosis in one direction is balanced by the hydrostatic head attempting to drive solvent in the opposite direction through the membrane, at which point solvent movement ceases. The differential pressure between the two sides at this equilibrium point is the osmotic pressure of the solution (which varies with the original concentration of the solution).

If, now, an extra pressure, greater than the osmotic pressure, is applied to the solution side, solvent will begin to move in the reverse direction through the membrane. This is the principle of **reverse osmosis**.

Osmotic pressure. [*See Osmosis*]

Osprey press. The first of the **compression belt presses**, developed by A Klein in the early1960s, had a thickened **cake** laid down on a porous belt and covered by an impervious belt, which squeezed the cake, first through a compression zone and then a shearing zone, before the cake was discharged. These effects were achieved by means of a series of small rollers on either side of the twin belts.

Ostwald ripening. In **agglomeration** or **coagulation** processes, this is the growth of large particles at the expense of smaller ones.

OSUF-Z test. A test procedure developed at Oklahoma State University for use with lubricating and hydraulic oil **filters**.

Outlet ports. Channels integrally formed in a **filter plate** to allow **filtrate** to escape from the filter **chamber**.

Overall effectiveness. [*See Screen effectiveness*]

Overall penetration. A measure of the effectiveness of a **separation** process, given by the ratio of the mass of solids passing through the process (i.e. that recovered from the **overflow**, or **filtrate**, or **centrate**) to the mass of solids in the **feed**. This is equal to 1 – (total efficiency).

Overdrain classifier. [*See Hardinge classifier*]

Overflow. The product stream from a **sedimentation** operation (e.g. a **clarifier** or **thickener**) that contains a lower concentration of solids than the feed. In the case of a clarifier, this should be a completely clear liquid.

Overflow limit. In terms of the **pusher centrifuge**, this is the maximum throughput beyond which the feed slurry flows over the **cake** and out of the **bowl**.

Overflow weir. The rim around the top of a **sedimentation** tank, over which the clarified liquid flows into the discharge channel. The weir may be serrated in order to cope with varying throughput rates.

Over-pressure device. A control system for **filter presses** that are hydraulically closed. It regulates the closing pressure to a pre-determined level, so that a higher pressure can be achieved at the end of a cycle, to overcome any excess pressure generated by the **filtration** process.

Overproduct. A term sometimes used for the product collected from above a **screen**; the term **oversize** is preferably used.

Oversize. In a **sieving** separation, the oversize is that fraction of the mixture fed to the sieve that is retained by the **apertures** of that sieve.

Oversize particles. Particles in a **suspension** with a considerably greater diameter than the average value for the whole suspension, which appear sporadically. Because they could easily interfere with a particular **separation** process (such as blocking feed channels in a **membrane** module), they usually have to be removed in a **pre-filter**.

Owl. An instrument that can determine the particle size of a **monodisperse** cloud of spherical particles, by means of the colours generated by the scattering of light by the cloud.

Oyster filter. A development of Bokela's Hi-Bar **filtration** technology. It is a **rotary drum filter** mounted in a pressurised vessel, whose lid opens like a shell, to give easy access to the whole filter assembly.

Ozone. The triatomic molecule of oxygen, which is chemically active, and the presence of which in **electrostatic precipitators** is to be avoided. However, it is beneficial as a disinfectant in fresh water production, not causing tastes or odours like chlorine.

P

Pachuca tank. A simple vertical cylindrical vessel, with a conical hopper bottom, in which a **slurry** of solids in a **leaching** solvent is agitated by air bubbles to improve the leaching characteristics.

Packed bed. A mass of discrete particulate materials, such as mineral granules or fabricated rings or saddles, assembled in a confined space, to act as a **filter medium** for liquids or gases. Such systems are also used for other **mass transfer** operations.
[*See also Deep bed filter, Trickle filter*]

Packing density. In any **packed bed** or other porous mass, the ratio of solid matter volume to the total volume of the contained solid is the packing density.

Pad. A thick sheet of (usually) **nonwoven** material to be used as a **filter medium**.
[*See also Panel, Sheet*]

Paddock cloth. A term sometimes employed for an open **weave** fabric, woven from coarse **yarn**, which is used either as a **drainage** layer or as a tough support for a more delicate **filter medium**.

Paddle flocculator. A large tank in which a set of paddles, rotating about either a horizontal axis or a vertical one, agitates a **suspension** to which a **flocculating agent** has been added, in order to give both the necessary time for **aggregation** and the maximum of liquid movement before the **shear** would be too great and start to break up formed **flocs**.

Paddle washer. A countercurrent **washing** system, for granular solids, which has an array of semi-circular basins in a line rising from the low point where the solid enters the system. Each basin, which sits at right angles to the line of the array, is fitted with a paddle, whose function is both to mix the solids with the **wash liquor** in the basin, and to move the solids uphill to the next basin. Wash liquid is introduced to the uppermost basin and flows downhill to leave at the lowest point, enriched with the solution washed off the solids.
[*See also Reslurry washing*]

Filter panels

Panel. A flat **filter element**, rectangular in shape, which is fitted into a dividing wall to provide **filtration** for the air taken in to the working or living space inside the wall. The panel consists of a frame holding an assembly of **filter medium** material, which can be a **pad**, or some **pleated** material arranged to fill the space in the frame, or a set of filter **pockets** with their mouths filling the frame (and then extending inwards towards the clean space). Sometimes termed a cassette.

Pan filter. A **gravity** or **vacuum filter** consisting of one or more broadly rectangular pans, each with some **filter medium** sitting on a false bottom at the base of the pan. The pan may be fixed, in which case, once **filtration**, **washing** and **dewatering** are complete, the **filter cake** must be removed, either by hand, or by means of some kind of solid discharge device. Typically, the pan is supported in such a way that, for cake discharge, it may be turned upside down, so that the cake falls out into a collector below the pan.

The pan filter can have just one pan, or a set arranged in a circle around a central shaft (which carries the main **filtrate** offtake pipe), or a set in a straight line, either set **indexing** in turn from feed station to cake discharge.
[*See also Table filter*]

Pannevis filter. The version of the reciprocating **tray filter**, made (among other types of filter) by Pannevis. This has a single **belt**, of tough porous material, which drags the set of **vacuum trays** along with it when it moves. The belt moves continuously, the trays intermittently.

Paper. A more or less porous sheet material made on what can be regarded as a gravity **filter**, by the drainage of water away from a thick **slurry** of **fibres**.

A typical and conventional definition of paper is that it is a substance made from **fibrous cellulose** material, such as rags, wood or bark, treated with various chemicals and formed into thin sheets for writing, printing, wrapping and a wide variety of other uses. This has been the interpretation of the word paper for centuries, but a major change has occurred in the last half-century, namely the manufacture of fibres of other materials that can be formed into paper-like sheets by adapting the conventional paper-making process; the outstanding example of this is the variety of **glass fibre papers**, which are of major importance in **filtration**.

In essence, the paper-making process comprises dispersing fibres to form a **suspension** in water, and then filtering this through a **wire mesh** to produce a thin mat, which can be compressed and dried. Whilst any fibrous material can potentially be processed in this way, the resultant sheet will only have sufficient strength to be usable if the fibres bond together, either because of their intrinsic properties or by impregnation of the sheet with a suitable adhesive or resin.

The preparation of the fibre suspension is of crucial importance and typically involves a sequence of mechanical and chemical treatment stages to ensure that the original cellulose fibres are well separated from each other, and also that the structure of each fibre is partly disintegrated so that it surface is **fibrillated** (i.e. hairy). The possibility of achieving this state is apparent from the typical multi-layered structure of cellulose fibres; the fibres are relatively coarse, about 30 μm in diameter, but the fibrils are very much finer, their dimensions and numbers depending on the extent of the chemical and mechanical treatment.

By variation of this pretreatment process, and of the nature of the fibrous raw materials, the structure of paper made from cellulose fibres can be controlled to give a wide range of products of different **permeabilities**, **porosities** and strengths. The **strength** may be further enhanced by impregnating the paper with a suitable resin, especially for use under wet conditions, because absorption of water reduces the strength of untreated cellulose.

Multi-layer papers of different grades, possibly combining different materials (e.g. **membranes**) or including chemical reagents for specific functions, can be produced by lamination using a variety of binders and adhesives. An alternative approach uses a single manufacturing operation to produce multi-layer graded density papers, which combine high **dirt holding capacity** with low pressure drop characteristics.
[*See also Dry-laid, Filter paper, Wet-laid*]

Paper disc stack filter. A **cartridge filter element** made of a stack of circular discs of paper, held together on a central former by a spring at one end. The edges of the paper discs form an apparently complete cylinder, with **filtrate** flow from outside the cylinder through the narrow interstices between adjacent discs.

Paper roll filter. A **belt filter**, operating usually by gravity, but sometimes also vacuum assisted, in which the **filter medium** is a roll of paper, which is used once. The paper is set initially in the filtration zone, and then is pulled across it when fully loaded, to place a fresh part of the roll in place, while the dirty paper is rolled up at the used end, for ultimate discharge.

A continuous roll of paper can also provide the filter medium for a **vacuum drum filter** with the roll folded over an external roller to discharge the formed **cake**.
[*See also Flat bed filter*]

Parabolic rate law. [*See* $t/V = f(V)$ *method*]

Parallel connection. An arrangement of items of processing equipment in which the individual items (or groups of items) are mounted across the direction of flow of the main process stream. They are all connected to the same delivery point of the **feed** flow, which divides into as many (usually equal) streams as there are units. Parallel connection is used either when the individual units are unable to accept the full flow of feed fluid (because they cannot be made large enough, or because the items cannot be scaled-up in

design), or where it is necessary to stop one item (for example to empty or clean it), and yet maintain reasonably full flow through the whole system. In the latter case, the items would all have different operational cycles, with each stopping in turn.
[*See also Serial connection*]

Paramagnetic. Any material that is attracted along the lines in a magnetic field towards a point of greater field intensity is termed paramagnetic.
[*See also Diamagnetic, Ferromagnetic*]

Partial efficiency. A measure of the efficiency of a **clarification** or **classification** operation, which considers recovery only within a specified range of particle size. **Recovery index** and **rejection efficiency** are partial efficiencies.

Particle. The basic unit of a mass of finely divided solid. The range of particle sizes may be wide, but the term is normally used of units of a few millimetres or less in size (down to the limits of identification). The particle may have very different shapes: from spherical to plate- or needle-like.
[*See also Granule*]

Particle attachment, Particle detachment. **Clarifying filtration**, usually in a **depth filter**, relies upon the success of some kind of attachment mechanism between the particle and the **filter medium**, and a correspondingly low rate of detachment. Attachment can be achieved by any one of seven or eight different capture mechanisms.

Particle bleeding. The passage of fine particles right through a **filter medium**, which obviously is an unwanted occurrence. It may occur in **cake filtration**, before **bridging** is complete, and also at the start of a **precoat** process. It may also happen if **particle detachment** occurs, following, perhaps, a sharp increase in fluid velocity. (It is not the same as **shedding**.)

Particle classification. This term refers to the unwanted effect of the **separation** of particles by size or density in flow towards the **filter medium**, or even in fluid distribution over the surface of the medium. This can cause selective deposition of solid particles on the medium, and the formation of an uneven **cake**. **Filter** design should thus be such as to minimise the possible classification in the **feed slurry** where this is not the purpose of the separation.
[*See also Particle segregation*]

Particle collection. A term used in the theory of **classification** in a **sedimenting centrifuge**, referring to the separation of a particle into the **underflow**, the solids discharge. Once removed from suspension into the accumulated solids, a particle is said to be collected, and a collection efficiency can be assigned.

Particle count. The number of particles present in a given sample of gas or liquid **suspension**, determined by volume or mass. Normally qualified by the size parameter measures, and by the technique used for the measurement.

Particle counter. An item of test equipment used to effect an actual count of a set of particles in order to give a comparison between, say, a feed stream and a **filtrate** of a **filter**, so that the efficiency of **filtration** can be determined.

Particle fraction. A component of a mixture of particles that is defined by some particular property, such as chemical nature, density, shape or average size.

Particle migration. The formation of a **filter cake** may be affected by irregularities in liquid flow, but it can also depend on the way in which fine particles migrate within the formed cake to block flow passages.

Particle Peclet number. A **dimensionless group** used to characterise the **washing** regime for a **filter cake**.
[*See also Peclet number*]

Particle polarisation. A particular form of **concentration polarisation**, which refers to the increase in the concentration of particles in the approach to the **filter medium** in a **cross-flow filtration** process, and which hinders the diffusion of liquid in that region.

Particle retention. A term referring specifically to the entrapment of a particle by a **filter**, by whatever mechanism of **capture** is being employed by the **filter medium**.

Particle segregation. A term used generally to refer to the **separation** of a mass of particles of different sizes into different **size fractions**. This may be an unwanted phenomenon, such as when a suspension may have its particles segregated during the flow of the suspension, such that when it reaches a **filter medium** it is no longer a homogeneous mixture, and

so may separate abnormally on the **filter**. It can also refer to the separation of smaller size particles at the surface of a **cross-flow membrane**, with the finer particles bridging the pores of the membrane, to form a secondary membrane, while the coarser particles flow on in the **concentrate**.
[*See also Particle classification*]

Particle separation. A very general term referring to the **separation** of particles from **suspension** in a fluid, or their separation one from another, in some kind of **classification** operation.

Particle shape. An important characteristic of particles, affecting rates of **settlement** or **filterability** of particles. Roughly characterised by descriptive terms such as **rounded**, **angular**, **flaky**, and **acicular** – all shapes deviating from the truly spherical to a greater or lesser degree. A **shape factor** can be developed to relate actual shape to the spherical equivalent.

Particle size. A particular geometrical characteristic of a **particle**. This is normally thought of as the equivalent to a diameter, were the particle actually spherical in shape. Most real particles are non-spherical, departing more or less from this ideal shape. The real particle will thus need to be described by two size factors in order to characterise the particle, such as the length and cross-sectional diameter in the case of a **fibre**. Particular diameters often used are: volume diameter (diameter of sphere of same volume as particle – the basis of measurement by the Coulter Counter); surface diameter (diameter of sphere of same surface area as particle); projected area diameter (diameter of circle of same projected area as particle, when viewed in a direction perpendicular to a plane of stability for the particle); sieve diameter (the width of the minimum square aperture of a sieve through which the particle will pass – the basis of **particle size analysis** on a set of sieves).
[*See also Feret's diameter, Martin's diameter, Sauter mean diameter, Stokes' diameter*]

Particle size analysis. Strictly this refers to the process of determining the factors defining the composition of a mixture of particles in terms of a size characteristic, although the term is also used for the description resulting from the determination process. The basic method is the separation of the mixture into **fractions**, each limited by a small range of the characteristic in question (by **elutriation** or **sieving**, for example), followed by weighing or otherwise assessing the quantity of each fraction. The analysis would then be expressed in terms of the weight of each fraction listed in the order of the size characteristic.

Particle size determination, measurement.
[*See Particle size analysis*]

Particle size distribution. The spectrum of particle sizes in a mixture of particles, as determined by a **particle size analysis**. This is normally expressed as the quantity of the mixture whose component particle sizes lie between two limiting size characteristics, such as the weight percentage of the mixture that lies between 1 mm and 2 mm in effective size. An alternative form of expression uses cumulative quantities, such as the total weight percentage of the mixture whose size is above 1 mm.

The distribution is usually described either by a table of such fraction limits, or by a graph showing the frequency distribution.
[*See also Normal distribution*]

Particle-stopping power. One of the models of the behaviour of a **suspension** in flowing through a **filter medium** is concerned with the way in which particles are captured, or stopped, by the medium. The term is related to the **equivalent pore size** of the medium, but is effectively the same as the efficiency of filtration.

Particle velocity. The speed of movement of an individual particle, as it moves with or through a fluid, in **sedimentation** under the influence of gravity or **centrifugal force**, or through a porous mass on its way to being **captured** from suspension.

Particulate. Relating to minute and separate particles.

Particulate settling. The **sedimentation** regime in which each particle settles individually, unaffected by the proximity of other particles.
[*See also Hindered settling*]

Partition curve. A **grade efficiency curve** with the particle size plotted on a logarithmic axis.

Paste. A very thick **sludge**, although still pumpable, and from which moisture can still be expressed. A **filter cake** can be describes as pasty if it is of this nature.

Patch test. A test primarily concerned with determining the degree of contamination of a fluid, by passing the fluid through a fine **filter medium**, under standard conditions, and observing the intensity of the stain, or patch, made on the filter medium by the contaminants in the fluid. The procedure can also be used to determine the **efficiency** of a **separation** by carrying out patch tests on **feed** and **filtrate**, and comparing the intensities of the patches.

Pathogen. Any body capable of transferring disease to humans. Usually refers to **microbes**, especially **bacteria** and **viruses**.
[*See also Pyrogen*]

Pattern repeat. A term relevant to **woven fabrics**, describing the way in which the pattern of the **warp** and **weft** threads repeats itself across the width and length of the fabric. Thus, the simplest **weave** pattern, **plain weave**, has the weft thread passing over one warp thread, under the next and over the third, to give a pattern repeat of 2 in each direction.

PDMS. [*See Polydimethylsiloxane*]

Peclet number. A **dimensionless number** of value in the analysis of the settling of an **aerosol** as it passes through the **pores**, or over the **filaments**, of a **coalescing filter**. The Peclet number is then:

$$Pe = ud/D$$

where u = gas velocity
d = fibre or filament diameter
D = a diffusion coefficient calculated from the Langmuir equation

It is also a number that characterises mass transfer in a porous mass, and so is used in the correlation of **filter cake** washing test data by characterising the **wash curve**. In this case the symbols are:

u = velocity of the wash liquid in the pores
d = particle diameter
D = diffusivity of the solute in the wash liquid,

and the Peclet number can be called the cake Peclet number

The Peclet number is also employed in describing capture mechanisms in the flow of a liquid suspension through a **packed bed**. Now the Peclet number can be called the particle Peclet number:

u = velocity of the particle
d = particle diameter
D = diffusion coefficient for the particle

Peeler centrifuge. A cyclically operating **centrifugal filter** that has a perforated **drum**, covered on its inner side with a **filter medium**. The slurry to be separated is introduced at one end of the drum, spreads across it and forms a **cake** of solids inside the drum, as the liquid drains away through the medium. Periodically, the feed of slurry is stopped, and the accumulated cake of solids (normally after being washed) is removed by lowering into it a **peeling knife**, which cuts the cake away, usually at the same rotational speed as that used during separation. The knife removes almost all of the cake, apart from a thin layer, or **heel**, close to the medium, which is left there

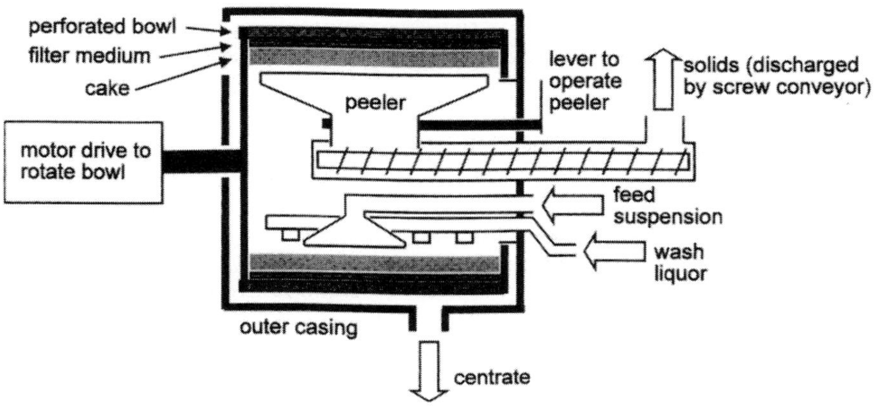

Peeler centrifuge

to avoid damage to the surface of the medium by wear from the knife.

Peeling disc. A **centripetal pump** used to remove a liquid from a **centrifuge** that takes advantage of the rotational energy of the liquid and converts it into a pressure sufficient to drive the liquid into and through an external pipe system. It usually takes the form of two parallel discs, with curved flow channels built into the space between them. The channels lead the liquid from the periphery of the disc to its centre, from which the liquid leaves at high velocity. The disc sits on, but at right angles to the axis of the centrifuge, and is used in most **disc stack centrifuges**, and quite a number of **decanter centrifuges**.

Peeling knife. A device used in **peeler centrifuges** to remove the accumulated **cake**. The knife blade may stretch the length of the centrifuge drum, or it may be shorter than the drum, in which case it has to reciprocate across the length of the drum to remove all of the cake. The shorter knife requires less effort to hold it in the cake. When sufficient cake has been dislodged, the knife is retracted towards the axis of the drum, out of the way of the next filtration stage.

Peeling pipe. A discharge device for the removal of clarified liquid from the **bowl** of an **imperforate basket centrifuge**. In the case of a liquid /solid separation this would be done as part of the separation cycle, but for liquid/liquid separations, it would operate continuously. In either case, the pipe will be lowered into the layer of clarified liquid while the centrifuge rotates, and the inertia of the rotating liquid is converted into pressure head to drive the liquid out of the centrifuge and into a collection system. The pipe thus acts effectively as a **centripetal pump**.

A very differently acting device, with the same name, is used in some **peeler centrifuges** that are equipped with a **peeling knife**. The pipe sits behind the knife and as the cake is cut away it is sucked pneumatically up the pipe and so out of the centrifuge.

Peg. A protrusion from the top of a **plate** in a **filter press**. Several pegs stand out from the plate, and on them is hung the sheet of **filter medium**.

Pegging. The blockage of a **screen** by particles that get lodged in the **apertures** and are not removed by the screen motion.
[*See also Plugging*]

Pelleting flocculation. A process for producing large compact **flocs**, with low **moisture contents**, in which **feed** and **flocculant** are introduced at the top end of an inclined **drum**, which rotates slowly. At the top end the flocs roll on each other. A single lifter transfers the pellets and liquid into a lower **decanter** section where the liquid flows out through slits. In the final consolidation section, further **compaction** is provided by mechanical contact.

Pelletised sludge blanket clarifier. A **clarifier** for water treatment in which the **sludge blanket** is made up of firm **agglomerates** or pellets, so as to give much higher throughput rates. The pellets are formed by the creation of **flocs** with a carefully chosen **inorganic flocculant**, followed by agitation and further flocculation by means of a **polyelectrolyte**, with more agitation after that.

Pellet reactor. A **sludge blanket clarifier** containing seed solids on which the particles to be removed are deposited.

Pendular (moisture). Any moisture remaining at the contact points between particles after a **cake** has been **dewatered** is described as pendular moisture.
[*See also Funicular*]

Pendulum centrifuge. A batch **basket centrifuge**, rotating about a vertical axis, mounted upon an oscillating housing, supported by three telescopic sprung legs, in order to eliminate the need for a vibration resistant mounting. Solids discharge can be by means of a **peeling knife**, or by removable bag. Such centrifuges operate in the speed range of 200 to 3000 RPM, and have drum diameters from 30 cm to 2 m.

Also used of a **centrifuge**, usually a basket centrifuge, suspended on a shaft passing through the centre of the basket, and supported and driven from above the basket, with the basket free to move as a pendulum in the event of vibrations.

Penetration. [*See Overall penetration*]

Peptisation. Generally used of the production of a **colloidal** solution of a substance (i.e. the opposite of **coagulation**). More correctly used for a chemical means of achieving **dispersion**, in which the colloidal particles are **stabilised** by the **adsorption** of charged **ions**.

Percolating filter. [*See Trickle filter*]

the form of a horizontally rotating tube. The wall of the bowl is retracted intermittently for cake discharge.

Pipefitting screen. The name given to a simple in-line **strainer** consisting of a piece of **wire mesh**, either conical or cylindrical in shape. The screen is mounted in a housing that can be connected directly into a pipeline or easily removed from it for cleaning. Versions can be supplied to fit into straight pipes, elbows, tees, etc. **Disc strainers** and pipefitting screens are often used as temporary strainers to remove coarse debris from a newly installed line before start up.

Pipeline filter, strainer. Also known as an **in-line filter**, this is a coarse filter or strainer held in a relatively small housing, which is then mounted in the middle of the pipeline carrying the fluid to be filtered. It is most often used as a **pre-filter** to protect some other processing unit downstream of it.

Pipe manifold. An array of pipes that splits a single flow into a number of equal flows, or which collects a group of fluid flows into a single line for onward flow.

Pipe module. A relatively uncommon arrangement of a **membrane** medium as a single tube housed in a piece of pipe. The ratio of separation surface to process volume is very low.

Piston press. A laboratory scale **filter** used for measuring the compressibility of a **filter cake**.

Piston-spring model. A theoretical model used to describe the behaviour of a **compressible** mass of solids saturated by a liquid, such as a **filter cake**. If such a mass is subjected to a compressive force, then the structural pressure within the mass will rise to equal that of the force, whilst the pressure within the liquid will fall from that of the compressive force to zero at the equilibrium state.

Pitch. The distance between the centres of two adjacent holes in a piece of perforated material, such as a **screen**. This distance should normally be as small as possible, consistent with the required strength of the finished plate. It is usually the case that as many holes as possible be created in the plate, and, if the holes are circular, then this nearly always results in a hexagonal array of holes, with each equi-distant from six others.

In a very different usage, the pitch is the axial distance between two adjacent turns of the flight of a **screw**, as in any screw conveyor, or in particular of the flight of the **scroll** in a **decanter centrifuge**.

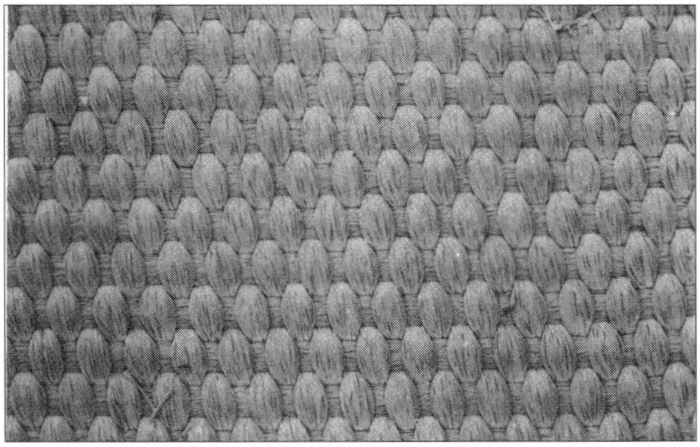

Plain weave multifilament fabric

Placer jig. A **jig** used for the recovery of heavy mineral materials, usually deriving from surface or placer deposits.

Plain weave. The simplest of the types of **weave** relating to **woven fabrics**, in which the **weft** thread passes alternately over and under adjacent **warp** threads across the whole width of the material. The next weft thread does the same, but goes over the thread that the previous weft went under. The third weft then repeats the pattern of the first, and so on. Plain waves look the same from front or back, and represent the tightest and strongest way of interlacing warp and weft.

Plain weave is also called calico or muslin weave when the threads are of cotton, cloth weave for wool, and taffeta for silk. If two or more weft threads pass over and under two or more warp threads, the resultant pattern is called rib- or basket weave.

Planetary centrifuge. A **centrifugal filter** idea, proposed by Alfa Laval, in which a number of perforate baskets are arranged in a circle. Each basket rotates around its own vertical axis, and the circle of baskets rotates about its axis. The system aims to use **cross-flow filtration** in parts of each basket: by a suitable choice of rotation speeds, cake can be caused to collect on the parts of the baskets furthest away from the centre of the circle, while cross-flow filtration occurs on the other parts.

Plastic. Specifically, this term refers to any deformable material, but it is used generically of all man-made polymeric materials, of petrochemical or other origin, but, in this context, is usually confined to **thermoplastic** materials.

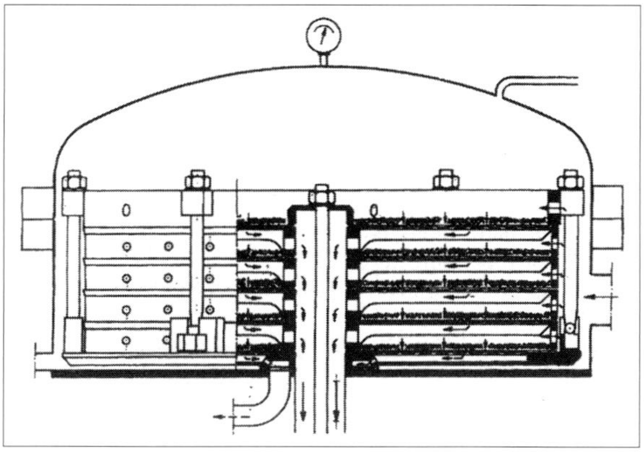

Horizontal plate filter

Plasticity index. A parameter used in defining the flow characteristics of a **filter cake**.

Plastic limit. [*See Atterberg limits*]

Plastic sheets. Nowadays one of the major forms of **filter media**, with a **porosity** imposed in a great variety of ways: most **membranes** are basically plastic sheets.
[*See also Filter media*]

Plate. Has two major connotations in filtration technology, either as a component of a (plate-and-frame) **filter press**, which holds the **filter medium** and the formed **cake**, or as a form of filter **leaf**, i.e. a rigid flat structure, with filter medium fixed to one of the flat sides only (and with that side usually facing upwards).

Plate-and-frame filter press, Plate filter press. [*See Filter press*]

Plate (and frame) modules. [*See Flat sheet modules*]

Plate inclination angle. The angle to the horizontal at which the plates of a **lamella separator** are mounted. The inclination angle and the **plate separation** are the main determinants of the separation efficiency of a lamella separator.

Plateless press. A **chamber filter press** made by Kurita, consisting of a series of alternate **frames** that either hold **filter medium** on a thin net support, or a rubber **diaphragm**. The cake is **dewatered** using the rubber diaphragm, and is not normally washed or gas blown. The filter is more compact, lighter and cheaper than conventional **diaphragm presses**. However, it requires uniform feeding of all the chambers with no pressure imbalance between them and this restricts the maximum size.

Platen press. A batch **expression press**, open at the sides, with the bags of material that are to be pressed stacked on a series of plates or platens, as in a **box press**. Although normally vertical, it may also be horizontally oriented or inclined. The platens can be heated to aid the expression, and they usually have gutters to collect the expressed liquid.

Plate press. A version of the **filter press**, developed by Krauss Maffei, the plate press consists of a set of circular vertical chambers each fitted with a protruding rubber collar, which, under compression, forms a seal. The chambers are filled with **slurry** at low pressure, then a hydraulic ram is used to compress the chambers together, and so compress the **cake**, while the collars slide back. The cake can be gas blown but not washed. Within each chamber is a metal disc to which the cake attaches itself. The chambers are opened simultaneously, by withdrawing the ram, and the discs lowered for the cake to be removed by scraping.

The term is often used, unwisely, as synonymous with **filter press**, and can also refer to a **sheet filter**.

Plate settler. [*See Lamella separator*]

Plate spacing. The distance between adjacent plates in a **lamella separator**. To minimise the physical size of the separator, the spacing should be as small as possible, consistent with the avoidance of blockages of the spaces by overlarge particles.

Plate strainer. A **metal edge filter cartridge** made from a stack of flat rings with spacers to provide a precise gap for **filtration**.
[*See also Plate-type strainer*]

Plate tower. A **gas cleaning** system, in which a vertical cylindrical tower is fitted with a number of horizontal perforated plates. Water flows downwards and the gas upwards.

Plate transfer mechanism. A feature of a **filter press** in which an automatic system transports the **filter plates**, one at a time, in either direction, along side suspension bars, or along an overhead beam, so that they may be cleaned.

Plate-type strainer. A **strainer cartridge** made up as an assembly of machined circular plates, mounted on a central shaft. The **separation** action is achieved by the narrow axial spaces between the plates. Between the plates also is a complementary set of cleaning blades mounted on a shaft that rotates around the outside of the separating surface, to scrape the outer rim of the cartridge and dislodge any particles trapped by the plates. Typified by the Auto-Klean Plate Strainer, this device is mainly used for lubricating oil straining.

Pleat, Pleated, Pleating. A pleat is a single sharp fold across the width, or along the length, of a piece of material, creating a narrow, V-shaped protrusion from the bulk of the material. The whole material is usually pleated, with a series of parallel, accordion-like folds formed into the entirety.

The pleating of a sheet of **filter medium** creates a very much larger filter area within the dimensions of the outside edge of the pleated sheet, an area increase that is the greater the deeper the pleats. The consequence of this is that a much greater filter area can be contained within a given volume of filter **housing**.

Pleated media can be used flat, mounted in a supporting frame, or made into a cylinder, as a **cartridge** element.

Pleat spacers. Any of a number of devices included in a **pleated filter medium** as used in a **cartridge** or **panel**, so as to prevent the collapse or squeezing together of the pleats, and hence a loss of effective **filtration area**. Spacers may be installed within each pleat, or a sheet of support material, such as **wire cloth**, can be folded together with the filter medium.
[*See also Pinched pleat*]

Pleide ultrafilter. An early type of **ultrafiltration** system using **membranes** as flat **sheets**.

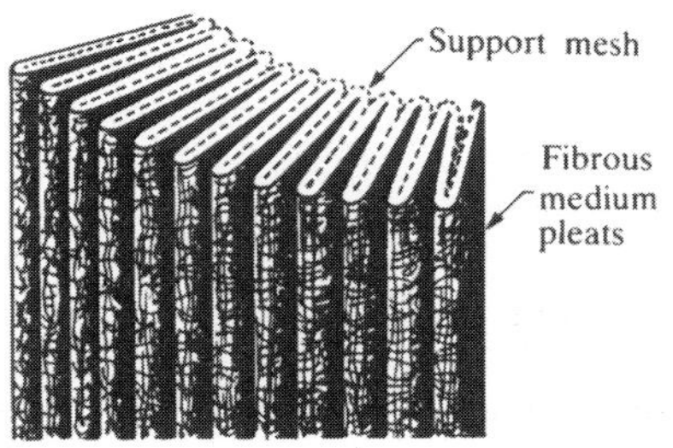

Pleated filter medium

Plenum chamber. A compartment formed in a gas treatment system to ensure even distribution of gas flow into subsequent flow channels or equipment items. Thus, a system of several **filters** operating in parallel will normally have a plenum chamber at their common inlet.

Plough. A term often used for the **knife** used to discharge collected **cake** from a **centrifugal filter**.

Plough tile. In a **decanter centrifuge**, **tiles** are fixed to the tips of the **conveyor flights** to protect the flight tip from wear by the solids being scrolled. Sometimes these tiles have a plough-shaped profile, so that they will lift and rotate the cake while scrolling.

Plug flow. If a mass of fluid passes through a space in such a way that its front surface moves at the same velocity throughout, then it is said to move in plug flow. This is not a completely achievable state because the resistance to flow at the walls of the space will slow the fluid flow adjacent to the walls – but, if the space is reasonably wide, then plug flow can be considered as reasonable description.

The same term is also used for flow through a mass of porous solids, when the front moves through all of the pores at about the same speed. This is the expected behaviour where the pores are all of the same size, but it cannot be expected where there are large channels in the mass.
[*See also Channelling*]

Plugging. When a particle enters a pore or an aperture in a **filter medium** it will block the hole if it is too big to pass through. The hole is then said to be plugged, and will take no further part in the separation process. Eventually the entire medium becomes plugged (or enough of it to cause an unacceptable rise in driving pressure or reduction in fluid flow), and it has to be cleared by a reverse flow of fluid to blow the plugging particles out of the holes in which they are held. Also called **pegging** when referring specifically to the plugging of a **screen**.

Plugging factor, index. A term used to characterise **fouling** in a **cross-flow membrane filter**.

Plunger. In some cases of **bag filtration**, the hold up of **filtrate** in the centre of the bag may be disadvantageous. A closed, impervious hollow cylinder, or plunger, can be lowered into this space to reduce the hold-up of liquid. The diameter of the plunger is matched to the nature of the liquid and also to the filtration process: too close a fit, for example, will increase the pressure drop across the filter.

Ply. When individual **yarns** are twisted together (usually in the opposite direction from the twist in the yarn) to make a composite yarn, the product is described, according to the number of yarns used, as two-ply, three-ply, etc.

PM. An abbreviation for "particulate matter", which has come into use with concern over air quality, and especially relating to diesel engine emissions. PM10 now refers to any suspended solids with a mean size of 10 μm or less, and PM2.5 of 2.5 μm or less, these being two accepted limit values for different kinds of urban environment. The ability to **filter** diesel engine exhausts to these limits will be a problem for filter and engine system designers in the near future.

Pneumatic cell. A **flotation** cell with no mechanical agitation, relying entirely upon bubble action.

PNEUROP classification. A system of standards and recommendations for pneumatic and compressed air circuits, now part of the **ISO** system.

Pocket. A form of **filter bag**, which has a flat oval cross-section (rather than circular), and a rectangular external shape. Often used mounted side by side with others in a **panel** frame for air conditioning use.

Point bonded. In the production of **spun bonded** materials, a random layer of extruded **fibres** is laid down on a moving belt. This layer is then thermally bonded to give cohesion to the material. If this bonding is done at isolated points over the material, it is said to be point bonded.
[See also Area bonded]

Pocket filter module

Point of Zero Charge (PZC). The pH value in a **suspension** at which there is no charge on a particle.

Poiseuille's law. The law and its associated equation relate flow rate, in the **laminar** flow regime, fluid viscosity and pressure drop to the diameter of a cylindrical **pore**. Poiseuille was a French physician and physiologist, whose interest in the circulation of blood led him to conduct a series of experiments on the flow of liquids in narrow tubes, from which he determined the law that bears his name. (Discovered independently by Gotthilf Hagen, a German hydraulic engineer, this relation is thus also known as the **Hagen-Poiseuille equation**.)

The equation gives the average velocity of a fluid, u´, moving in laminar flow through a straight circular capillary of length L and diameter D:

$$u' = D^2 \Delta p / 32 \mu L$$

where Δp = pressure drop along the capillary
μ = fluid viscosity.

Although this law is used in the derivation of the **Kozeny-Carman equation**, **Darcy's equation** is more useful in **filtration**.

Police filter. [*See Polishing filter*]

Polishing. In **filtration** terms, polishing refers to the final stage of any filtration process that may be necessary to achieve the required degree of clarity in the **filtrate**. All practical filters will allow a little of the **influent** suspended solids to pass through (unless there is a very sharp **cutoff** in the **particle size distribution** of the **suspension**) and some filter systems operate most efficiently with a main filter to remove the bulk of the suspended solids, allowing a little through, to be caught in a second stage.

Polishing filter. The final stage in a processing system, intended to remove suspended solids to a high degree of clarity. These may be solids that have passed through an earlier **filter medium** because they were too small to be trapped, or those solids which pass by a **filter** before a **pre-coat** layer becomes fully effective.

Polyacrylamides. The most common form of **polyelectrolyte**, much used in the treatment of organic **sludges**, especially those from municipal sewage treatment.

Polyamides. Thermoplastic polymers, made by a condensation reaction, and most commonly known as **nylon**. They make good **filter media**, as **monofilament** or **membranes**.

Polyaramids. Synthetic polymeric aromatic amides, the best known being Nomex, much used for **fabric filter bags**, because of their relatively high temperature resistance.

Polycarbonates. An important engineering series of **thermoplastic polymers**, of note in the **filtration** business as the main material used for **track-etched membranes**.

Polydimethylsiloxane. A synthetic silicone material, of increasing interest as a membrane material for **nanofiltration**. In common with all silicones, it features the –Si-O-Si- link.

Polyelectrolyte. A generic term for synthetic **polymers** used for **flocculation**. They are long-chain linear polymers, with ionisable groups distributed along the chain, and are soluble in water (although not necessarily easily). The polar groups attach to a **particle** in suspension and so link them together to form the **agglomerates**.
[*See also Polymeric flocculants*]

Polyesters. A major type of **thermoplastic polymer**, much used as **fibre** in a wide range of products, including **filter media**. PET (polyethylene terephthalate) is the best known, as "Terylene".

Polyethers, Polyetherketones, Polyether ether ketones (PEEK). Thermoplastic polymers of increasing importance as **membrane** materials.

Polyethylenes. Thermoplastic polymers of ethylene, available in a range of grades (high density, low density, etc) and widely used, mainly as film, in industry and commerce. Not so good as **polypropylene** for making **fibre**, and so much less common as **filter media**.

Polyimides. A group of engineering thermopolymers, available as **fibre**, and of quite good thermal resistance.

Polymer. A large molecule formed by the joining together of a large number of relatively simple molecules (**monomers**), and therefore made up of a large number of repeating structural units. An addition polymer is formed by the breaking of a carbon-carbon double bond in the monomer, and the resulting structural units are very similar in composition to the original monomer. On the other hand, condensation polymers are formed by reaction between monomer molecules in such a way as to eliminate a simple molecule (usually water). The structural units are now no longer similar to the monomer.

The word is also used loosely to refer to the **polyelectrolyte** used as **coagulant** in **centrifuge** applications, especially the **dewatering** of sewage **sludge** in a **decanter centrifuge**.

Polymeric flocculants. Chemical materials with high molecular weights, formed by polymerisation, which act to form **flocs** from suspended solids. They may be ionic (anionic or cationic) or non-ionic, and form their flocs in different ways, accordingly. They are used in solution, usually in water, but their dissolution is very difficult, and special units of equipment are normally required to effect the solution.
[*See also Polyelectrolyte*]

Polymer utilisation baffle. A disc fitted to the **conveyor hub** of a **decanter centrifuge** just down stream of the **feed zone**. It dips slightly in the **pond** and prevents the **flocculant** solution from skimming over the surface of the pond with resultant poor utilisation.

Polyolefin. Any **polymer** made from an olefin molecule as **monomer**, such as ethylene, propylene, butylenes, etc.

Polypropylene. A synthetic polymeric material widely used in **fibre** form as a **filter medium**. It has a good range of chemical resistance, and a better temperature range than some other synthetic fibres.

Polystyrene. The main **polymer** of styrene, although elastomeric copolymers are also very important (acrylonitrile-butadiene-styrene and styrene-butadiene-rubber). Polystyrene is not **thermoplastic** and has only limited application in **filtration**, other than as a **foam**.

Polysulphones. A rapidly growing range of thermopolymers for **membrane** applications.

Polytetrafluoroethylene (PTFE). A major material source for **filter media** because of its resistance to chemical attack, and its much higher operating

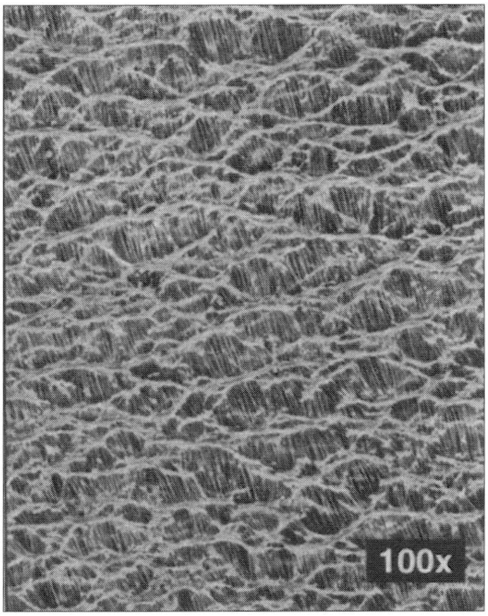

Micrograph of ePTFE membrane

temperature range. In its expanded form (ePTFE) it is now an important **membrane** material as well.

Polyurethane. A synthetic **polymer** easily formed as a **reticulated foam**, and used as such for the **filtration** of air in **HVAC** systems.

Polyvinylidene (di)chloride, (di)fluoride (PVDF). A **thermoplastic** fluoropolymer, with many of the properties of PTFE.

Pond. The layer of liquid (**suspension** and **centrate**) held in the **bowl** of a **decanter centrifuge**, between the **front hub** and the **beach**. Terms used for the characteristics and behaviour of the pond, using up and down in their gravitational sense, now refer to distance from the axis of the centrifuge. Thus the pond level is its radial depth, and the pond surface is the cylindrical boundary between liquid and the gas above it.

The layer of liquid in a **mechanical wet classifier** is also called the pond.

Pore. The individual flow channel formed within a block of porous solid, or between the particles of a mass of particulate material. In either case the shape of the pore may be quite complex, with a great deal of division and merging of channels. The prime characteristic of a pore from the point of view of **filtration** is that it should run from one side of the

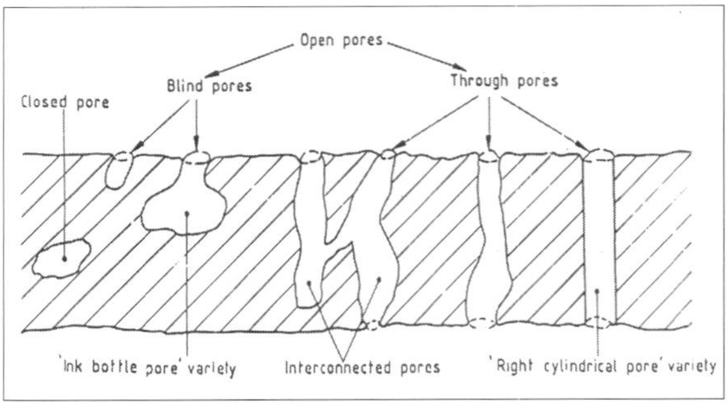

Types of pore

filter medium to the other, no matter how convoluted its path. A closed off pore is of no value to filtration – but contributes to the apparent pore volume by most methods of measurement of porosity.

The ideal pore would have a circular cross section, but most are a long way from ideality, especially those through a mass of granular particles. Nevertheless, average dimensions of pores are determined, such as by the **bubble point method**, and such values can be very valuable in modelling the behaviour of a filter medium.

Pore bridging. [*See Bridging*]

Pore diameter. [*See Pore size*]

Pore flow. The flow of a fluid through the pores of a porous mass such as a **filter cake** or **filter medium**. The mean rate of such flow can be related to the rate of fluid flow though the empty space (in the absence of the porous mass) by use of the mean **porosity** of the mass. The wall effect on pore flow is obviously very much greater than in an open pipe.
[*See also Kozeny-Carman equation, Gupte equation*]

Pore size. A pore in a porous mass is usually characterised by a size term, which is derived as though the pore were cylindrical in shape, and is then the diameter of the equivalent cylindrical pore.

Pore size determination. The physical process used in measuring the average pore size in a porous mass, such as the **bubble point method**.

Pore size distribution. Apart from those materials in which pores are created by regular mechanical processes (such as punching or drilling), almost all real porous materials have pores exhibiting a wide range of sizes. The range of sizes and the frequency related to any particular size can be determined, and is then quoted as the pore size distribution of the material.

Pore size distribution index. A measure of the distribution of pore sizes in a porous mass, especially as used in **gas blowing** predictions.

Pore size span. [*See Span*]

Pore velocity. [*See Interstitial velocity*]

Poro-edge. A brand name for an automatically cleaned **filter**, with a set of **elements** made from **wedge wire** (but with filtering flow from the wider end of the **aperture** to the narrower), the elements being stationary and the cleaning system as a rotating arm, which also provides the **backwash** exit.

Porosimetrics. The measurement processes by which **pore size**, **pore size distributions** and **porosities** of porous masses are determined. Two major methods exist: filling the pores with a liquid, and blowing air through pores completely filled with liquid. In **mercury porosimetry**, the porous mass is evacuated free of entrained air, and then liquid mercury is forced into the spaces under pressure, the volume taken up at a particular applied pressure being indicative of the totality of pores of a particular size. In the **bubble point test**, air pressure is gradually increased upstream of a porous mass that is fully saturated with water or some other test liquid, and the point at which a bubble is first seen downstream is indicative of the largest pore.

Porosity. The property of a porous mass that characterises the closeness of the packing of the **fibres** or **particles** that make up the mass. It is a measure of the volume of the space, within a mass of solid material, which is made up of **pores** or the **void ratio**, expressed as the ratio of the void volume to the total volume of the mass. If the volumes respectively of the void spaces, the solid making up the mass, and the overall external volume of the mass are V_v, V_s and V_t, then the porosity

$$\varepsilon = V_v/V_t = V_v/(V_v + V_s)$$

It is also given by (1 – dry bulk density/true particle density), and is frequently reported as a percentage.

Porous. A mass of solid material is said to be porous when it contains an array of open spaces or **pores** within it, and, in particular, when this array of pores is interconnected, in such a way that fluid could flow through the mass from one side to the other under an imposed pressure.

Sintered porous plastic

Porous metal. A **filter medium** made by **sintering** metal powder or fibres.

Positive pressure respirator. A **respirator** to which filtered air is supplied at a rate higher than needed by the wearer, so that the pressure in the face-piece or hood is always slightly higher than the ambient pressure. Inleakage of ambient air is thereby prevented.

Post-treatment processes. Although sometimes used for the operations carried out on a **filter** immediately after **cake formation** – e.g. **washing** and **dewatering**, the term more usually refers to the treatment of the dewatered cake after it leaves the filter, including thermal drying or stabilising, or even incineration.

Potential vortex. A feature of **cyclones** (also known as the Helmholtz vortex), which rely upon a spinning mass of fluid to create a **centrifugal separating force**. In a potential vortex (of which the **primary vortex** in a cyclone is an example), the flow velocity increases towards the centre of the vortex. The relation:

$$vr^m = \text{constant}$$

where v = tangential velocity of fluid
r = radius of vortex at which that velocity is measured
m = an index whose value is unity for frictionless flow, and close to 0.5 for frictional flow of **suspensions**

Pot press. A batch **expression press**, similar to a **box press**, in which a stack of open cylindrical pots are pressed by a hydraulic or mechanical ram pushing downwards. Loose feed material is filled into every pot, with a **pad** of porous material or a **screen** at the base of each pot, and on top of the filling. The ram's plunger compresses the top pot, while the bottom of each pot acts as plunger to the one below it. Pressures of up to 400 bar are used, and the pot press has a higher throughput than most other types of batch press.

Prayon filter. A rotary **tipping pan filter**, marketed by Bird.

Precipitation. The formation of **particles** within a fluid, which have then to be separated from it. One particular meaning of the term comes from the reaction between two materials dissolved in a liquid, to produce a substance that is insoluble in the liquid, and so forms small particles that can be separated from the liquid. Another meaning refers to the formation of water droplets in air, either as a mist or as rain.
[*See also Crystallisation*]

Precipitator. The equipment in which a controlled precipitation process is carried out, and which is frequently a form of **nutsche filter**, enabling the recovery of the precipitate once the reaction is finished.

Can also refer to a **sludge blanket clarifier**.
[*See also Electrostatic precipitator*]

Precoat, Precoat layer. A layer of inert and insoluble material, granular or fibrous, that is laid down on the surface of a **filter medium**, prior to the commencement of a **recovery filtration** process. The precoat layer becomes the filter medium, upon which the **cake** collects, thereby ensuring that none of the solids suspended in the slurry to be filtered passes though the filter medium while the filter cake is being established. The precoat also protects the filter medium from being fouled by sticky material or very finely divided solids.

The most common materials used as precoat are **diatomite** and expanded **perlite**, replacing **cellulose** (although the latter still has wide usage for particular applications). Precoat materials have a high specific surface, so as to be able to set up a suitable filtration layer quickly.

Precoat drum filter. A **drum filter** used for **clarification** rather than solids **recovery**, in which a thick layer of **precoat** is laid down on a drum, and slowly removed, with its trapped solids, by an **advancing knife**.

Precoat filter, Precoat filtration. By the laying down of a **precoat layer** on a relatively open support, a good **filter medium** can be established, to match the **filtration** application. Precoat filtration is used to achieve a high degree of **clarification** of a feed liquid, and is frequently employed in the beverage industries. It would normally be undertaken in a **batch filter**, but can also be used in a range of continuous **vacuum filters**. Occasionally, the precoat layer can act to treat the **filtrate**, as, for example, in the decolorisation of the liquid.

Prefilt. A term sometimes used for the feed **slurry** to a **filter**.

Pre-filter. Any item of **filtration** equipment that is situated upstream of another **filter** (or indeed, of any other kind of equipment), whose function is to remove oversize particles from the flow of fluid, that would otherwise damage, or interfere with the operation of, the downstream item, by, for example, plugging nozzles or distribution flow spaces. The pre-filter can itself be quite a finely separating device, for example in the use of a **microfilter** or even an **ultrafilter** to protect a **reverse osmosis** unit. On the other hand, most coarse filters and **strainers**, mounted as **in-line filters**, act to protect mechanical equipment such as pumps and valves.

Press. A term widely used in the **filtration** industry for any device with a filtering action supplemented by mechanical or fluid pressure.
[*See also Belt press, Filter press, Screw press*]

Press belt. A flexible strip of material that is pressed down on top of a separated wet **cake** lying on a porous belt beneath it (the combination being known as a **double belt press**). The press belt is forced onto the cake by a set of rollers and the cake is then **dewatered** by **compression**. If the press belt seals the edges of the cake, then an applied gas pressure may also be used to achieve dewatering.

Press filtration. A rather imprecise term referring to any **filtration** process where the **filter cake**, once formed, is subject to **compression**, either by the supernatant **suspension**, or by some ancillary mechanical means, such as a **belt** or a squeezing **diaphragm**. Press filtration can only be effective if the cake is **compressible**, which would be the case where the particle size is small (say, below 10 μm), where the particles are non-rigid or are themselves flocculated. Pressing forces range from a bar or two, up to 150 bars in some cases.

Press plate. The plate in a **filter press** from which the **wash filtrate** is collected during **through washing**.

Press roller. A rotating cylinder carrying a belt that moves across the roller, propelled by its rotation, and an important component of a **belt filter**. The roller presses the belt into the layer of solid on it, and the belt may wind through a series of rollers to achieve the required degree of dewatering. Two rollers may be mounted close opposite one another with a pair of belts nipped between them to give additional compression.

Pressure difference, Pressure drop, Pressure loss. The difference in fluid pressure across a porous mass that causes a fluid to flow through that mass – a **filter cake** or **filter medium**, for examples. The differential can be created by applying a vacuum downstream of the mass, but the term almost always applies to a reduction in pressure from a value higher than the ambient pressure.

Less specifically, a drop in pressure occurs whenever a fluid flows through a pipe or vessel, because of friction at the walls. The terms also relate to a difference in pressure observable between two points in a flow system.

Pressure disc filter. One of a range of **disc filters**, either **indexing** or **rotary**, which is totally enclosed in a pressure vessel, and operated at a pressure well above **ambient**. Models exist to match the range of **vacuum disc filters**, but the mechanical construction is more complex.

Pressure drum filter. A **drum filter** operating under super-ambient pressure as the **filtration** driving force. Less common than the equivalent **vacuum drum filter**, because it is mechanically much more complex.

Pressure effect. Refers specifically to the operating (fluid) pressure in a **cake filtration** process, in which the higher the pressure the more effective the **filtration**, for a completely **incompressible** cake. However, all cakes are **compressible** to some degree, and so a higher pressure may cause **compression** of the cake, and so a reduction in its **permeability**.

Pressure filter. Any **filter** in which the driving force necessary to cause the fluid to flow through the **filter medium** is a fluid pressure above the ambient level, as distinct from vacuum, gravity or **centrifugal force**.

Pressure flotation. A **dissolved air flotation** process in which the flotation air is dissolved under pressure and the cell operates at atmospheric pressure.

Pressure leaf filter. A **filter** consisting of a pressure vessel containing one or more **filter leaf** type **elements**. The **feed slurry** is introduced to the vessel under pressure, and filtration occurs through the **filter medium** on each side of the leaf. **Filtrate** flows to and through the inside of the leaf, and out of the vessel via the filtrate offtake system, linked to each leaf. The pressure vessel is usually cylindrical in shape, and may be mounted horizontally or vertically. In order that accumulated **filter cake** can be easily removed from the leaves, these are usually hung vertically.

Filter cake can be removed in the wet state or after **dewatering**, and this is usually an automatic process. Removal can be achieved by scraping or vibration or blowback of filtrate, after which the solids fall to the bottom of the vessel for later removal. Such a filter can be operated with quite a short cycle time.

Pressure plate filter. A **pressure filter** with a stack of circular **plates** accumulating **filter cake** on their upper surfaces.

Pressure pot. A small test **filter** used to predict the performance of a range of **pressure filters**.

Pressure separations. Upstream fluid pressure is one of the major driving forces in **filtration**. Pressure separations have the important characteristic that they usually have a smaller overall size than other methods, and so are more easily fitted into a working space. Fluid processes are also very often undertaken at pressures above ambient, so inclusion of a **pressure filter** is easily done. Pressure separations make the control of vapour emissions easy – but create great difficulties in the discharge of collected solids to the outside.
[*See also Pressure filter*]

Pressure vessel filter. Any design of **filter** in which the **separation** equipment is enclosed in a pressure vessel – from the simple **leaf filter** to the complex **rotary drum filter**. (Also known as a tank filter.)

Pre-thickener, Pre-thickening. The concentration of a **suspension**, by the removal of as much suspending fluid as possible. This is done by relatively simple equipment, ahead of a much more complex separation stage. Pre-thickening may just be for the reduction of the separation load on the subsequent equipment, or it may be necessary in order that the next stage should work at all. Thus, a **pusher centrifuge** will need a certain minimum slurry concentration fed to it, or the suspension will wash out over the edge of the **filtration** zone in the **drum**. Pre-thickening is often achieved by gravitational **sedimentation**, and frequently is accompanied by **agglomeration** to maximise the **separation efficiency**.

Pre-treatment chemicals. Chemical additives used to condition a **suspension** ahead of a **separation** process to improve that process's efficiency. They include acids or alkalis added to adjust pH values, and **flocculants** intended to **agglomerate** the particles of the suspension.

Pre-treatment processes. Pre-treatment involves changing the properties of a **suspension**, by physical or chemical means, in order to improve its **filterability** (or **settling** behaviour). The processes therefore are the conditioning steps applied to a suspension prior to **filtration** or other **separation** process. These normally include pH adjustment and charge neutralisation, and **flocculation**, but may also cover **thickening**, heat treatment or even **freeze-thaw** processing. Other physical processes can include size **classification** (especially the removal of oversize particles), aeration, agitation, ultrasonic or mechanical vibration, magnetic treatment or ionisation.

Primary coagulant. A **coagulant** that is used to aid the performance of a subsequently applied **flocculant**.

Primary particles. A term used to refer to the basic particles in a suspension that will be **agglomerated** into **flocs** by a **coagulation/flocculation** process.

Primary treatment. The first major stage in the treatment of municipal sewage, which mainly involves the removal of particles larger than 100 mm from the flow of waste water (sand and grit, and larger detritus), usually by **sedimentation**. May also include the precipitation of soluble components and the removal of oils and floating particles.

Primary vortex. The main swirling motion in a **cyclone**, resulting in the **separation** of the suspended solid particles. This vortex (which is a **potential vortex**) forms between the tangential input to the cyclone, and the thickened slurry outlet nozzle at the cyclone's throttled apex. At the apex, the vortex is reflected back up the axis of the cyclone, as a **secondary vortex**, inside the primary one, to discharge, as a clarified fluid, through the **vortex finder**, whose diameter decides the **cut size** for the separated particles.

Probability screen. A **screen** for **classification**, specified on the basis that particles moving at right angles to the screening surface are likely to pass through when their size is greater than about half that of the aperture. Used especially in the selection of **sieve bends**.

Process filter. A **filter** (or **centrifuge**) whose operation is essential to the process in which it is used, and whose operating characteristics are dictated by the nature of that process.
[*See also Utility filter*]

Process parameters. The physical (and possibly chemical) characteristics that define the **separation** process. These may include the density, viscosity and surface tension of the liquid, and the particle size, **particle size distribution** and solids concentrations in **feed** and **filtrate**.

Projected diameter. The diameter of a circle having the same projected area as a particle resting on a horizontal surface, when viewed from above.

Pseudo-equilibrium period. In the operation of a **membrane process**, **permeate flux** reduces as a function of time, as the membrane surface becomes **fouled**, until only a very low flux is achievable. If, however, steps are taken to minimise fouling, permeate flux will still fall, but more slowly, and will reach a pseudo-equilibrium value, when the depth of the fouling deposit has stabilised.

Pseudomonas diminuta. One of the most common **bacterial** species used in **challenge tests** for **filters**.

Pseudoplastic. An old term, no longer in correct use, for those non-Newtonian fluids whose viscosity decreases with increasing shear.
[*See also Shear thinning*]

Psi factor. A parameter used in the design of **decanter centrifuges** to be used for **thickening** work, to indicate the control position between the extremes of high dryness, poor **centrate**, and wet cake, good centrate.

PTFE. [*See Polytetrafluoroethylene*]

Pull-action centrifuge. A proprietary design of batch **basket centrifuge**, made by Ferrum, which has a flexible liner supported on a perforated **drum**, which rotates about a vertical axis and has an open bottom. The liner acts as the **filter medium**, and when the batch has been filtered, the drum is slowed, and the liner pulled downwards to release the accumulated **cake** out through the bottom. This results in complete discharge of the cake, without leaving a **heel**.
[*See also Inverting basket centrifuge*]

Pulp. Another term for **slurry** or **suspension**, but used specifically for suspensions in the paper industry, and in some parts of the mineral processing industry.

Pulp density. Refers to the bulk density (and sometimes solids concentration) of a **suspension**, and is used mainly in the paper and mineral dressing sectors.

Pulse-jet cleaning. The removal of collected **filter cake** from the outside of a filter **element** by means of a rapid pulse of compressed air (or **filtrate**) applied inside the element. This is usually applied to the elements in succession, so that the bulk of the **filtration** flow can continue. Mostly used in **fabric filters** for **gas cleaning**, but liquid filters can also be cleaned in this way.

Pulse-jet filter. Any **filter** using **pulse-jet cleaning** of its filter **elements**, but mainly used of large **fabric filters**, employed in dust removal from gases.

Pump. [*See Centripetal pump*]

Pumpable cake. A **filter cake** that is fluid enough to enable it to be handled in a pump and conveyed in a pipeline to the next processing stage.

Purification. Specifically used to refer to the separation of traces of one liquid from its suspension in another liquid. This can be done by means of a **disc stack separator**, but is more often undertaken by a **filter-coalescer**.

Purifier. A word used generally of any device that purifies a fluid stream by removing contaminants from it, and so usable of almost any of the equipment covered by this encyclopaedia. Sometimes used specifically for equipment in which **purification** is achieved.

Purolator filter. A long-established **filter** for lubricating oils, in the form of a **cartridge** made from a continuously wound steel **ribbon**, wound flat turn after turn, with the turns kept apart by spacer projections from the flat surface of the ribbon. In fact, the ribbon is of a long wedge shape, with its outer edge wider than the inner, to form the filtration surface. A scraper device is rotated around the outside of the cylinder to keep it clean.

Pusher, Pusher centrifuge. An important type of **centrifugal filter**, widely used for the separation of crystals and other coarse solids (with particle sizes of 80 μm or above). It is usually mounted with its axis of rotation horizontal, in cantilevered form with drive, gear box and bearings all at one end, and the **drum** open at the other. The **slurry** is fed continuously onto the drum at the end nearest to the drive, and the formed **cake** is then pushed across the drum towards the open end, by an axially reciprocating annular plate. The drum is formed from a screen of parallel **wedge-wire** bars, mounted in line with the drum axis.

One major variant has multiple stages in the form of separate sections of drum wall, moving with respect to one another, such that the solids from the first stage tumble onto the start of the second and so on. Another has a symmetrical, two-drum form with drive at the centre, and the reciprocating pushing of the solids moving them away from the centre in both directions (double acting pusher).

A pusher centrifuge can achieve very good washing of the filter cake, but needs a fairly concentrated feed slurry (at least 10%) so as not to wash straight out of the drum. They normally operate in the G-force range 200 to 2500, and can range in size from a drum diameter of 15 cm up to 1.5 m.

Pusher plate. The axially mounted plate in a **pusher centrifuge** that pushes the accumulated

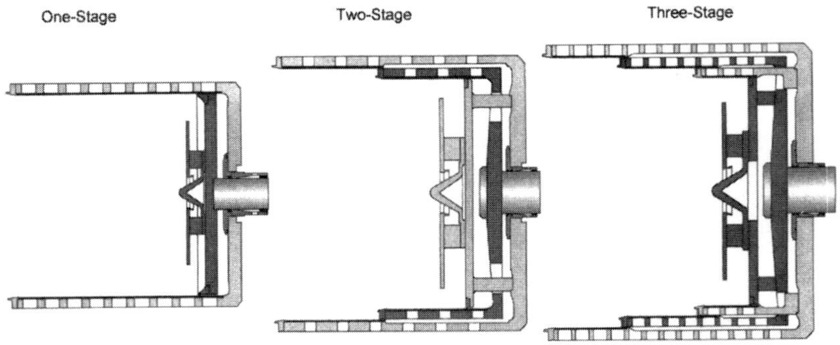

One-, two- and three-stage pushers

solids along the drum and over its far edge. The plate is as close a fit as possible to the screen. The plate movement does not cover the whole length of the drum, relying instead upon the pushing force being carried by the dewatered solids.

PVDF. [*See Polyvinylidene (di)fluoride*]

Pyrogen. Any of a group of materials that, upon ingestion by an animal, cause a rise in body temperature.
[*See also Pathogen*]

Q

Queen plate. The end plate of a filter press, which moves as the press is opened or closed.
[*See King plate*]

Quick filter. A term sometimes used for a **deep bed, sand filter**, more usually termed a **rapid filter**.

Quick-start centrifuge. A batch laboratory **centrifuge** which is able to reach its operating speed very quickly (in 1 to 2 seconds), and to slow down just as rapidly. This enables it to reproduce under test conditions the behaviour of **centrifugal filters** that have short residence times, such as a **pusher** or some **conical screen** machines.

R

Race track. A close fitting **collector** for the discharging solids, fitted around the solids discharge holes at the top of the **beach** of a **decanter centrifuge**.

Radial control head. [*See Control head*]

Radial flow. The flow of a fluid or a particle suspended in a fluid in a **separator** that is rotating, in order to achieve **centrifugal** behaviour. In a **hydrocyclone**, the direction of this flow changes along the axis of the device, and the direction of flow becomes an important design parameter.

Radial velocity. A parameter in the design of any **centrifugal** system, either **centrifuge** or **(hydro)-cyclone**. The radial velocities of fluid and solid must be different, or there would be no separation.

Raffinate. In an liquid/liquid **extraction** process, the raffinate is the depleted **solution** or **suspension**, after treatment of the **feed** with the extracting **solvent**.

Ragged profile. A situation in a **cake washing** process that produces an uneven **interface profile** because **wash liquid** flows through some **pores** in the cake much faster than through others.
[*See also Plug flow*]

Rake. A device installed at the base of a **sedimentation tank**, especially a **mechanical classifier**, which moves slowly across the bottom to pull or push the collected solids to the **underflow** (solid discharge) point. It must move slowly, so as to avoid stirring up the solids and in that way return them to the suspended state. In a circular tank, the rake will move continuously around the circle to pull the solids to a central discharge. In a rectangular tank, the rake will move from end to end and then back to the start again, acting in one direction only, or effective in both directions, with a discharge at each end.

A completely different use for the term occurs in a **decanter centrifuge**, relating to the angle between the tip of the **conveyor flight** and the **bowl wall**. The rake is the angle the flight subtends to a perpendicular from the axis. It can be positive, when the action of the scroll is to lift the solids (then sometimes called cant), or it can be negative, when the solids will be pressed against the bowl wall. In some designs the flights are canted in the beach section, to provide zero rake on the beach; i.e. the flight tips are perpendicular to the beach. **Tiles** are designed to give the conveyor flight tips a cant to improve scrolling efficiency.

Rake classifier. A **mechanical classifier** in which the solids discharge from the **pond** is by means of an array of **rakes** that drag the **underflow** up a long slope. The rake is actuated by an eccentric motion so that after each pull it is lifted up and returned to its starting point. Multiple rake designs are available with some rakes dragging while others are being returned.

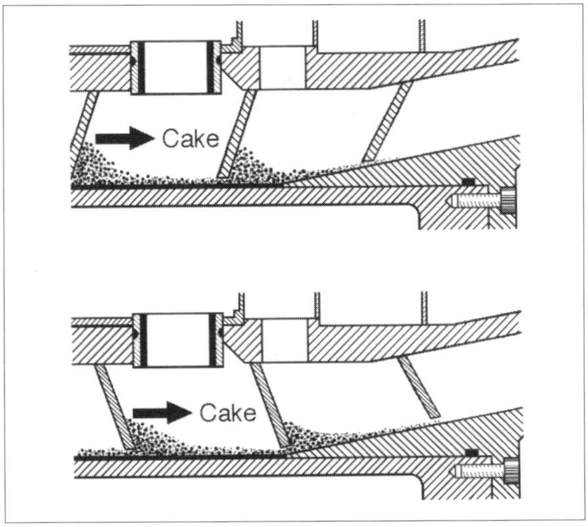

Flights with negative and positive rake

The motion of the rake causes some liquid movement, which makes **classification** sharper than in simple tanks, without any **reslurrying**. **Wash liquid** can be added above the level of the pond to wash **fines** or **mother liquor** off the coarse solids. The underflow may be allowed to **dewater** on the beach above the pond. [Note that a **drag conveyor** has a similar action, but the blades of the rake are attached to a belt and move continuously through the pond.]

Random fibre media. **Filter media** can be very broadly divided into two classes, those with some structure (such as **woven media**) and those with little or no structure (such as **nonwoven** materials or **bonded** particles). The second group have a structure that is close to completely random. Random fibre media, either **felts** or **spunlaced** are of increasing importance in the **filtration** business.

Rapid gravity filter, Rapid sand filter. A **deep bed filter** with down flow of fluid at a relatively fast rate, cleaned by **backflushing**. Also known as a quick filter, the rapid sand filter is the most common form of filter in the processing of fresh water, and in the treatment of waste waters.

It is a filter, using a deep bed of sand or some other inert granular material as the **filter medium**, and operating at high throughputs, of the order of 10 m^3/m^2h. The driving force is gravity, working on the hydrostatic head of feed water.

The solid in the feed water is removed by **adsorption** within the bed, which must eventually be regenerated by backflushing with a stream of clean water, to expand the bed and so release the trapped solids (the backwashed slurry is more concentrated, and must be filtered on another filter). **Filter aids** may be used to assist the separation.
[*See also Sand filter*]

Rapid settling clarifier. [*See Reactor clarifier*]

Rapid settling suspensions. A broad class of suspensions, usually of coarse and/or dense particles, larger than about 50 μm. The designation is used by some authors to classify the types of equipment best suited, economically and technically, to achieve the required separation for such suspensions – which often involve **classification** by particle size or density.

Rated pore size. An indication of the effective pore size of a **cartridge filter** or **membrane**, quoted by the manufacturer, but rarely of use in the comparison of the products of different manufacturers.

Rate of filtration. The speed with which a **filtration** process occurs, almost always used of **cake filtration**, and hence is governed by the rate of **cake formation**.

Rate thickness product. A parameter derived from the equation giving the **cake thickness** in a **cake formation** process. This parameter, the product of the **filtrate** flow rate, q, and the cake thickness, L, can be used to describe the **filtration** characteristics of different materials. There is usually a logarithmic relation between Lq and the filtration pressure.

Rayon. Originally a brand name, referring to regenerated **cellulose**, especially in **fibre** form. Much used as **filter media** for a while, but now largely replaced by better materials.

Reactor clarifier. A **sedimenting clarifier** that is designed to give high throughput rates by careful design of the **flocculation** section. Also known as high rate or rapid settling clarifiers, the reactor clarifier (equivalent to a **high-rate thickener**) has a **feedwell** containing a set of paddles or some other mixing device to ensure good mixing of the **flocculant**.
[*See also Sludge blanket clarifiers*]

Reagent. A general term used for any chemical substance added to a system in order to change its behaviour by a physico-chemical reaction. Can therefore be used of **coagulants** and **flocculants**.

Rear end, Rear hub. In a **decanter centrifuge**, the rear end is conventionally the end of the decanter where the **cake** discharges. The rear **hub** is therefore the flange, or end plate, which encloses the **bowl** at the cake end, and includes the spindles that support the rear main and rear conveyor bearings.

Rebecel. A bulking **filter aid** material based on **cellulose**, derived from reneweable sources. It is produced as pellets, which are dispersed in water before being added to the slurry to be filtered.

Receiver. Generally refers to any tank following a separating device, into which the **filtrate** or **centrate** is transferred as intermediate storage before moving on to subsequent processing or discharge. A receiver can also act as a **degassing** (deaerating) stage for the removal of the air that becomes entrained in the processed liquids, especially in the **casing** of a **centrifuge**.

Recessed plate press. [*See Filter press*]

Reciprocating screen. A **screen** shaken by means of a reciprocating motion along one axis. This would normally be a large, fairly coarse screen, used in **minerals processing**.

Reciprocating tray filter. A **vacuum belt filter** in which the belt moves continuously over a set of trays, which each fasten on to the underside of the belt and travel with it for the duration of the **filtration cycle**, then to be released and return to the start position. The fastening and releasing are effected by the intermittent application of vacuum to each tray.

Recirculation clarifier. A **continuous gravity clarifier** with internal recirculation of the solids to improve **flocculation**. Some of the separated solids are fed back internally (or occasionally externally) to mix with the **feed slurry**, in order to accelerate the rate of flocculation and promote the capture of fines by interception. The recirculation clarifier is smaller than an equivalent one pass clarifier (or, for a given size, gives better overflow clarity), but it cannot be stopped easily nor operated with fluctuating loads.

Recovery. A general term referring to the removal of suspended solids (usually) from a liquid; it implies that the solids are wanted, not wastes. It also implies that the **suspension** is fairly concentrated.

It can also mean the percentage of the suspended solids in the **feed** that is discharged as accumulated cake, especially when the recovered material is a valuable one.
[*See also Harvesting*]

Recovery index. A measure of the efficiency of particle **separation** in **screening**. It is defined as the mass of that part of the fine product that is finer than the **cut size**, divided by the mass of the **feed** solids finer than the cut size.
[*See also Rejection efficiency*]

Recovery plate. A horizontal **filter plate** located in the bottom of the vessel of a **pressure leaf filter**, used to filter the remaining slurry in the vessel at the end of the main cycle through the leaves. Also called a **scavenger filter**.

Recrystallisation. Most **crystallisation** processes are controlled by a rate of **diffusion** through a **mother liquor**. This may not give a sufficient degree of purity the first time, and so separated crystals are washed, redissolved and recrystallised. There is, however, a particular use of the term in the behaviour of a **filter cake** at the **filter medium** surface, when it comes to **cake release**. It is possible for recrystallisation of the solute to occur in the space between a particle of the cake and a fibre of the filter medium, thereby forming a solid bridge between the two, and so making cake release more difficult.

Rectangular basin, Rectangular clarifier. A simple **continuous gravity clarifier**, in the form of a long rectangular tank, usually having a mechanical **scraper** to move solids towards the **underflow** outlet. The flow of liquid through the clarifier is approximately **plug**-like, the residence time being sufficient to achieve the required degree of **separation**, with or without prior **flocculation**.

Reduced efficiency, Reduced mass recovery. In a **separation** process whose performance has been tested in the laboratory, separation efficiencies and mass recoveries may be determined relatively easily. Where the ratio of **underflow** to **feed** is high, however, the performance is diminished, because some fine material will appear in the underflow, because of its high relative volume. Under these circumstances the separation efficiency is changed to a reduced efficiency:

$$E_r = (E_t - R_f)/(1 - R_f)$$

where E_t = total efficiency
R_f = the underflow/feed ratio

A reduced **grade efficiency curve** could then be plotted, and a reduced **mass recovery** would be calculated as total efficiency minus the underflow/feed ratio.

Reemay. Originally a brand name for the then new **spun-bonded** materials developed by du Pont. Now a brand of BBA.

Re-entrainment. The picking up again by the flowing fluid of particles that have already been separated from the fluid, as might happen if the flow velocity were to increase, or if the **filter** should get vibrated.

Refreshing knife. An **advancing knife** for a **precoat drum filter** that is periodically retracted to its starting position, in order to allow a fresh layer of **cake** to build up.

Regeneration. A term used specifically of **membrane filtration** processes to refer to the various ways in which the decline in **permeate** flux can be prevented, perhaps even with a return to higher flux rates – when the membrane is said to be regenerated.

Regulator. A term used specifically in **flotation** processes to refer to a chemical substance added to the **suspension** to control flotation by activating or depressing the **attachment** of the particles to the bubbles.

Reichert cone. A **continuous gravity concentrator** in the form of a wide, shallow, inverted cone, into which the **slurry** is fed at the periphery. **Separation** between lights and heavies occurs as the slurry flows over the lower surface of the cone, with the heavy fraction taken off through a slot away from the centre, and the light fraction at the centre. Two or more cones may be stacked one above the other to increase the separation efficiency.

Rejection. A term used specifically in describing the action of **ultrafiltration membranes**, which are generally characterised in terms of **solute** rejection, namely the fraction of material of a specified size (usually measured in terms of **molecular weight**) in the feed that is retained on the membrane, i.e. that is rejected from passing through it.

Rejection efficiency. A measure of the efficiency of a **screening** process, defined as the mass of that fraction of the coarse product coarser than the **cut size**, divided by the mass of the **feed** particles coarser than the cut size.
[*See also Recovery index*]

Reitema model. A basic model of **hydrocyclone** behaviour based on residence time. It assumes that, for a 50% efficiency, a particle at the centre of the inlet nozzle must travel outwards (a distance of half of the diameter of that nozzle) to reach the inside of the hydrocyclone wall in the time that it is inside the hydrocyclone.

Relative flowrate. A term used specifically to refer to the **permeability** of a **filter aid**. Manufacturers of these materials often define the permeability of their finest grade as 100, then give a figure of 200 for a grade with twice the permeability, and so on.

Relative permeability. The parameter defining the rate of flow of one fluid phase (gas or liquid) when a **filter cake** contains both phases.

Relaxation. A term generally referring to the changes in a material when an applied pressure is removed, and specifically referring to the expansion of a compressed **cake** that has been **dewatered** under applied gas or mechanical pressure when that pressure is released.

Rellumit filter. [*See Fipoca filter*]

Remanent moisture. [*See Residual moisture*]

Remashing. A term, specific to the alcoholic beverages industry, meaning the **reslurrying** of a previously dewatered mass of solid particulate material, for reprocessing.

Remoisturization. A phenomenon observable at the end of a **dewatering** process for a **filter cake**. If the dewatering driving force is stopped before the cake is discharged, then some moisture may be sucked back by **capillary attraction**, into the cake from the **filter medium** or adjacent parts of the filter.

Renewable element. A **filter element** that can be treated, by mechanical or chemical cleaning, or in some other way, so as to return it to its original condition (or close to that condition), in terms of flow resistance.

Residence time model. [*See Rietema model*]

Residual liquor, Residual moisture (content). The liquid remaining in a **porous mass**, such as a **filter cake**, that cannot be removed any further by mechanical means. This will normally be **capillary** or **interstitial moisture**. The amount of liquid so retained has to be determined by the weight change following thermal **drying**, and is normally expressed as a percentage of the weight of the moist mass. Determination of moisture content by infra-red absorption is also an effective technique.

Residual product layer. [*See Heel*]

Residual saturation. [*See Equilibrium saturation*]

Residual volume filtration. In **batch filtration** processes, especially, for example, in **candle** or **leaf filters**, where the **slurry** to be filtered occupies a vessel surrounding the **filter elements**, there will always remain a quantity of unfiltered slurry in the vessel at the end of each batch. This must be removed before the **filter cakes** can be properly **dewatered**, and this presents the problem of what to do with it. Where the next batch is the same in composition as the previous one, then the residual volume can be recycled to the feed tank as part of the next batch.

Where, however, the next batch is of different composition, then some ingenuity is required to filter the residual volume, especially if it is valuable. This may require a small **filter** mounted in parallel with the main one (whose residual volume, in turn, may be small enough to be drained away), or it may be possible to spray the residual material onto the formed filter cake. [*See also Recovery plate, Scavenger filter*]

Resilience. The ability of any material to absorb physical shock without permanent deformation, and so applicable in **filtration** terms to refer to, say, the properties of a **filter bag** that is regularly cleaned by the sudden expansion caused by a **reverse pulse**.

Resin. A high molecular weight material, which softens at moderate to high temperatures, and is used as a **bonding** adhesive. It can be natural (rosin or shellac) or synthetic (polymeric – usually phenol-formaldehyde, polyurethane or epoxy).

Resin bonded, Resin impregnation. The treatment of a **fibrous** material with an adhesive resin (usually a synthetic one) to give it added mechanical strength, by bonding together the particles or fibres of a **filter medium**. This is done with **paper**, especially to give it wet strength, in **pleated** paper **cartridges**, and also in the forming of **bonded depth filter** cartridges. In the case of paper, about 30% by weight of the finished product is made up of resin.

The resin may be added as a solid in the medium formation stage, or injected as a liquid, in either case being set at higher temperatures.

Resin-wool. A **filter medium** used in gas cleaning, and made from the resin impregnation of carded **wool** fibres. The resin used is a rosin modified phenol-formaldehyde.

Resistance, Resistance to flow.
[*See Flow resistance*]

Resistance force. When a fluid flows past a particle, the effect on the particle is a sum of the pressure and friction forces. The component of this total force in the direction of the flow is called the resistance force (and that in the perpendicular direction is the dynamic buoyancy).

Reslurry plate. A particular **plate** in a (plate and frame) **filter press** that is designed to allow the accumulated **cake** to be reslurried.

Reslurry washing. Some **separation** equipment enables the separated solids to be washed thoroughly inside the equipment. In other cases, the **filter cake** is removed from the **filter** or **centrifuge** and resuspended in **wash liquor** (usually fresh water) in a stirred vessel, and is then passed to a second stage of separation. The whole process can be repeated as necessary, in order to achieve the required degree of purity in the final cake.

Resonance screen. A **vibrating screen** that is designed to operate at its resonant frequency, and so to minimise energy requirements.

Respirator, Respiratory filter. A respirator is a device that enables the wearer to breathe clean air, when working in an environment where the ambient air contains dusts, or noxious or toxic gases. The inhaled air passes through a **filter**, which, in the case of a dust respirator, is a highly efficient particle filter. For gaseous impurities, the respirator will be fitted with an **activated charcoal filter**, to adsorb the impurities. Increasingly, **combination filters** are used, which are able to do both jobs.

Resuspension. [*See Re-entrainment, Remashing, Reslurry washing*]

Retentate. The thickened slurry leaving a **cross-flow filter**, especially a **membrane separation** system, a portion of the suspending liquid having been removed as **permeate**. (Also known as concentrate.)

Retention, Retention efficiency. Used in a similar way to "capture", to refer to the entrapment or the holding back of suspended material by a **filter medium**, and so to the ability of a **filter** to retain particles of a given size and greater. The proportion of a **feed suspension** that is retained by the filter gives the efficiency of retention, which is the same as total efficiency.

In terms of **membrane separation** processes, it has a rather more precise meaning, being expressed in terms of concentrations, with the retention efficiency equal to (1 – the concentration of the species in the feed) divided by the concentration in the **permeate**.

Retention capacity. [*See Dirt holding capacity*]

Reticulated (foams). Foams that are **impermeable** in the form in which they are made can be rendered **permeable** by chemical or thermal methods that

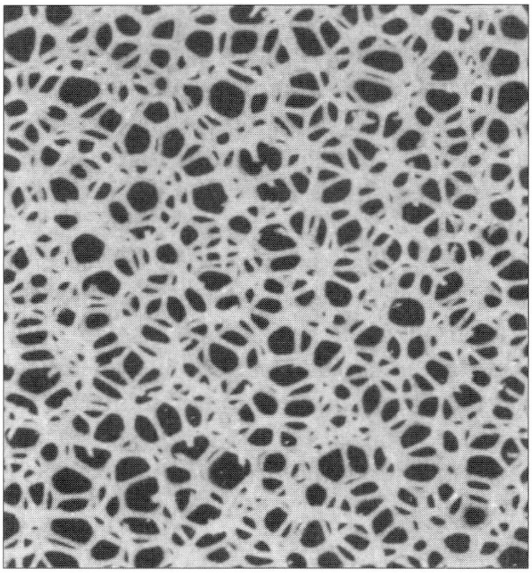

Reticulated polyurethane foam

erode the cell walls to create pores through the material – the process being called **reticulation**.

Reticulation. A process for the treatment of polymeric **foams** that breaks down a high proportion of the cell walls within the original foam to make a series of interconnected pores, thereby converting the foam into a format whereby it will enable fluid flow through it. Reticulation is usually achieved by thermochemical processing, whereby the wall material shrinks back to the nodes of the original bubble network.

Return line filter. A **filter** in a hydraulic system that is positioned in the circuit after the last operating unit, and before the oil reservoir, so that it can remove all of the dirt produced in the system. Here it is unlikely to suffer from any of the pressure surges that might occur in the circuit.

Revamping. The re-engineering of an existing process system to increase the output of product or its quality (or both). This may involve anything from slight internal adjustments, through to the exchanging of a major system component for a better model.

Reverse flotation. A **flotation** process used to clarify a dilute **suspension**, in which the waste material is transferred to the **froth**.

Reverse jet cleaning. The cleaning of a **filter element**, especially a **filter bag** or **filter sleeve**, by the action of a jet of air on the side of the **filter medium** opposite to that upon which the **filter cake** is deposited. This can be done whichever side of the medium is being used for filtration, frequently by means of jets from a blow-ring that travels up and down the sleeve or bag.
[*See also Pulse jet filter*]

Reverse osmosis. A major **separation** process, whereby a solvent is recovered from a solution. It is employed largely for the generation of pure water from saline or brackish waters, and was the first of the **membrane separation** processes. The impure water is presented to one side of a **semi-permeable membrane** under a hydraulic pressure greater than the natural **osmotic pressure** of the impure water. The pure water then passes through the membrane, and the impurities are retained and discharged as **retentate**. The membrane is impermeable to ionic and most molecular species in solution.

The driving force for reverse osmosis is a very high fluid pressure, and the membrane for use in the process needs to have a high ratio of **filtration area** to **housing** volume. The process operates in **cross-flow mode**, so that the movement of the **feed**/retentate flow helps to keep the membrane surface free of collected solids.
[*See also Osmosis*]

Reverse pitch. Refers to the **conveyor** of a **decanter centrifuge**, which has an inner **flight** as well as the main outer flight. The inner flight has opposite pitch to that of the main one, so as to scroll surface cake towards the **front hub** to increase its residence time in the **bowl**.

Reverse plain Dutch weave. A **plain weave** with the **monofilament weft** threads close together, showing a wavy form as they cross the **warp** threads, which remain straight. A common form for **wire mesh**.

Reverse pulse cleaning. A short pulse of fluid, usually the clean **filtrate**, in the reverse direction from that of the main **filtration** flow, used to clean the surface of the filter **element**.

Reverse pulse filter. Any **filter**, for gas or liquid flow, where the **filter medium** is cleaned by a reverse pulse of fluid. The **cake** blown off the element must settle quickly into a cake reception zone, so as not to be re-entrained in the flow of fresh feed.

Reversible filter. A **filter** with an array of flat filter **elements**, each capable of filtering a **slurry** flowing in either direction. Thus, with flow in one direction, **cake** is built on the upstream side of the element, to its optimum thickness. Then flow is stopped and reversed, whereupon the previously formed cake is blown off, and that **filtrate** is rejected. After a short rejection period, full filtrate flow proceeds, until the cycle is repeated. The filter is used for **clarification**, and may employ **precoat**.

Reversible tray filter. [*See Reciprocating tray filter*]

Reynolds. Osborne Reynolds was born in Belfast in 1842. After a mechanical engineering apprenticeship, he graduated in maths from Cambridge University in 1867. A year later, he became the first professor of engineering at Owens College in Manchester, a position that he held until he retired in 1905. Although primarily known for his work in fluid flow, his first major contributions were in heat transfer, leading to improved boiler and condenser design. He made an early estimate of the mechanical equivalent of heat.

He contributed significantly also to the development of turbine pumps, formulated the theory of lubrication, and then developed the standard mathematical framework for the representation of fluid turbulence. His 1883 paper on the law of resistance to flow in parallel channels led to the formulation of the first useful "**dimensionless number**", by which he is now best remembered.

Reynolds number. The most important of the various **dimensionless numbers** used to describe physical behaviour, the Reynolds number is the criterion by which it can be shown if the flow of fluid through any channel is absolutely steady (streamlined or **laminar**), or moves with small unsteady fluctuations (turbulent). If the Reynolds number is well under 2000, flow in a pipe is generally laminar, whereas at values well above 2000, flow is usually turbulent. The transition between laminar and turbulent flow does not occur at a specific value of the Reynolds number, but over a range beginning between 1000 and 2000, and extending upward to between 3000 and 5000.

In 1883 Osborne Reynolds, a British engineer and physicist, demonstrated that the transition from laminar to turbulent flow in a pipe depended upon the ratio of the product of the average flow velocity, the diameter of the pipe and the density of the fluid to the absolute viscosity of the fluid. This ratio, a pure number without dimensions, soon became known as the Reynolds number, and it was subsequently applied to other types of flow that are completely enclosed or that involve a moving object completely immersed in a fluid.

Thus the Reynolds number represents the ratio of inertial forces to viscous forces:

$$Re = Lu\rho/\mu = Lu/\nu$$

where L = a characteristic length (usually a pipe or pore diameter)
u = linear flow velocity
ρ = fluid density
μ = dynamic (or absolute) viscosity
ν = kinematic viscosity

Rheology. Strictly speaking, this term means the study of the flow and deformation properties of matter (solid, liquid or gas), including **elasticity**, **plasticity** and **viscosity**, but is often used in phrases such as "the rheology of the suspension", implying the specific values of these properties.

Rheopexy. A variant of **non-Newtonian** flow behaviour. The dynamic viscosity of a rheopectic solution increases with the increasing duration of the shear forces acting upon it.
[*See also Shear thickening*]

Rhodes model. A model, developed 30 years ago, to describe the **washing** of a **saturated filter cake**, which assumes that the concentration of **solute** in the **wash liquor** leaving a cake is directly proportional to its concentration in the cake, and that the two change equally (and oppositely) with time. This gives an exponential decay equation relating the change of solute concentration in the wash liquor with time.

Rib. One form of drainage surface below the **filter medium** on a **filter plate** consists of a number of ribs, protruding from the drainage surface, to hold the medium away from the surface and allow the flow of **filtrate**. The drainage grid thus consists of parallel straight ridges.

In a **decanter centrifuge**, ribs are often welded to the **beach**, or part or all of the **bowl** in a direction parallel to the centrifuge axis, in order to aid the scrolling of the collected solids. Used sometimes in place of **grooves**.

Ribs are also used across the water flow down a **sluice** or **launder** to form a **riffle**.

Ribbon. A continuous strip of material wound flat in the form of a spiral, so that its outer edges may be used as a **filter medium**. The apertures between these edges are formed by indentations on the flat

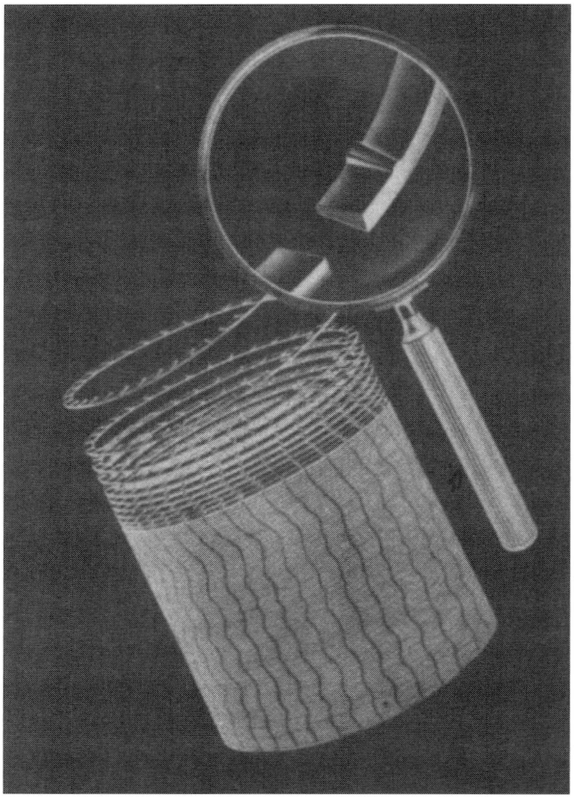

Ribbon cartridge

surface of the ribbon, which act as spacers between consecutive turns of the ribbon.
[*See also Edge filter*]

Ribbon flight. A **decanter centrifuge** may have a flight of small height, welded to pillars or vanes attached to the **conveyor hub**, instead of the full height of **scroll**. This causes less disturbance to the liquid in the **pond**.

Rib weave. [*See Plain weave*]

Richardson & Zaki equation. A relationship developed to describe the behaviour of **suspensions** under the conditions of **hindered settling**, where particle size and shape lose their influence on settling velocity, and the **concentration** of particles becomes the decisive factor. Richardson and Zaki's work (in 1954) produced a relationship of the form:

$$u = u_s (1 - c)^n$$

where u = actual settling velocity
u_s = an ideal settling velocity of a single isolated particle of average size
c = volume concentration of particles in the suspension
n = an empirical index.

Richardson & Zaki found a value for n of 4.65 for spherical particles, resulting in a rapid decrease in **settling velocity** with increasing values of the concentration.

Riffle. A riffle is an inclined surface down which a **slurry** flows, carrying solids of different densities. Across the surface (at right angles to the slurry flow) are a set of raised bars or grooves, such as on a shaking table. Into the space behind the bars, or into the grooves, settle the heaviest solid fraction (as would the granules of gold in a prospector's gold mine).

Rigid belt filter. A **belt filter**, similar to those with a **rubber belt**, but fitted instead with a set of rigid open ended **trays**.

Rigidised media. The term used specifically to describe a range of **filter media** made from stiff polymeric material, formed by moulding into shapes resembling **bags** or **pockets**, but essentially rigid in their final form.

Rigidity. The stiffness of a **filter medium** material, a term descriptive of one of its essential mechanical properties.

Rigid porous medium. A general term descriptive of a range of **filter media** made from metal, ceramic or plastic. This definition usually excludes woven metal, however rigid, i.e. it includes only foams and sinters.

Filter element in rigidized medium

Ring dam. [*See Weir plate*]

Ringelmann charts. A series of charts each carrying a blackened area, of increasing intensity of blackness (indicated by the numbers 0 to 5), intended for measuring the density of black smoke.

Ring shear cell. A particular piece of test equipment used for measuring the flow characteristics of a **filter cake**.
[*See also Jenicke shear cell*]

Ring stacks. A filter **cartridge** made from a series of flat rings, with suitable indented spacers, stacked one above the other, around a central core, so that the outer surface of the stack forms a **filter medium**, with circumferential apertures, precisely defined in width by the ring spacers.

Rinse chamber. In a **decanter centrifuge**, this is a compartment in the **hub** of the **conveyor**, behind the **feed zone** into which the rinse or wash water is applied before spraying onto the **cake**. A buffer chamber is sometimes interposed between rinse and feed chambers. The **floc** chamber is sometimes used as a rinse chamber.

Rising channel. A series of holes, one above the other, in each of the discs in a **disc-stack centrifuge**. The holes create a channel for the flow of one of the liquids in a **separation** process involving two immiscible liquids of different densities. The positioning of the channels is dictated by the volume ratio of the two liquids in the mixture to be separated, and also by the degree of clarity sought for either component.

Rising knife. A means for the discharging of the accumulated solids from a **basket centrifuge**, using a short **knife** that moves through the **cake** in a single traverse.

Roller discharge. A special design of **filter cake** removal device for **rotary drum filters**, for use where the cake is sticky or pasty in character, and where a **knife discharge** would not be fully effective. It consists of a small diameter roller, held against the width of the main **drum** of the filter, to which the cake sticks, and so is removed from the drum. The cake can then be cut from the roller more easily by a knife (which may have a straight blade, or a blade that is toothed, so as to leave an irregular surface to the residual cake on the roller, which then more easily picks up fresh cake in the next revolution).
[*See also Transfer roller*]

Roller elutriator. A solids **classification** device, which consists of a single vertical tube, terminating in a **thimble filter** intended to trap the finest particles. Different sizes of particles are separated by varying combinations of tube diameter and air velocity.

Roller mill. A machine for achieving the **expression** of oils and juices from vegetable matter, in which the **feed** material is fed directly between adjacent pairs of rotating rollers.

Rollfilt. A version of the hot **filter press** made by Bertrams.

Roll filter. A simple **filter** used for the cleaning of air and other gases, in which the **filter medium** is on a continuous roll. In one type, the roll is made of a coarse fibrous **mat** or of **wire mesh**, and is a continuous **belt**, one roller of which is immersed in a bath of oil, which both cleans and re-oils the medium. The other type uses a large roll of medium, such as **paper**, which is presented section by section to the filtration zone, with the dirty medium rolled up on the other side for eventual disposal.
[*See also Flat-bed filter*]

Rollgoods. Any material produced (and sold) in rolls, for conversion to suitably shaped **filter medium**, including both **woven** and **nonwoven** materials (although not usually **wire cloth**).

Room air filter. Any **filter** used to clean the air entering (or recirculating from) a living or working space. The demands on such a filter are steadily

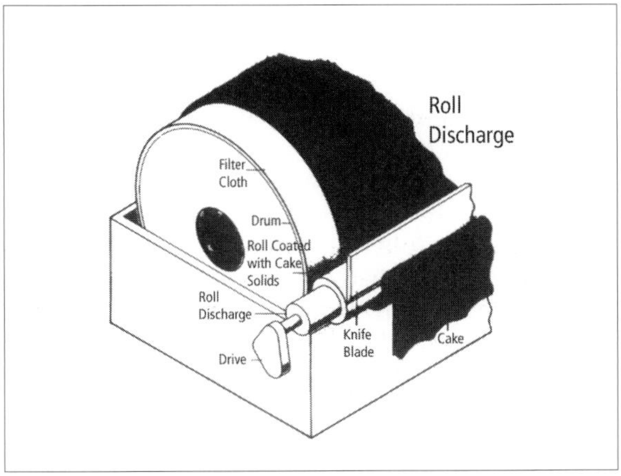

Drum filter with roller discharge

increasing as breathing air standards improve, and as more is known about the problems of air quality.
[*See also Indoor air quality*]

Rope discharge. A term descriptive of the discharge of solids from the nozzle of a **hydrocyclone** that is operating at high solids concentrations.

Rosenmund filter. A type of mechanised **nutsche filter**, supplied by the Rosenmund company.

Rosin-Rammler plot. A graphical method of showing **particle size distributions** that enables the linearization of the data. Such a plot can be used to facilitate the prediction of the **residual moisture content** of a **filter cake**.

Rota-plug thickener. A gravity-powered thickener in which a **sludge** is **dewatered** by being rotated in a fine **mesh trommel**, and mainly used for **flocculated** sewage sludge. The wire mesh of the drum is covered with a **monofilament plastic mesh** (usually of nylon). Flocculated **feed sludge** is pumped in at one end and **filtrate** flows out through the medium. The rotation of the drum causes the solids to build up as a plug, which acts as the means for compressing fresh sludge. Excess solids discharge from the thickener by building up over an annular weir at the end opposite to the feed. The discharged solids may be fed to a subsequent **drum filter** fitted with **compression** rolls.

Rotary barometric leg filter. A **rotary vacuum disc filter**, used in the paper industry, in which the vacuum is achieved by means of a drop leg on the **filtrate** exit line, which is long enough to create a vacuum at the filter.

Rotary disc/drum filter. *See* pages 211–13.

Rotary disc strainer. A continuous **gravity** operated **strainer** for use in raw water treatment. It consists of a large disc rotating about a horizontal axis, with a set of buckets at its periphery, and with its lower part immersed in the raw water flow. Suspended solids in the water are captured and held in the buckets, from which they are washed at the top of the revolution. It serves a similar purpose to the **travelling screen strainer**, but is used for finer solids. (Also known as a rotary wheel screen.)

Rotary drum filter. [*See Rotary disc/drum filter*]

Rotary extractor. A horizontally rotating table used for **percolation leaching**. Similar in concept to the **tipping pan filter**, it has wall compartments that form radial sectors, each with a perforated floor. The machine rotates so that each sector passes through a feed point; under a series of sprays that are usually arranged to give countercurrent flow; a drainage section, and a discharge section where the floor opens.

Rotary filter. Any of a number of continuously operating **filters** in which the **filter elements** or **filter medium** rotate at a constant (and fairly low) speed on a hollow axis or shaft, which usually carries the **filtrate pipe**(s). The filter surface is progressively dipped into a bath of the **suspension** to be filtered, with a **cake** forming on it, which is then raised out of the bath to be dewatered, washed and removed. Normally operating under vacuum, they can also be pressure driven, and they include **rotary disc** and **drum filters**.
[*See also Table filter*]

Rotary hydrocyclone. A **hydrocyclone** with rotating walls, offering lower shear stresses than conventional hydrocyclones. It is thus a cross between a **centrifuge** and a hydrocyclone. It is cylindrical in shape, and **feed** enters axially rather than tangentially, at one end where it passes through rotating blades. Both products of the separation are removed at the opposite end, the light product passing through an axial tube and the heavy product through a radial orifice.

Rotary magnetic drum filter. A rotating **drum** having a high magnetic flux over its surface, and thereby able to separate magnetic materials from a liquid flowing through a trough at the lower part of the drum. A machine tool coolant, for example, carrying ferrous **swarf** will flow below the rotating drum, in the opposite direction to that of the drum's rotation. The swarf is attracted out of the coolant, and then removed at the top of the drum's rotation, by means of a non-ferrous knife.

Rotary screen. [*See Trommel*]

Rotary siphon cup. A device fitted to a **peeler centrifuge** that creates a liquid head of a few centimetres outside the centrifuge **drum**, and so causes a small vacuum effect on top of the centrifugal force to draw the **filtrate** through the collected **cake**. As soon as the first section of cake is free of moisture, gas flow through the pores collapses the siphon, which is not re-established until filtrate flow starts again.

Rotary vacuum drum filter

Rotary disc/drum filter. The rotary drum vacuum filter is almost as important a **process filter** as the **filter press**, having widespread industrial uses. It is easily the most popular form of continuous **vacuum filter**, although it is losing market share to the **vacuum belt filters**. It is certainly the most important member of the class of rotary disc and drum filters, which include pressurised versions as well as vacuum ones.

The first patent for a rotary drum vacuum filter was granted in 1872 to the Harts, father and son, and, although many variations on the basic design have been developed over the ensuing century or so, that design remains in use largely unaltered. The operating principle involves a horizontally rotating **drum**, covered on its outside by **filter medium**, dipping into a trough of fairly thick **slurry**. The **filtrate** is sucked through the medium by a vacuum imposed from inside the drum, and a **cake** of collected solids forms on the outside of the medium.

The drum is made up of a set of flat **cells** or compartments, divided parallel with the axis of the drum, so that the cross-section of the periphery of the drum is actually polygonal, rather than circular. Each cell is independent of the others, with its own vacuum and filtrate drain pipes, connected to a rotary valve at one end of the drum. As the drum slowly rotates, a cell will dip into the slurry and the vacuum be applied to it. **Cake formation** occurs on the filter medium all the time that the cell is submerged, and the relatively clear filtrate passes through the medium on its way to the rotary valve.

When the cell has left the slurry, the cake drains under the influence of the vacuum, and is then washed for a period and finally dried, still under vacuum. The cake is then removed from the filter medium, usually by a **knife** stretching the width of the drum, normally aided by a reverse blast of air to lift the cake away from the medium, although there are several other discharge methods in common use. A "belt" of strings or chains can be formed around the drum, on which the cake forms, to be carried off the drum when dry enough; the strings and cake then pass over a roller of a much smaller diameter, to break the cake away from the strings, which return to the filter drum before it dips again into the slurry (passing through a **comb** on the way, so as to maintain the alignment of the strings). Alternatively, and especially for sticky cakes, a small diameter roller can be pressed against the cake on the drum: the cake transfers to the roller, from which it can be cut without any risk of damaging the filter medium on the drum. An obvious extension of the string discharge is to take the filter medium off the drum and pass it around a smaller diameter roller to release the cake – a design with the marked advantage of allowing the filter medium to be thoroughly washed before it returns to the drum. The Coilfilter forms its cake on a medium made from a double layer of tightly coiled springs, the two layers of which separate as they leave the drum, and pass over different rollers, where the springs open a little to release the cake, before the springs return over a re-alignment roller.

The individual cells are usually about 30 cm wide (in the direction of the drum circumference), so that larger diameter drums have more cells: 12 for a 1 m diameter drum, up to 24 for a 4 m drum. The depth of the cell may be around 25 mm. It is normal for a rotary drum vacuum filter to turn at less than 1 rpm, perhaps as slowly as 10 revolutions per minute. It is normal also for the drum to be submerged for about 30-35% of its surface (some designs submerge as

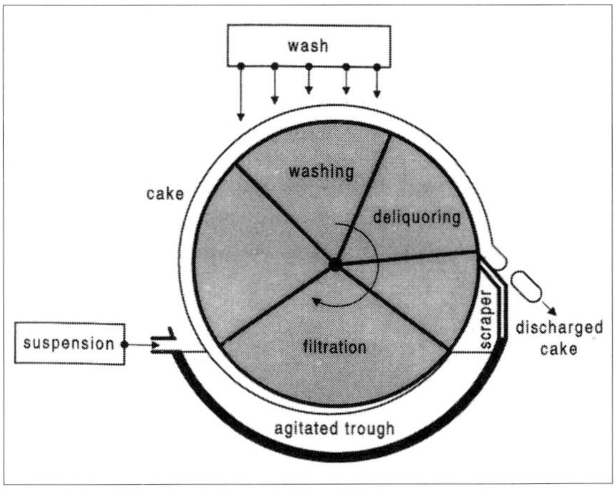

Rotary drum filter cycle

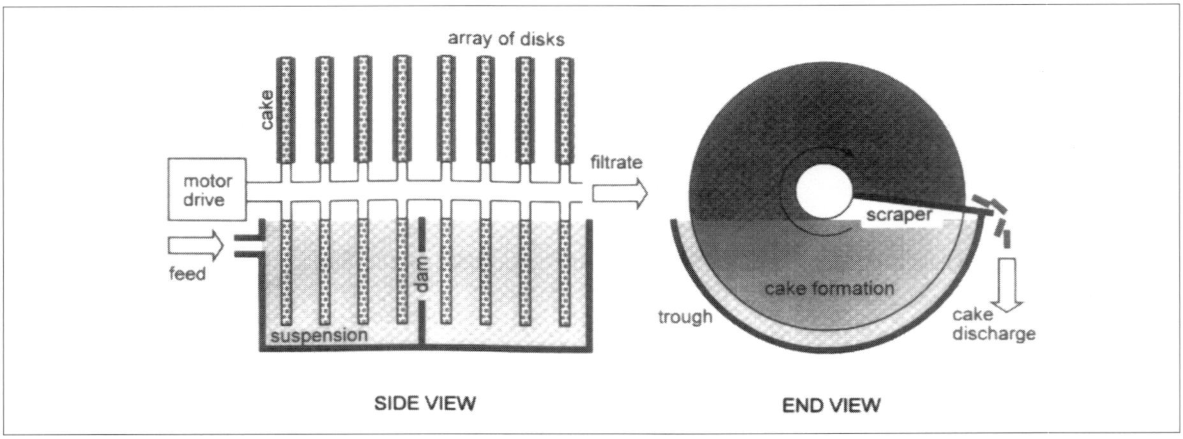

Rotary vacuum disc filter

much as 75%, but this creates sealing problems, since the central shaft is then also submerged).

The first major variant to the multi-cell version, dating from almost as long ago, had the whole drum under vacuum, with no cell structure at all. This does away with the complex internal piping and the rotary valve. The filtrate collects in the bottom of the drum, from which it is pumped through a single pipe. A close fitting shoe on the inside of the drum covers the cake discharge sector of the rotation, which can thus run at ambient pressure, and use reverse gas flow to blow the cake from the medium if wanted. Another shoe can be used to collect **wash liquor** if it is necessary to keep this separated from the filtrate. A popular design for many years, the single cell version has mostly given way to the multi-cell, with the exception of the Bird-Young filter. This operates with a thin cake, so that production rates are high.

There is a risk that a spray of wash water onto a drained cake will simply run down the outside into the slurry trough. This can be prevented by a **drag blanket**, a piece of fabric that floats over the surface of the cake, spreading the wash liquor and closing any large cracks in the cake. A more complicated device is the **compression belt**, a continuous porous belt held onto the cake by a set of rolls across the washing region, with wash liquor sprays onto the belt.

The designs described so far are bottom fed, i.e. the slurry trough is at the lowest part of the drum's rotation. A major variant is the **top-fed** rotary drum filter, of multi-cell construction but in which the feed slurry enters the filter from a head box, at or near to top dead centre of the rotation, onto a cell that is momentarily horizontal. The head box has seals to the drum on either side and behind, so that a pool of slurry forms on the top, to be filtered, drained and dewatered as the cell descends. This design is suitable for very rapidly filtering materials (for which feed sprays across the top of the drum may be sufficient), but is less effective for cake washing than the bottom-fed designs.

The rotary filters discussed so far have all used drums as their **filter element**. A major alternative is to employ disc-shaped elements, which have the important advantage of being able to pack much more filtration surface into a given area of factory floor. The rotary vacuum disc filter has an array of large circular discs mounted on a horizontal shaft, which carries the vacuum and filtrate pipes, as well as being the means whereby the whole assembly rotates. A rotary valve, similar to that used in a rotary drum filter, sits at one end of the shaft to control the application of the vacuum and the filtrate withdrawal. The discs have filter medium on both sides, and the lower part of each disc dips into a feed slurry trough.

Each disc is actually made from a number of pie-slice-shaped segments, separately connected to the vacuum source and filtrate discharge. As the shaft rotates, each disc dips a segment into the slurry and vacuum is applied, so that filter cake forms on each side of the disc, and the filtrate is sucked through to the interior of the segment. When the segment leaves the slurry, it drains and dewaters under the influence of the vacuum, and is finally discharged, by means of a reverse air blast and a scraper.

The discs are normally about 30 cm apart, with up to 12 on a single shaft. They are made in sizes from 50 cm in diameter, up to 5 m, so a large filtration area is possible. The major disadvantage of the rotary disc filter is that washing of the cake is not easy, even if possible.

The discs of a rotary disc filter are normally submerged to just below the point at which they join the central shaft, but versions do exist in which the shaft is also submerged, although this is mechanically more complicated. The discs turn at somewhere around 2–5 minutes per revolution.

The rotary drum and disc filters are normally found driven by vacuum. However, versions exist driven by super-atmospheric pressure, which are much more complicated machines, although offering higher filtration efficiencies because of the higher driving force. Mostly, they consist of the equivalent equipment as for a vacuum filter, but now completely enclosed in a pressure vessel, with a special mechanism now necessary to extract the filter cake through the wall of this vessel.
[*See also Precoat drum filter*]

Rotary strainer. A pressure **strainer** consisting of an array of porous cylindrical elements arranged in a circle. The whole assembly rotates so that each element passes through a zone with an outlet open to the atmosphere. This allows each element to be backwashed in turn. The elements can be **edge filters**, coarse **mesh** or perforated sheets.
[*See also Poro-edge*]

Rotary vacuum filter. Term usually applied to **rotary drum** or **disc filters** but also includes **table** and **rotary tipping pan filters**.

Rotary wheel screen. [*See Rotary disc strainer*]

Rotating knife. A device for discharging accumulated solids from a **basket centrifuge** with a **knife** blade that operates over the full length of the basket.

Rotating siphon peeler centrifuge. A **peeler centrifuge** fitted with a siphon through which the **filtrate** passes on its way out of the machine, this siphon tube rotating with the **basket**, and providing a little more driving force for the **separation**.

Rotating table filter. [*See Table filter*]

Rotoklene strainer. A **wedge-wire** wound **cartridge filter**, in which the **screen** rotates driven by a geared rotor meshing with grooves on the rear of the **element**. As the teeth and grooves meet, a small jet of liquid flows back though the screen to clean it.

Roto-Shear filter. A type of **dynamic filter**, developed by Kotobuki, which has a set of rotating **discs** on a central shaft, and a corresponding set of annular stators between the discs. The spaces between discs and stators are small, so that the rotating discs cause **shear** at both surfaces, all of which are porous. **Filtrate** is taken off through a pipe within the rotating shaft, and at an outlet from each stator.
[*See also Artisan filter*]

Roughing filter. A coarse **filter** designed to act as protection for another process downstream of it, to remove **tramp material** that might otherwise cause damage to or block that process.
[*See also Prefilter*]

Roving. A **yarn** treatment process, which imparts a slight twist to the yarn as well as compressing it. Also used to refer to the yarn produced in this way.

Rozka filter thickener. A **clarifier** in the form of a high rate **thickener** made up of an array of **inclined tubes**, above the outlet of which floats a **packed bed** of polystyrene spheres, held in place by a screen above them, to act as a **polishing filter**.

Rubber belt filter. A continuous **vacuum belt filter** in which an endless light filter belt is supported on a stout rubber belt, also endless. The rubber belt has a drainage-aiding surface and holes that allow **filtrate** to flow through the belt into a longitudinal vacuum box. The top of the box forms a moving seal with the bottom of the rubber belt. The box may be partitioned to separate the filtrates for countercurrent **washing** applications. Both filter and support belt can be cleaned in the return run. Rubber belt filters are more expensive than many other types of belt filter, but are more suited to large heavy duty applications. (Also known as the travelling belt filter.)

Rugosity. The reciprocal of **sphericity**, and hence a measure of the irregularity of a particle.

Rupture strength. A mechanical property of a **filter medium** material, determined by standard tests, and used as equivalent to **bursting strength** and **tensile strength**. In a typical test, the material is stressed until it ruptures, and the load on it at the point of failure, divided by the original cross-sectional area gives the rupture strength.

S

Sabre-shaped cell. A special design for the **filter cell** of a **disc filter**. A frequent problem with conventional disc filters is an uneven **cake** thickness resulting from the differences in immersion times in the **suspension** of different elements. To counter this effect, the sabre cell is bent at the **cell foot** in the direction of rotation of the disc array, so that the cell emerges from the suspension not just with the innermost point of the **forward edge**, but with the entire cell.

Salt cloud test. [*See Sodium flame test*]

Sand. A coarsely granular natural material, used commonly in **deep bed filters**. In the mineral processing industry, there are defined grades of sand: fine sand is 60–200 μm, medium sand is 200–600 μm, and coarse sand is 600–2000 μm.
[*See also Grit, Silt*]

Sand cone. A term frequently used of a **deep cone separator**, since the substance most frequently separated is **sand**.

Sand filter. A **deep bed filter** using a packed bed of sand as the **filter medium**. The sand may be of uniform size throughout, or it may be graded in size through the depth of the bed, with the coarser particles at the top, facing the incoming liquid. Sand filters are widely used in the filtration of fresh waters (and to a lesser extent, waste waters).

Sand filters normally operate batchwise, accumulating dirt from the incoming water, until the maximum dirt load is reached, whereupon the main flow is stopped, and a little of the filtrate is returned in the opposite flow direction, to flush the dirt out of the sand. There are two types of this batch filter, the **slow sand filter**, which uses a surface layer as the actual filter medium, and the **rapid sand filter**, which uses the entire depth of the sand bed as a depth filter medium. Modern developments have taken the rapid version into continuous operation, such as the Dyna-Sand filter.

Sanding. The deposition of coarse particles on the base of any tank through which a liquid flows, but especially the feed tank for, say, a **thickener** or **drum filter**.

Sand trap. [*See Grit trap*]

Sateen weave, Satin weave. Satin weave is one of the three main types of **weave** in **woven fabrics** (the others being **plain** and **twill**). Satin is an extension of twill, but without the regularity of pattern that characterises twill. It has a sheen produced by exposing more **warps** than **wefts** on the right side of the fabric. The exposed warps are called floats. In the sateen weave the process is reversed, and the exposed wefts form the floats. The amount of twist in the yarns and the length of the floats produce variations of visual effect.

Satin weave fabrics are smooth faced, with an unbroken surface made up of long floating warp yarns. A true satin must have at least five warp and weft yarns in each complete weave repeat. Most satin fabrics are made of smooth, lightly twisted yarns that heighten the effect of light unbroken by visible crosswise bindings. Satins, having long floats, are susceptible to the wear caused by rubbing and snagging and are, therefore, generally regarded as luxury fabrics, and have only a very specialised place as **filter media**.

Saturation, Saturation degree. The measure of liquid content in a porous mass, and in particular in a **filter cake**. If all the voids in the cake were full of liquid, then the cake would be described as fully saturated, while if only some of the voids contain liquid, it would be partially saturated. The saturation degree is the ratio of the actual liquid volume to the total void volume:

$$S = V_l/V_v$$

where V_l = volume of liquid contained in the pores
V_v = total void (pore) volume.

For a fully saturated mass, S = 1, while for a fully (thermally) dewatered mass, S = 0
[*See also Capillary pressure curve*]

Sauter mean diameter. The ratio of the volume of a perfect sphere to the area of its surface is d/6 where d is its diameter. This relationship leads to one method of expressing a mean value for the diameter of a mass of particles: the total volume of the mass of particles is divided by the total surface area of the mass. The Sauter mean diameter is then defined as:

$$d_{32} = \Sigma d_i^3 \Delta n_i / \Sigma d_i^2 \Delta n_i$$

where Δn_i = number of particles with (effective) diameter d_i

Another way of expressing it is as the diameter of the sphere having the same equivalent specific surface as the particle in question.

Saveall. A term used in the paper industry for a **rotary filter**, operating under gravity or vacuum, to recover good fibres that have passed through the paper machine, so that they may be returned to the paper making process. The filter must be able to handle large volumes of very thin **slurry**, although any accumulated solids can usually be washed off the drum and pumped back to the paper feed tank.

Scale factor, Scale-up factor. A numerical factor used to uprate performance data derived from laboratory tests to a larger item of **separation** equipment, to be used for production.

Scale-up. The projection of results from laboratory or pilot scale tests to the full scale design size of equipment.

Scalping. The removal of a small quantity of **oversize particles** from a **slurry** or a mixture of particles.

Scanning electron microscope (SEM). An electron microscope able to give an apparently three-dimensional picture of the surface of a material. Its resolution is in the range 10–20 nm, and it is a very useful tool for determining surface structures of **filter media** in general and **membranes** in particular.

Scavenger filter. A small **filter** in the base of a **tank filter** to treat the **slurry heel**, after the end of the main **filtration** cycle. It may be a curved **plate** matching the curvature of a horizontal cylindrical tank.
[*See also Recovery plate*]

SCFT. [*See Standard cake formation time*]

Scheibler filter. A batch operated, **vertical leaf filter** of unusual leaf design. The leaves consist of a loose **bag** fitted on a collapsible frame, with the bag some three times the width of the frame, thus offering increased surface area within the pressure **housing** – which may be cylindrical or rectangular.

Schenk filter. A horizontal **pressure plate filter**, with circular plates mounted one above the other on a central shaft, inside a vertical cylindrical vessel. **Filter cake** forms on top of the plates, and when they are full, the central shaft rotates spinning the plates, and discharging the cake. If just **centrifugal force** is enough, then the cake is discharged dry, but in some designs, the cake is washed off the plates as they spin.
[*See also Funda filter*]

Schmidt number. A **dimensionless number** used in the scale-up of **membrane** processes. It is defined as:

$$Sc = \mu/\rho D$$

where μ = fluid viscosity
ρ = fluid density
D = diffusion coefficient

Schmutzdecke. An upper layer on a **slow sand (deep bed) filter**, which operates with downwards flow of raw water. A biological layer forms on the surface of the bed, and this layer acts both to filter solids from the water, and also to purify it free from dissolved organic contaminants by bacterial action.

Schone elutriator. A **particle size classifier** using **elutriation** in a vertical column of liquid, with different fractions recovered by using different liquid velocities.

Schultze-Hardy rule. A rule established in the **coagulation** of particles by means of inorganic **electrolytes** with multivalent ions. The rule states that the minimum ionic concentration necessary to produce coagulation is approximately proportional to the inverse of the sixth power of the counter ion charge.

Scoop pipe. [*See Peeling pipe, Skimmer pipe*]

Scouring. The mechanical removal of the dirty layer on the top of a **deep bed filter**, but can also mean the action of the liquid flow in a **cross-flow filter**, to keep **fouling** of the medium surface to a minimum. Air scouring is the use of a jet of compressed air as part of the backwash process for a **rapid sand filter**.

Scraper. A device intended to remove accumulated **filter cake** from the surface of the **filter medium**. Similar to a **knife**, but made of wood or rubber in the form of a blunt knife, so as to do no damage should it come in contact with the medium.

Scraper discharge. A blade covering the full width of the **filter medium** on a **drum** or **disc filter** to scrape the cake from the medium (or to act as a deflector if the cake is blown off with compressed air). [*See also Knife discharge*]

Screen. A major type of separating equipment, used for relatively coarse **separations**, and usually made with some precision as to the shape and size of the openings through which the solids pass. The **filter medium** is made out of coarse **wire mesh**, or **perforated plate**, or arrays of parallel bars. Screens can be used singly, for example to prevent onward passage of oversize particles, or they can be used in series, producing a number of closely sized **fractions** from a mixed **feed**.

Screen media formats

A particular form of screen as used especially in **centrifuges**, where the separated solids slide over the screen surface, employs small wedge wire bars mounted on support rods in a direction parallel to the intended movement of the solids. The screen is made in the shape of the centrifuge, most usually as a cone.

Screens can be operated horizontally, or sloping, or they can be made into a cylindrical shape that rotates as a drum. They may be stationary (with solids washed over them), or vibrating (including a gyratory motion). The largest versions are those used in the mineral processing sectors, which tend to be made from toughened metals, but the smaller sizes can be made of plastic.

Screening can be done in the dry state, to produce the separate fractions similarly dry, or a screen can process a slurry, producing two **slurry** products containing differently sized solids, or leaving a relatively dry coarse solid on the screen, with all the liquid passing through the screen.
[*See also Screen mesh, Sieve*]

Screen-bowl decanter centrifuge. A normal **decanter centrifuge** has a cylindrical **bowl** section in which the suspended solids are settled out to the bowl wall, along which they are conveyed by the **scroll**, with a short conical section or **beach** at one end, decreasing in diameter away from the cylindrical section, up which the solids are then conveyed – draining as they do so. The solids are then discharged over the edge of the smaller diameter end of the beach.

The time spent on the beach may be too short for adequate **dewatering** of the solids, or for **washing** them. Under these conditions, the bowl is extended beyond the end of the beach by means of a short cylindrical **screen** (usually made of parallel, **wedge wire bars**). The scroll continues to move the solids across this screen section, for discharge at its end. The screen bowl decanter can handle solids of a wide particle size distribution, and is a very good separator for solids that settle easily but need washing.

Screen changer. In the extrusion of molten **polymers** to make continuous **filaments**, it is vital that the melt be free of solid contaminants that might block the **spinneret** nozzles. This freedom is achieved by special **filters** mounted in the delivery line to the extruder, which are usually made from **sintered metal powder** or **wire mesh**. Each screen is fixed in a holder that can be pushed into the polymer flow when clean, and can be removed when used. The screen changer is the mechanism for achieving this interchange.

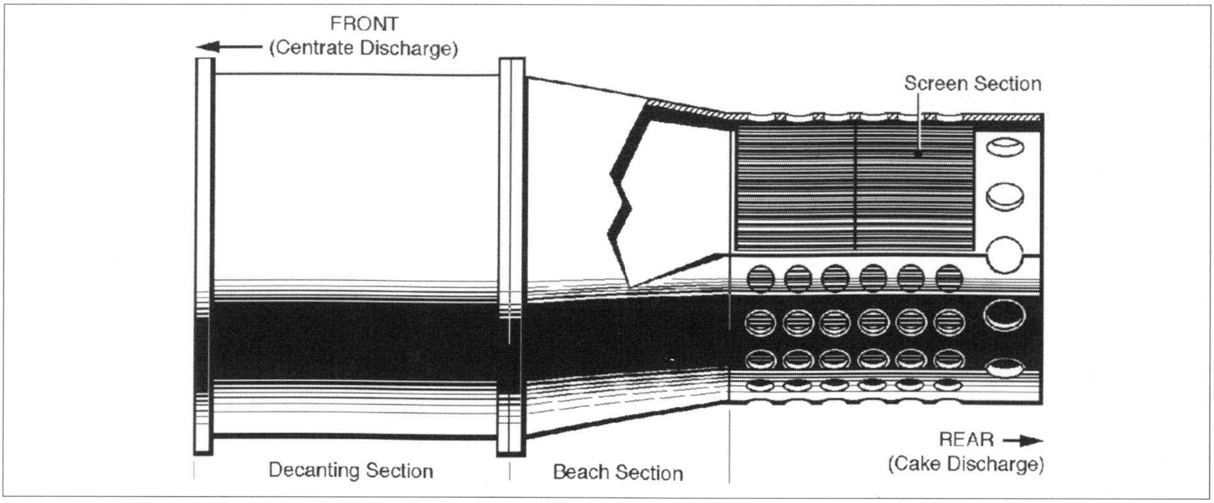

Screen-bowl decanter centrifuge

Screen effectiveness. The effectiveness of **separation** (i.e. **classification**) by a **screening** process can be based on the **oversize** or the **undersize**. For the oversize it is the (overflow/feed ratio) times (mass fraction of particles in the overflow divided by that in feed). A parallel definition for effectiveness based on undersize uses underflow instead of overflow. The overall effectiveness of the screen is the product of the two definitions.

Screen efficiency. A performance index for a **screen separation**, which assumes that the fine product is free of any particles coarser than the screen **aperture** size.

Screening. A **filtration** process employing coarse **media**, possibly for **separation** of a mixture of solid particles by particle size, or for **pre-filtration** (protection) applications, wet or dry. It is the process that comes closest to **surface filtration**.
[*See also Screen*]

Screen mesh, Screen size. The measure of the size of the opening in a **screen** – an **aperture** in a **woven mesh** or in a **perforated plate**, or the space between two adjacent bars. For the coarser sizes the dimension is usually stated as the distance between adjacent wires or bars, or the diameter of the hole. For finer separations, the size is quoted as the number of openings per unit length. The solid products produced by **screening** are also characterised by the same dimension.

Screw. In **filtration** terms, a screw is a device for moving masses of particles from one place to another, by means of a helical blade arrangement fastened to a central shaft, and running in some kind of cylindrical container or a trough, as is done in some **mechanical classifiers**. The simplest version is the screw conveyor, intended just to move the solids over a distance (flat or elevated), such as that used to discharge the accumulated solids from a **filter** or **centrifuge**.

The decanter centrifuge uses a screw conveyor, or **scroll**, to move the settled solids out of the supernatant liquid and up the beach to the discharge point. Some types of conical basket centrifuge use a screw conveyor as a means of regulating the discharge of collected solids from the basket. A **screw press** has a screw at its centre, forcing the wet solids against and past a perforated surface, to achieve dewatering of the solids.

Screw press. A continuously operating machine used to squeeze liquid from a mass of fibrous solids, such as in the recovery of animal or vegetable oils, or to dewater a pasty filter cake. It consists of a **screw** rotating inside a perforated or slotted **shell**, so that the mass is pressed against the inside of the shell. The solids must be coarse or stiff enough so that they do not pass through the holes in the shell, while the expressed liquids are collected in a surrounding **casing**. The press can be operated with its axis vertical or horizontal. The design of a screw press tends to be specific for one product, so that it is not usually suitable for multiple products. Pressure is increased along the length of the press either by reducing the screw pitch, or by increasing the shaft diameter or by reducing the diameter of the shell, and can be controlled by varying the size of the outlet. The shaft may be hollow to permit steam heating, and **wash liquid** can be added along the screw to increase recovery.

Scrim. A strong simple **woven** material, with **yarns** well separated, laid between the webs of a **felt** to give it tensile strength.

Scroll. A **screw** conveyor, used to move separated solids through and out of a range of **separators**.

Scroll discharge cone screen centrifuge. A **cone screen centrifuge**, with a shallow cone angle, such that the dewatered solids will not slip across the screen and discharge unaided. A **scroll** filling the conical space within the screen is then used to discharge the cake.

Scroll discharge (decanter) centrifuge. [*See Decanter centrifuge*]

Scroll-screen centrifuge. [*See Screen bowl decanter or Scroll discharge cone screen centrifuge*]

Scrubber (wet/dry). A **gas cleaning** system in which the dirty gas (carrying dust or gaseous impurities or both) passes upwards through a (usually cylindrical) tower where it contacts either a spray of liquid or a **packed bed** of **granular** solids. The dry scrubber acts by **impingement** of the dust particles on the material of the bed, and is therefore a true **filter**. It may have no effect at all on any gaseous impurities, and suffers from the risk of **re-entrainment** of particles.

The wet scrubber may have a spray of liquid droplets in a relatively large space, or a film of liquid with some droplets falling through a packed bed, when the dust particles or gaseous impurities are absorbed by the liquid. The wet scrubber transfers the problem of a dirty gas to one of a dirty liquid – which may be as serious a disposal problem.

Scrubbing. The cleaning of a dirty gas stream by passing it through a **scrubber**, either dry or wet.

Seam. The place(s) at which materials are joined to make non-flat **filter media** structures, especially bags. The seam is a region where the **porosity** of the filter medium may be very different from the bulk of the medium, so possibly creating a weakness – hence the search for **seamless** construction of bags, etc.

Seamless. Filter elements made from flat materials into three-dimensional structures, such as **bags** or **pockets**, would normally contain a **seam**, where the material is joined. This creates an irregularity in the structure, and was a cause of the loss of market share by filter bags. The ability of filter bag makers to manufacture them without a seam has given a new stage of life to the filter bag, especially in solid/liquid **filtration**.

Secondary air. A flow of air through a **vacuum filter**, either intentionally through a valved pipe, to control the level of vacuum applied to the **suspension**, or unintentionally through cracks or channels in the filter cake. The latter is to be avoided as it can cause the vacuum to collapse.

Secondary membrane. In any **membrane** process, a **fouling** layer rapidly forms on the membrane surface, and effectively becomes the **filter medium**. Its thickness is minimised as far as possible, but theories of membrane **filtration** assume that this secondary membrane is the active **separation** medium.

Secondary treatment. The second major phase in the treatment of municipal sewage, which essentially involves the bacterial oxidation of organic materials in solution and suspension, but is followed by the **sedimentation** and/or **filtration** of residual suspended solids, mainly in the size range 1 to 100 μm.

Secondary vortex. The inner vortex in a **cyclone**, carrying the fluid from the solids discharge apex back to the feed end of the cyclone, where it is "located" by the **vortex finder**, through which the fluid escapes. The **undersized** solids suspended in this vortex remain in it and are discharged with the fluid.

Sector. A circumferential division of a **drum** or **disc filter** connecting to a single **filtrate** outlet and undergoing the same part of the **filtration cycle**.

Sedicanter. A proprietary version of the **decanter centrifuge**, manufactured by Flottweg, to provide a thick **sludge** from a difficultly separated **suspension**. The solids being scrolled up the beach of the decanter have to pass a barrier, under the effect of the hydraulic head of liquid in the bowl, itself controlled by a **peeling disc**.

Sediment. Any mass of solid particles that has separated out of a **suspension** by **sedimentation** (under gravitational or **centrifugal forces**). A sediment consists of solids that are more dense than the fluid, and will be completely saturated with the fluid.

Sedimenting, Sedimentation. *See* pages 221–2.

Sedimentating centrifuge. Any **centrifuge** in which the **separation** mechanism is that of **sedimentation** (rather than **filtration**). These can be batch-operated or continuous, and include the **tubular bowl centrifuge**, the **imperforate basket centrifuge**, the **disc-stack separator** and the **decanter centrifuge**. Some sedimentation centrifuges operate with very high centrifugal forces, to achieve settlement of the very finest particles, using G-forces of up to 10,000.

Sedimentation equipment. Any item of equipment or equipment system in which the process of **sedimentation** is undertaken. This can be gravity sedimentation or **centrifugal sedimentation**.

Sedimentation front. Describes the definite surface marking the top of the sedimenting layer below the clear supernatant above the sediment. Below the sedimentation front the particles are settling at a very similar velocity within the **hindered settling** zone, which occurs once a critical **suspension** concentration has been reached.

Sedimentation potential. A small electric potential that is created when charged particles settle through a liquid.

Sedimentation tank. [*See Gravity thickener*]

Sedimentation rate, velocity. The velocity of a particle settling out of a **suspension**, under the influence of gravity or **centrifugal force**. In the case of an isolated particle, under **laminar flow**, the velocity is described by **Stokes' law**. Once a thicker sediment is formed, and **hindered settlement** begins, the velocity of the **sedimentation front** is described by **Richardson and Zaki's** equation.

Sedimentator. A proprietary (Bokela) counter-current **washing** device employing several sets of **re-slurry** units, mounted above one another in a cylindrical tower. Suitable for easily separated solids, but can offer washing under higher pressure and temperature conditions.

Sedimenter. In principle, any item of equipment used for **sedimentation**, but is mainly used of simple **gravity settling** equipment.

Segregation. In general terms, this refers to the **separation** of a mixture of particles of different particle sizes into a set of closely sized **fractions** (for example, by **sieving**). It also refers to the unwanted separation of the particles in a suspension that can easily occur in a number of items of separation equipment. It is very likely to occur where **sedimentation** is the main separation process, or where there is a large volume above a horizontal **filter medium**, such that the **sediment** or **filter cake** will form first from the coarser (and so fastest settling) particles, followed by progressively finer particles in higher layers. This may result in a very fine, and easily clogged layer at the upper surface of a filter cake.

Selectifier screen. A coarse **screen** used in the preparation of stock in the paper industry, made by Black-Clawson, which has a vertical cylindrical screen, with solids collected on the inside. The screen is kept clean by an aerofoil-shaped blade that is rotated inside the screen, the liquid flow around the aerofoil being enough for the cleaning action.

Selective agglomeration.
[*See Spherical agglomeration*]

Selective flocculation. A **separation** process between two solids in **suspension**, whereby one solid is **flocculated** and the other is not, so that a subsequent **separator** will cause one to **filter** or **settle** out of the suspension, and the other will stay suspended.

Selective flotation. A **classification** achieved by selection between suspended particles, such that some particles attach to the gas bubbles while others do not and so stay in the **suspension**.

Selectivity index. A term sometimes used for the efficiency of **separation** in a **classifier**.

Self-cleaning filter. Any type of **filter** that operates in such a way as to remove accumulated solids from the **filter medium** by an automatic process, based either upon elapsed **filtration** time, or on the pressure drop across both the accumulated **cake** and the medium. A common form of such a filter would have two filter **elements** mounted in a **duplex** housing, such that one can be cleaned, while the other remains on line. The flow is then switched from one

Sedimenting, Sedimentation. One of the two basic mechanisms of fluid/solid **separation** (the other being **filtration**), sedimentation is the process whereby dense solids settle through a less dense suspending fluid, under the influence of gravitational or **centrifugal force** fields. This is a very different separational mechanism from filtration, and sedimentation is also characterised by being able to handle very large quantities of relatively dilute suspensions.

Two broad classes of sedimentation equipment are recognised, with a very imprecise boundary between them. The first concerns the settling of sands, or relatively coarse particles, which occurs rapidly, to give a clear supernatant liquid and a layer of **sediment**. The other involves much finer particles, and is correspondingly much slower, producing a thickened **sludge**.

In a sedimentation process there are four distinguishable stages: **free settling** (where the particles do not interfere with one another), **hindered settling** (where the movement of one particle is considerably affected by the presence of other particles settling with it), **consolidation** (the formation of the sediment) and **compression** (the compaction of the sediment under its own weight). The process is modelled by **Stokes' law** for free settling, and by a number of approaches to hindered settling, including that of **Richardson and Zaki**.

There is a major division among sedimentation equipment types: between those driven by gravity, and those in which separation occurs under the influence of **centrifugal forces**. **Gravity sedimenters** are usually large open tanks, whereas those operating centrifugally are much smaller, contained devices.

Sedimentation equipment can be reviewed based on this division, gravity driven versus centrifugal. The gravity range covers:

- **deep cone** thickeners, with a variety of ways of controlling solids discharge, and
- circular and rectangular **settling tanks**.

The settling tanks can be designed for **clarification** or **thickening**, with the latter intended either for simple separation or for **classification** of a mixture of solids. These all have quite complex mechanisms for solids collection within the settling tank, such as rotating **rakes**, or **drag belts** or **screw conveyors**. The classifiers can be either **mechanical** (with drag belt or screw conveyor), or **hydraulic**, operating effectively by **elutriation** out of a fluidised bed.

A very different type of gravity settling device employs the short settling distance principle, supplied by the inclined parallel plates of a **lamella separator**. A similar settler uses inclined tubes, both of these types occupying considerably less ground area than the normal settling tank, and being as much used for **liquid/liquid separation** as for the separation of solids.

Sedimentation under centrifugal forces can be achieved either by:

- inertial movement of the fluid, consequent upon the way in which the fluid enters the device, or by:
- rotation of the equipment, driven by a motor.

The inertial devices are the **cyclone** and **hydrocyclone**, the cyclone being used for the removal of solids from gases (such as in process exhausts or at the delivery end of a pneumatic conveyor), while the hydrocyclone separates solids from liquids, in particular acting as a classifier (for instance, in starch processing). The suspension is introduced tangentially to a cylinder and the resultant swirling motion of the fluid creates the necessary centrifugal field.

Sedimentation in a driven rotating device is the province of the **sedimenting centrifuge**, in which an imperforate bowl is rotated rapidly to create a force field several hundred (if not several thousand) times that of gravity. There are four main types of sedimenting centrifuge:

- the **tubular bowl centrifuge**, which has a long, relatively thin cylindrical bowl, rotating at very high speeds, and mainly used for liquid/liquid separations;
- the **imperforate basket centrifuge**, which has a drum rotating about a vertical axis, able to separate two liquids, but also accepting solids in the mixture;
- the **disc stack centrifuge**, with a roughly spherical bowl, containing a stack of conical discs (the centrifugal equivalent of the lamella separator), developed initially for liquid/liquid separation, but able to handle relatively high solids concentrations in the feed in some of its variants; and
- the **decanter centrifuge**, which has a cylindrical bowl containing a screw conveyor able to move collected solids up an inclined surface at the end of the bowl (the centrifugal version of the mechanical classifier).

There is, finally, another class of sedimentation equipment, operating under the influence of gravity, in which the material to be separated is caused to float upwards from the bulk of the suspending liquid. This is the technique of **flotation**, in which bubbles of gas (usually air) are attached to the suspended solid particles, giving them the required buoyancy for them to rise upwards, to the top of the flotation cell as a surface layer of froth containing the solids. This layer is then skimmed off, and washed to recover the solids. The process is much used in the mineral processing sector, and is a successful method of separation of one component from a solids mixture – by appropriate chemical dosing to achieve selective bubble attachment.

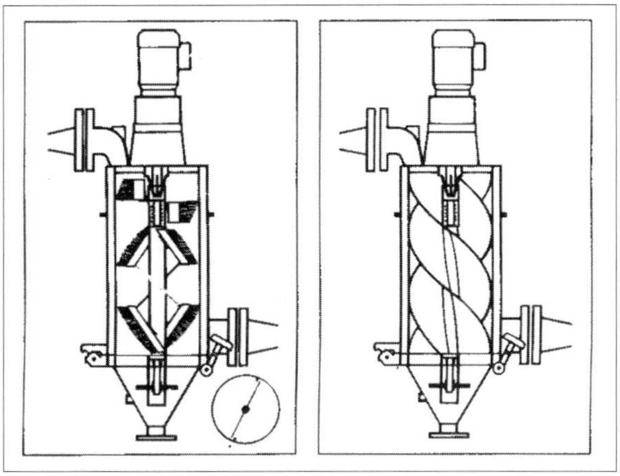

Self cleaning filters (brush and screw)

element to the other, once it has been adequately cleaned.

The cleaning process may involve the scraping or brushing of the element from the upstream side, or it may be achieved by flushing it with clean **filtrate** or water through from the downstream side. In some cases, the scraping or brushing of the surface of the filter medium is continuous throughout the operation of the filter. This maintains simple operation of the filter, which can then comprise a single element, but increases the risk of pushing some solid particles through the medium. The mechanical cleaning processes are best matched with filter media made from metal or plastic screens, to resist the wear from the cleaning mechanism.
[*See also Automatic filter*]

Self-cleaning separator. A **disc-stack centrifuge** is commonly used for the separation of two immiscible liquids. If either or both of these liquids contains a small quantity of suspended solids, then these solids will settle into the widest part of the separator's **bowl**. A self-cleaning separator is one that has an automatic means for the discharge of any accumulated solid, and this can be achieved by continuous flow through small **nozzles**, or through a set of **valved nozzles**, or by an **opening bowl** mechanism (in which the two halves of the bowl separate from a short time). The opening of the valves or the bowl is controlled by the build-up of solids at the bowl wall.

Self-suctioning drum filter. In a **drum filter**, a low vacuum can be applied by connecting each filter **cell** to a **filtrate** pipe that runs on the inside of the **filter medium** in an arc of the drum's circumference, and in the direction of the drum's rotation. The liquid in this filtrate pipe creates a hydrostatic head sucking on the filter cell, sufficient to drive the **filtration** of easily settled solids.

Self-transporting centrifuge. A term describing a range of continuous **basket centrifugal filters**, whose basket is in the shape of a cone, with its axis of rotation horizontal, and whose diameter increases with distance from the feed point. The **filtrate** drains quickly through the conical basket and the solids move outwards under the balance of **centrifugal force** and **friction force** at the drum's surface, a balance dependent upon the moisture content of the separated solids. Such a centrifuge will only work satisfactorily with a narrow range of solids and solid concentrations, for any one cone angle, so there are available a number of devices to control the rate of movement across the drum – either to hold it back, or to speed it up.
[*See also Cone screen centrifuge, Scroll discharge centrifuge, Tumbler centrifuge*]

SEM. [*See Scanning electron microscope*]

Semi-included angle. A dimension that is half of the apex angle of a cone, and which is used to define the shape of a cone, in a **cone basket centrifuge**, in a **cyclone**, in a **deep cone** settler, or in the **beach** of a **decanter centrifuge**.

Semi-permeable. A material is said to be semi-permeable if it allows passage through it to part of a **mixture** or **solution**, and retains the remainder. In principle, therefore, any **filter medium** can be regarded as semi-permeable, but the use of the term is usually restricted to the function of a **membrane** in allowing

permeation by the fluid, and a very limited part of the suspended or dissolved materials.

Separating force. The driving force used to achieve a **separation**: gravity, vacuum, pressure or centrifugal force.

Separation. *See* pages 224–5

Separation control head. The control valve that forms the transition piece between the rotating part of a **cell-less drum filter** and its stationary components. Through it flows all of the **filtrate**, and any entrained air.

Separation efficiency. The separation efficiency of a **filter** is a measure of how good the filter is at doing its job – removing solid from a fluid suspension. It can be defined as the number of particles captured by the filter as a ratio to the number in the feed slurry to the filter, and is usually expressed as a percentage. Its relationship to the **β-value** of the filter is that:

$$\text{separation efficiency} = 100(1 - 1/\beta)$$

Separation mode. It is usually accepted that there are two modes of **separation** – **filtration** (i.e. passage through a barrier) and **sedimentation** (i.e. relative settlement through the fluid).

Separation process. The complete **separation** process may include **pre-treatment** and **post-treatment** as well as the separation itself.

Separation selectivity, sharpness. A term that defines the efficiency of a solid particle **separation** process, in terms of the proportion of fine solid that is present in the coarse fraction, and of coarse in the fine. This can be seen in the fractional **grade efficiency curve**, where the steeper the curve, the greater the selectivity of the separation process. Thus, if x_{25} is the particle size in the original mixture of which 25% goes to the coarse fraction, and x_{75} is that for which 75% does, then the separation sharpness:

$$\kappa = x_{25}/x_{75}$$

In analytical separations the separational sharpness should be between 0.8 and 0.9, while a good technical separation would have sharpness figures between 0.6 and 0.8. Normal separations have lower values, between 0.3 and 0.6.

Separator. Although this term is used to refer to any device in which a **separation** occurs, it refers specifically to the high speed **disc stack centrifuge**.

Septic tank. A simple form of treatment system for foul sewage. It consists of a two-compartment tank, connected to a set of land drains filled with loose material. The first compartment contains some settled **sludge** and has a crust of sludge over it such that influent sewage is **digested anaerobically**. The digested sewage flows into the second compartment and the drains, where **aerobic digestion** finishes the work done in the first compartment. The first part will slowly fill up with settled grit and treated sludge, and eventually has to be pumped clean.

Septum. [*See Filter medium*]

Serial connection, Series connection. The combination of any set of items of equipment in such a way that flow of fluid through the set goes through each item in turn. The full flow thus passes through each item (unlike **parallel connection**, where the flow is split). In a **separation** scheme, a set might consist of a **pre-filter**, followed by the main **process filter**, and than a **polishing filter**. A set of test **sieves** is also operated in serial fashion.

Serrated weir. An overflow device in **decantation** operations, where liquid flows across a rim designed to control the flow (in, for example, a **clarifier** or **decanter centrifuge**). If the vertical part of the rim is serrated, then a regular flow can be achieved along the whole perimeter of the weir, even if it is not exactly horizontal. If the serrations are angled such that each slot is actually a V-shape, then the serrations will exercise some control over the liquid throughput, since the flow through the serrations will increase as the liquid level rises.

Serratia marcescens. A species of coccal **bacteria** used as a challenge test for sterilising **filters**. Easily identified by its blood-red pigment produced on all normal growth media (and especially potato), but needs to be handled with care, as it is not so harmless as was once thought.

Service life. The period of time during which a piece of equipment is fully available to carry out its expected function. This period is extended by proper and careful maintenance of the equipment. There is increasing demand for an understanding of the

Separation. In this one word is encapsulated the entire contents of this encyclopaedia. It is used here to mean all of those processes in which a two-phase mixture of distinct immiscible materials is divided into its component parts, by mechanical (physical) means, without a phase change being necessary. The last qualification differentiates the separation of the present context from those separations in which phase change is the major mechanism, such as fractional distillation, or absorption, or evaporation and drying, or liquid/liquid or liquid/solid extraction.

Such a sweeping definition is not, of course, without its minor inexactitudes: some three-phase mixtures are covered (as in, for example, the **disc-stack centrifuge** and the **three-phase decanter**); some processes within the scope of the definition do actually involve phase change (such as **pervaporation**); and some involve the separation of miscible species in a solution (**reverse osmosis**, **nanofiltration**, **ultrafiltration**).

Accepting that the demarcation line cannot be a sharp one, this definition basically includes the processes of:

- **filtration**, including **centrifugal filtration** and **screening**,
- **sedimentation** (or **settling**), including **centrifugal sedimentation** and **flotation**, and
- **magnetic separation**.

The systems upon which these processes are used include:

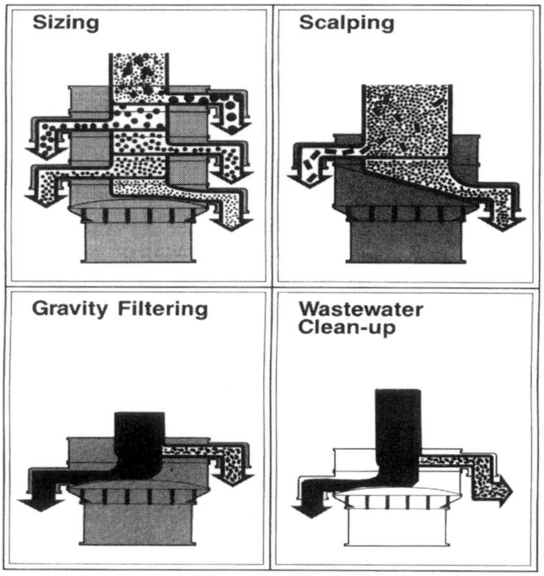

Vibrating separators

Membrane separation plant

- liquid droplets dispersed in a continuous gas phase,
- solid particles dispersed in a continuous gas phase,
- gas bubbles dispersed in a continuous liquid phase,
- droplets of a heavier liquid dispersed in a continuous liquid phase,
- particles of a single solid or a mixture of solid particles (mixed by particle size and/or particle density, and including magnetic or magnetisable solids) dispersed in a continuous liquid phase,
- small amounts of liquid dispersed through an effectively continuous, but porous, mass of solids, and
- a dry mixture of solid particles, perhaps including magnetic or magnetisable solids.

(To these can be added a gas/gas mixture, although not a mixture of immiscible phases, since some gas cleaning applications do include de-odorising and decolourising, which can be the dominant feature of the separation.)

Liquid droplet separation from gases is achieved by **demisting**, and gas bubbles from liquids by **degassing**. Solid particle separation from gases is achieved by a variety of gas cleaning/dust removal processes, including filtration and **scrubbing**, while liquid/liquid separations, which are often complicated by the presence of a small quantity of suspended solids, are undertaken in a variety of sedimentation equipment, including **lamella separators** and **sedimenting centrifuges**.

Solid particle removal from a liquid covers almost the whole range of filtration and sedimentation equipment, for **clarification** of the liquid and for **recovery** or **classification** of the solids. It is by far

Peeler centrifuge

the largest of the seven separation systems, although the demands of environmental and indoor air quality are magnifying the gas cleaning activities greatly.

Much of filtration leaves a wet cake, from which moisture must be removed, by centrifugal **drainage** or mechanical **compression**, the latter being necessary for the recovery of bound liquids (such as vegetable oils and fruit juices) from crushed vegetable matter.

The separation by size, density or magnetic properties of a mixed mass of dry solids can be achieved by screening, or differential sedimentation, or air **elutriation**, or **magnetic separation** if appropriate.

One problem with the term "separation" is its use in some areas as distinct from "filtration", rather than inclusive of it ("Filtration and Separation", "Filtrieren und Separieren", "Solid-Liquid Filtration and Separation Technology"). This usage has long history to it, but is better used if it be understood that the phrase should rather be "Filtration and other mechanical methods of Separation".

A final comment concerns efficiency of separation, since the objective of all separation processes must be to achieve the appropriate separative function with as close to perfect efficiency as possible: the removal of all dust from a gas stream, the recovery of all of the suspended solid in a thick slurry, the inclusion of all of the particles from a mixture that are above a certain size in one fraction. None of the processes described in this encyclopaedia are able to operate with 100.00% efficiency, so something less than 100% must be accepted – and the best process or the best equipment is that which achieves the closest to 100% in its implementation.

performance of an item of equipment throughout its life, in order to enumerate its life cycle cost – as a better measure of economic value that just initial cost.

Settling. [*See Sedimentation*]

Settling cone. [*See Deep cone thickener*]

Settling rate, velocity. [*See Sedimentation rate, velocity*]

Settling systems. The different components of the solid **settling** spectrum, including **dilute sedimentation**, **hindered settling**, **channelling**, and **compression**.

Settling tank. Any vessel through which a fluid passes and which is intended to capture solid particles that are suspended in the fluid. The particles are kept in **suspension** by means of the turbulent flow of the fluid in the pipe or channel leading to the settling tank, but are allowed to settle by the reduction in flow rate and less turbulent conditions once the fluid enters the tank. The term is used of any simple **gravity sedimenter** used for **clarification**.
[*See also Clarifier, Grit trap*]

Settling test. [*See Jar test*]

Settling time. The time taken for a **suspension** of solids to settle reasonably completely is an important design parameter for any form of solid/liquid **separation**. A good system of classification of separation equipment is on the basis of rapid, moderate or slow settling of the suspension.

Sewage, Sewage sludge. The treatment of foul sewage leads to the production of several different types of liquid/solid **suspension**, each presenting some challenging problems to the **filtration** industry.

Shaking screen. Any **screen separator** whose performance is improved by mechanically or electromagnetically applied motion, induced by vibrating, oscillating, reciprocating or gyrating the screen.

Shaking table. A **gravity concentrator** in the form of a sloping **deck**, with a **riffled** surface, down which the **slurry** flows. The deck is vibrated to improve the **classification** performance.

Shape coefficient. An empirical factor relating a real particle to an equivalent sphere. If the surface area coefficient is K_a then the surface area of the particle is $K_a d_e^2$, where d_e is the effective diameter of the particle. If the volume coefficient is K_v then the volume of the particle is $K_v d_e^3$. For spherical particles, $K_a = \pi$, and $K_v = \pi/6$. For real particles, the shape factors deviate further from these values, the less spherical the shape of the particle.

Sharpness of cut. A measure of the efficiency of a **classification** operation, which is usually derived from the **grade efficiency curve**. Specific definitions are usually those based on the slope of the curve at d_{50} or d_{75} / d_{25} or d_{90} / d_{10}.
[*See also Ecart probable imperfection*]

Shear. The action of a force across a surface such that the stress varies with distance, and the result is to distort that shape. When applied to a liquid, shear results in layers of the liquid slipping across one another.

Shear-drag force model. A model of the forces acting on a particle close to a **membrane** surface, when the **drag** on the particle by the **permeate** flow will pull it towards the membrane, while the **cross-flow** of **concentrate** will pull the particle across the membrane surface.

Shear filter. A term sometimes used for any **filter** in which fluid flow or mechanically induced **shear** is used to promote **filtration**.
[*See also Cross-flow filters, Dynamic filters*]

Shear rate. A **suspension** can be characterised by the behaviour of its shear stress (i.e. its flow or deformation) as a function of the shear rate applied. **Newtonian fluids** are the simplest to define, but there are several other types of fluid behaviour, some time dependent and more not.

Shear sensitivity. The ability of an already formed **floc** to withstand **shear** without breaking. This is especially important in the operation of **centrifugal separators** where there is a high level of shear in the **feed zone**.

Shear thickening, thinning. Characteristics of the behaviour of some **non-Newtonian fluids**, whose viscosity is dependent upon the applied **shear** stress. In the case of shear-thickening fluids, their velocity increases with increasing shear stress, while for shear-thinning fluids, the consistency decreases with increasing shear.

Shedding. An undesirable property of a **filter medium**, in which small particles or fibres are released by the medium under operating conditions (usually the drag of the **filtrate** flow), and enter the filtrate stream, and hence contaminate it. The term specifically relates to release of the material of the medium, and not to bypassing of material from the **feed suspension**. The ability to resist shedding under normal operating conditions is thus a valuable property of a material for use as a filter medium. Also known as media migration.

Sheet. A relatively stiff and thick piece of **filter medium**, usually in the shape of a rectangle, and strong enough to be self-supporting. It is made by the same wet-laying process as **paper**. Used for **clarification** by **depth filtration**, usually in a **sheet filter**. For a large part of its history the filter sheet was made mainly of asbestos, but since the realisation of the toxicity of asbestos it has been replaced by other fibrous or particulate minerals such as **diatomite** and expanded **perlite**.

Sheet filter. A batch **pressure filter**, with very long periods between successive shutdowns, used very largely as a **polishing filter** for liquid products that rely on high degrees of clarity for their marketability. It

Mobile sheet filter

consists of a set of filter sheets mounted in a cylindrical or rectangular **housing** with **suspension** flow into the space between adjacent pairs of sheets, and filtrate flow out through the centre space in each pair. The sheets are strong enough to give the filter its mechanical integrity, and they are thick enough to have an adequate solid loading before they need to be regenerated.

The regeneration of the sheets can be by back flushing of the whole filter, or, more likely, by the dismantling of the filter followed by cleaning or replacing of each sheet. The nature of its use implies that it will need to be sterilised, possibly by steaming, so the materials of its construction must be capable of resisting steaming temperatures.

Shell. A term used of the outer vessel of a **pressure filter**, or of the perforated cone of a **screw press.**
[*See also Casing*]

Shields curve. A relationship governing the movement of **particles** suspended in a liquid, as that liquid moves through an open channel. Much used in hydrodynamic modelling.

Shoe. A device used to distribute **feed** or **wash water** to a small area of a single compartment **drum filter**, or to collect **filtrate** from it.

Short tube test. A laboratory test used to size **gravity sedimenters**, especially for **classifiers** or rapid settlers such as **grit traps**.
[*See also Jar test*]

Shrinkage. The possibility of shrinkage of a **filter medium** as a result of many **filtration** cycles must be considered – this could result in tearing away from the supports, or changes in **porosity**. Filter media should thus be well tested for shrinkage before use on valuable products.

Shriver filter. A branded version of the **recessed plate filter press**, with patented opening and cloth **washing** mechanisms.

Shriver thickener. A continuous **pressure cross-flow filter**, constructed rather like a plate **filter press**, but with plates alternately acting as **concentrate** and **permeate** plates, with a porous **membrane** between them. The feed to each plate follows a spiral path in towards the centre of the plate.

Shriver tray filter. A horizontal **plate filter**, in which each plate has a rim all round it so that a layer of **slurry** or **wash water** can form on it. The plates are staggered so that liquor can overflow onto the plate below. The **heel slurry** may be blown up to the top plate for final treatment.

Shut down. The process of stopping the functioning of an item of **separation** equipment, either at the end of a batch, or for cleaning or replacing of the **filter medium**. This may be part of a routine procedure, or it may be necessary because of some malfunction.

Shute wires. A term sometimes used for the **weft** wires in a **woven wire** material.

Side bar. One of two supporting bars fitted to a **filter press**, from which the **plates** hang, and along which they move.

Side filtering centrifuge. A batch **centrifugal filter** in the form of a **drum** rotating about a horizontal axis with **slurry** feed at that axis. The drum has perforated or porous walls not only at its circumference but also at both ends. This is an attempt to increase **filtration** rates by utilising **cross-flow filtration**, which is the effective mechanism at the end walls.

Sieve. A simple **filter**, which operates almost entirely by **surface filtration**, with a **wire mesh filter medium** (or, less frequently, a plastic mesh). It can operate wet, to separate a quantity of solid particles from suspension, but the term is more often used for the **size classification** of a dry mixture of particles, especially under test conditions. It is also most often applied to a filter medium offering square **apertures**, of relatively small size, and so distinguished from the heavier **screens** of industrial use.

The term can also refer to the device that holds a **screen**, and thus enables the **screening** of solid particles.
[*See also Sieve sizes*]

Test sieve

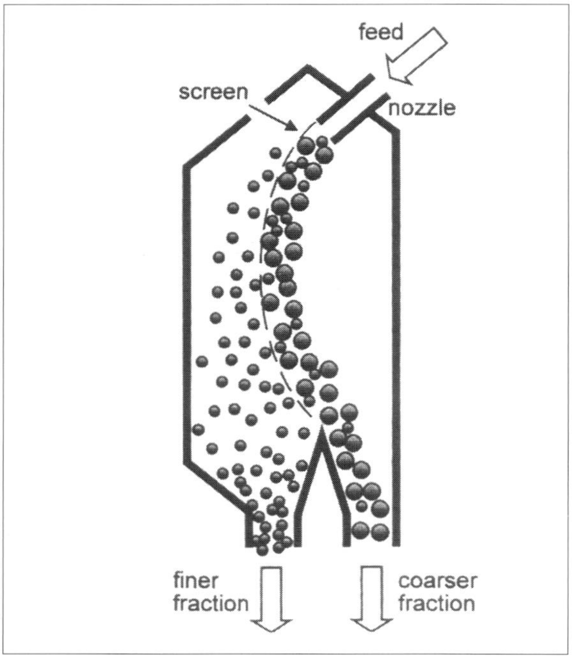

Sieve bend

Sieve analysis. A direct method of determining a **particle size distribution** by passing a mixture of particles of different sizes through a set of standard **sieves** and weighing the amount collected on each.

Sieve bend. A coarse, but high flux **filter**, made from parallel **wedge-wire bars**, lying across the direction of liquid flow, which is tangential to the filter surface. The filter is mounted in a vertical orientation, with the first section actually vertical (and, therefore, with the individual bars horizontal). It is to this top section that the feed slurry is introduced. The filter surface below its initial vertical section, curves gently to give a concave shape to the lower parts of the surface. A thickened slurry passes through the spaces between the bars, with the coarsest particles leaving through the upper spaces, and the finest through the lower ones. Originally developed by DSM, the sieve bend was marketed for a long time by Dorr Oliver.

Sieve diameter. An effective **particle size** allocated on the basis of the side of the square **aperture** in the **screen** through which the particle has passed.

Sieve sizes. Sieves are usually supplied with their **aperture** size fixed within a series of standard sieve scales – where each smaller sieve is a constant ratio in aperture size to the one before it (using ratios such as square root of 2, fourth root of 2, and tenth root of 10).

Sieving. The **separation** of solid particle mixtures into their **fractions** by using one or more **sieves** as a set. Sieving is usually used of dry screening (while **screening** is used of wet separation with a screen).

Sieving model. A model used for developing a theory for the transport of **filtrate** through **cross-flow filters**.

Sifter. Any device in which fine solids are separated, usually in the dry state, into different size **fractions**.

Sifting. Any process of separating solid particles by particle size, but usually refers specifically to the **separation** of fine particles, such as flour, in the dry state.
[*See also Sieving*]

Sigma enhancement. The clarification performance of a **decanter centrifuge** is described by **Sigma theory**, where Sigma is a measure of performance. Structural changes can be made to the **scroll** of a decanter centrifuge to improve its separation performance, and hence its Sigma value. These include the fitting of longitudinal vanes between the conveyor hub and the ribbon flight, at an angle to the conveyor radius. Alternatively, discs may be fitted instead of vanes.

Sigma factor. A parameter developed in **Sigma theory** for characterising the performance of a **sedimenting centrifuge**.

Sigma flow rating. Totally unrelated to the use of Sigma in **centrifuge** design, the sigma flow rating concerns a porous mass, and is that pore diameter such that 15.87% of the flow passes through pores larger than the identified diameter.

Sigma theory. A theory of behaviour in a **sedimentation centrifuge**, which considers the sedimentation of a small spherical particle to the bowl wall, and compares this with the movement of the element of liquid containing that particle to the liquid discharge point. It assumes **plug flow** of the liquid, and **unhindered settling** of the particle, which has reached its **terminal settling velocity** immediately on entry into the **pond**. The theory ends up with an equation between two sets of parameters – one set related to the process suspension, and the other to do only with the machine. Both halves of the equation are referred to as Sigma (Σ), which has the dimensions of an area. Sigma is then a scale-up factor for sedimentation centrifuges, indicating the **clarification** ability.

Silk. A natural material, produced as very long **filaments**, which have to be untangled. The filaments are spun into a delicate **yarn**, which has some very specific uses in **filtration**.

Silt. A specific term for a finely graded solid. As used in soil mechanics, it defines particle size: a fine silt is 2–6 μm, a medium silt is 6–20 μm, and a coarse silt is 20–60 μm.
[*See also Grit, Sand*]

Silt density index. [*See Silting index*]

Silting index. Silting is a problem in components in a fluid flow, which are made to close tolerances, where the deposition of fine particles or gelatinous materials could affect their operation. The silting index of a fluid measures its tendency to leave such deposits. The term also refers specifically to a parameter used to characterise **fouling** in a **cross-flow membrane filter**.

Silt Pac dynamic separator. An **inertial clarifier** in the form of a cylindrical open topped tank with a conical bottom. The feed to be clarified is introduced tangentially at the base of the cylindrical section, where it swirls around an inverted cone suspended in the centre of the tank. Solids settle to the bottom of the cone, and clarified liquid is removed at the top of the tank, through an annular **vortex finder**. It operates in a manner akin to the **hydrocyclone**, but it has a low pressure drop and shear, and so can be used with **flocculated** suspensions.

Simater filter. An early form of **moving bed granular medium filter**, used to clarify water, in which the bed granules are withdrawn from the bottom of the bed, air lifted to a cleaning zone and returned to the top of the bed. The bed is contained in a vertical cylindrical container with perforated or porous walls, through which the **filtrate** flows into an outer casing. The raw water also enters the bed at the top and flows downwards with the granules.

Simple washing. The washing of the **filter cake** in a **filter press**, in which the **wash liquid** follows the same path as the **feed slurry**.

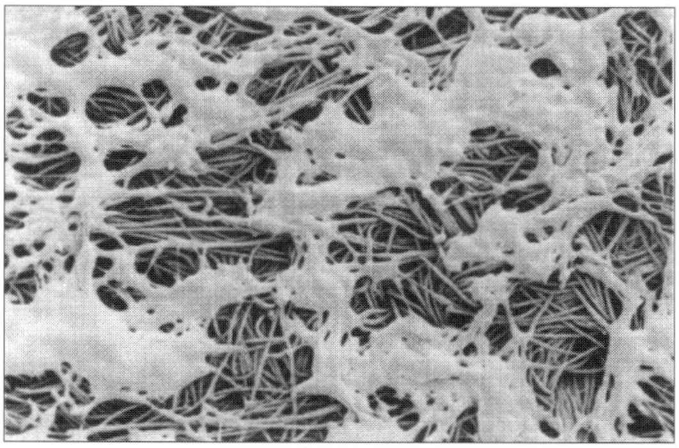

Singed medium

Singeing. The treatment of a material surface by a flame or by contact with a very hot surface, to cause partial melting of the material, and hence to change the surface **porosity**. It may also be used to remove a **nap** from the material surface.

Single bubble collision efficiency. A parameter in the design of **flotation** systems, related to particle and bubble size, particle and fluid density, and temperature.

Single compartment filter, Single cell filter. A continuous **rotary drum** (or, occasionally, **disc**) **filter**, without internal division into sectors and so requiring **shoes** to collect the liquids. The shoes are stationary and are in close contact with the internal surface of the drum. They are used for collecting **filtrates** from **cake formation** and **washing**, for applying gas for cake discharge, and for applying liquid for cleaning the cloth. This type of filter often operates at high speed to produce thin cakes. Accumulated cake discharge is by gas blowback and may be aided by a knife.

Single filter. A term sometimes used for a simple cylindrical **filter element** (**cartridge**) held in a cylindrical **housing**, and employed in a **clarification** process. It will eventually have to be taken off-line for cleaning (or the replacement of the element).
[*See also Duplex filter*]

Single particle sedimentation. The settlement behaviour of a single particle falling through a fluid in a wide container is easily studied, and leads to a relationship such as **Stokes' law** to describe its velocity. This behaviour can then be applied to that of dilute suspensions, with particles far enough apart so as not to interfere with one another's falling path.

Single pass test. A test procedure for **clarifying filters** or **filter media** that relies on the passage through the filter of a test **suspension** just once. The filter is fed with a uniform suspension of particles until a pre-decided **pressure differential** across the filter is reached, and the test is then stopped. During the test, flow rates, pressure differentials, and particle size and number are measured upstream and downstream of the filter.

Single pass through cascade. A means of arranging **membrane** filters to achieve a large **permeate** flow rate, by assembling several single **modules** in a cascade of decreasing permeate flow rates down the cascade. This is a way of scaling-up from the performance of a single module to a process of adequate size by using a number of the same size modules in a series of steps, of decreasing numbers in each step.

Single piece dam. A liquid discharge control device for the **decanter centrifuge** in the form of a one piece **weir plate**, which covers all **centrate** discharge points to the same level.

Single sheet filter. [*See Nutsche filter*]

Sink float process. [*See Heavy media separation*]

Sinter bridges. A feature of the **adhesion** of a particle to a surface as part of the **capture** mechanism of the particle in the **filter**. If the liquid around the point of contact forms a chemical bond between particle and surface, this is called a sinter bridge.

Sintered, sintering. The **bonding** of powders, fibres or meshes by heating under pressure, to fuse the material at the points of contact. Originally used of metals and ceramics, but now also of polymeric materials.

Sintered filter media. These are porous materials made by the close packing of finely divided materials followed by their sintering, whether into flat **sheets** or moulded shapes (such as tubes, thimbles or **candles**). The techniques of powder metallurgy have been rapidly assumed into the manufacture of a wide range of shapes as **filter media** or filter **elements**.

The basic material can be particulate or fibrous, or layers of woven mesh, and made from metal, plastic, glass or ceramic as appropriate. The final product's porosity and permeability can be changed by control over the size and uniformity of the basic powder, fibre or mesh.

Siphon (peeler) centrifuge. A version of the **peeler centrifuge** in which the **filtrate** is not discharged directly into the **casing** of the centrifuge, but is collected in a **rotary siphon cup** outside of the **drum**. This cup leads the filtrate into a siphon leg of pipe long enough to create a slight vacuum downstream of the drum, and so add a little vacuum to the centrifugal driving force. It is possible to design the siphon arrangement such that there is a reasonable seal around the **filter medium**, so that a slight overpressure can also be applied above the filtering slurry.

Size distribution. [*See Particle size distribution*]

Size fraction. The proportion of a mass of particulate matter that lies between two precise size limits. Usually expressed as a mass fraction, but can also be expressed as a number fraction.

Size grading. [*See Particle size analysis, Particle size distribution*]

Skewed Gaussian distribution. [*See Log probability distribution*]

Skid mounted. A **separation** equipment system is frequently supplied as a self-contained unit, with feed tank, pumps, piping, valves and controls, in a sturdy framework, which can be lifted as one by a crane (and transported probably in a container). The equipment is then said to be skid mounted.

Skimmer pipe. A movable pipe fitted into a **batch sedimenting centrifuge** for removing **centrate** or soft **cake** while the centrifuge is rotating.

Skimmer pump. A liquid discharge control device for the **decanter centrifuge** in the form of a special pipe attachment close to the **front hub**, which enables **centrate** to be skimmed off the **pond** surface, obviating the need for a **weir plate**. The skimmer can be adjusted while the bowl is at speed, which effectively alters the pond height. The pipe acts as a **centripetal pump** to convey the centrate away from the bowl.

Sleeve. A piece of **filter medium** formed as an open ended cylinder, which is slid over a **cage** or **core** to form a replaceable **filter element**, operated in a similar manner to a **filter bag**.

Sliding discharge centrifuge, Slip discharge centrifuge. The basic form of the continuous **cone-**

screen centrifuge, in which the separated solids slide over the surface of the conical **basket** and are discharged from its widest diameter into the solids collection zone of the **casing**. The flow of solids is caused by the component of the **centrifugal force** that is parallel to the surface of the centrifuge screen, provided that this is greater than the friction forces holding the solids on the screen. These latter change as the solids become progressively more dewatered, as well as with the characteristics of the screen. There is thus a limited range of **suspensions** suitable for **separation** on a given model of one of these centrifuges – but they are very cost-efficient when the match is good.

To ensure discharge of a range of solid cakes from a cone-screen centrifuge a variety of discharge control mechanisms are employed, such as vibrating the screen or giving it a tumbling motion. Discharge can also be effected with a screw conveyor.

Slimes. A term mostly used in the mineral processing sector for a **slurry** of fine particles.

Sliver. A loose, soft, untwisted rope-like strand of textile **fibre**, having a roughly uniform thickness. It is produced by the carding process, which separates raw fibres to prepare them for spinning, and may be the first step in the production of a spun **yarn**.

Sloping screens. Screens used for **classification** of a mixture of solid particles are often mounted at an angle to the horizontal. A combination of slope and vibration controls the rate at which solids move across the screen and hence the **screening efficiency**.

Slotted aperture (screen). A **screen** for **separation** that consists of parallel bars (usually of wedge wire), which may be arranged parallel or perpendicular to the **slurry** flow.

Slow-filtering slurries. A range of **suspensions** whose rate of **filtration** is among the lowest met in practice, typified by the filtering of pigments.

Slow gravity filter, Slow sand filter. A large **deep bed filter** used in the **clarification** of fresh water, which has a low water flow rate (0.05 to 0.1 m/h) downwards through the bed. It is often used with a biological layer on its top surface (the **schmutzdecke**), which operates both to destroy organic material in the feed water, and as the actual **filter medium**. To regenerate such a filter usually requires the complete removal of this upper layer.
[*See also Rapid sand filter*]

Slow-settling suspensions.
[*See Slow-filtering slurries*]

Sludge. A very thick **suspension** of solid particles in a liquid, in which the particles are close enough to touch one another, and possibly transmit imposed forces. It can vary in consistency from a still-pumpable state to an almost brick-like condition, depending on the residual moisture. A sludge is the usually intended product of any of the **dewatering separation** processes, and may be the feed to a subsequent processing stage, or it may be a waste, intended for disposal.

The most common sludge featuring in separation technology is the mixed sludge from waste water treatment, which must be dewatered prior to disposal. A wide range of equipment types competes for this application.

Sludge blanket. Where **flocculation** of a **suspension** is incomplete, then a **feed slurry** containing **flocculant** can be introduced near the bottom of a tank, from which clean water overflows from the top. As the water rises, small **flocs** will grow until they reach a size at which they remain stationary for a period, so forming a fluidised layer of sludge, or a sludge blanket. Eventually the flocs will grow too big and they will then fall to the bottom of the tank – being replaced by fresh smaller flocs. The blanket removes finer particles rising through it.

Sludge blanket clarifier. Sludge blanket clarifiers are solids contact **sedimentation clarifiers** to which the **feed** and added **flocculant** are introduced through a downward facing pipe into the layer of sludge at the bottom. **Coagulation** occurs in this sludge layer (the sludge blanket) which is fluidised by the flow. Clarified liquid overflows from a weir at the top. The blanket can be continuously bled off through a submerged weir and thickened if necessary. Alternatively the clarifier can be shut down periodically for removal of the solids. Sludge blanket clarifiers usually have square or circular hopper bottoms. Multiple clarifiers may be nested with no intervening walls above the hoppers.

Sludge compactor. A device for **thickening** a **sludge**, usually in the form of a perforated rotating **drum**, within which the sludge thickens and dewaters itself by its own weight as it rolls in the drum.
[*See also Rota-plug thickener*]

Sluice. An artificial channel down which a liquid flows. The term is much used in **mineral processing**

for a channel carrying a **suspension** of mineral material in water, during the flow in which some **separation** may occur.
[*See also Launder, Riffle*]

Sluice box. A long shallow **settling tank**, providing close to **plug-flow** of the **suspension**. It may have **riffles** along the bottom to prevent re-entrainment of the solids. A series of hopper bottoms can be provided to ease the removal of collected solids.

Sluicing. A method of removing collected **filter cake** in which the solids are washed off the surface of the **filter medium**, using jets of liquid above the medium surface.

Slurry. A **suspension** of solid particles in a liquid, where the solids concentration is high, and the solid particles can be easily seen.

Slurry concentration. The proportion of solids in **suspension** in a **slurry**, expressed either as a mass fraction, or as mass per unit volume.
[*See also Solids concentration*]

Slurry heel. The unfiltered **slurry** remaining in a **pressure** or **vacuum tank filter** at the end of the **cake formation** stage. If the next **cycle** is of the same material, then the heel can be left in place. If not, then either it must be drained away or blown from the tank, or filtered through a separate filter in the tank (a **scavenger filter**) or on the drain line.

Smog. A term deriving from a combination of "smoke" and "fog", to mean a polluted fog, containing water droplets, mostly condensed on to large dust particles.

Smoke. An **aerosol** of dust particles that are mostly below 5 μm in nominal diameter. Most smokes are of industrial origin, but some thick **hazes** can be produced naturally.

Smoluchowski equation. An equation developed to predict the rate of **flocculation** of a **suspension**.

Snake discharge. A description of the **underflow** from a **hydrocyclone**, which indicates a danger of impending blockage.

Snap blow valve. A device used in **drum** and **disc filters** for producing a sharp blast of compressed air for **cake removal** from the **filter medium**. Efficient cake removal requires a rapid build-up of air pressure below the filter medium, and the snap blow valve is fitted in the compressed air feed pipe to do just that.

Soak washing. A **filter cake displacement washing** process in which **wash liquid** is allowed to soak into the cake before pressure is applied.

Sock. A term sometimes used for the **bag** in a **fabric filter**.

Sodium aluminate. A material once widely used in the **coagulation** and **flocculation** of industrial waters. It dissolves in water to give an alkaline solution, and a fine **floc** of hydrated alumina, carrying a negative charge.

Sodium flame test. A major **challenge test** for high efficiency **air filters**, for **respirators** and for small test **filter sheets**. An aerosol of salt (sodium chloride) particles, of a mass median diameter of 0.65 μm, is generated under controlled conditions and carried through the filter under test. The mass penetration is determined by flame photometric measurements on the air flows upstream and downstream of the filter.

Softening temperature. The temperature at which a complex material begins to melt, such that points of contact fuse together. It is to this temperature that granular or fibrous materials are raised in order that they may be **sintered**.
[*See also Thermoplastic*]

Sol. A **colloidal dispersion** of a solid in a liquid.

Solid bowl centrifuge. Any of the range of **sedimenting centrifuges**, but usually refers to the **batch imperforate basket centrifuge**.
[*See also Decanter centrifuge, Disc stack separator, Tubular centrifuge*]

Solid bowl disc (stack) centrifuge.
[*See Solids retaining disc stack centrifuge*]

Solidity. A term used occasionally to characterise a porous mass. It equals (1 – **porosity**).

Solids concentration, Solids content. A measure of the quantity of solids present in a **suspension** of solid particles in a liquid. It can be expressed as a solids mass concentration, which is the ratio of the mass of solids to the total mass (solids plus liquid), and is normally expressed as a percentage:

$$c_m = 100m_s/(m_s + m_l)$$

or as a solids volume concentration:

$$c_v = 100V_s/(V_s + V_l)$$

where c = appropriate solids concentration as a percentage
m = mass of solids (s) or liquid (l)
V = appropriate volume.

The mass concentration is also referred to as the dry solids (or dry substance) content, although this can be a misleading term if the liquid contains a significant content of dissolved solids.

Solids contact clarifier. [*See Sludge blanket clarifier*]

Solids discharge. Any **separation** device that removes significant quantities of solids from **suspension** in a fluid (i.e. is involved in **recovery** rather than **clarification**) must be able to transport the solid from the separation zone to a solids receiving system. It is this need to be able to discharge the collected solids, often continuously, that has led to the complexity of many types of **filter** or **centrifuge**. Part of this complexity lies in the way in which a **filter cake** is separated from the **filter medium**, and part in the way that the accumulated solids are then pumped or conveyed away. For very dry solids, a simple chute may be sufficient, and a chute is the normal means for most centrifuges.

Solids ejecting disc (stack) centrifuge. Refers to any of the continuously operating disc stack centrifuges, such as the **nozzle**, **valved nozzle** or **opening bowl** types.
[*See also Solids retaining disc (stack) centrifuge*]

Solids flow. An obviously general term, which is used specifically to describe the flow properties of a **gravity thickener**. It deals with the settling properties in a thickener once **swarm sedimentation** has begun. The solids flow, which has dimensions of m^3/m^2h, is the product of **sedimentation velocity**, w, and **solids volume concentration**, c_v:

$$S = wc_v$$

The total solids flow is defined as the sum of this sedimentation flow and that in the **underflow**:

$$S_t = S + S_u = wc_v + V_u c_{v,u}/A$$

where V_u = sludge volume flow in the underflow
$c_{v,u}$ = solids volume concentration at the underflow
A = cross-sectional area of the thickener.

The sedimentation velocity and the solids volume concentration change, in opposite directions, over the height of the thickener, so that a critical minimum flow occurs at a certain position, and this is the major parameter in determining thickener size. The critical flow must be determined empirically.

Solids holding capacity. [*See Dirt holding capacity*]

Solids loading. A term used occasionally to refer to the quantity of solids reaching a **filter**, specifically the **concentration** of solids in the **feed** stream to a **clarification** filter.

Solids loss. When a suspension comes first to the surface of a fresh **filter medium**, then some small part of the entrained solids will pass through the medium until all of the larger **pores** in the filter medium are **bridged**, and the **filter cake** is established. This initial bypassing of the medium is termed the solids loss, and if no such loss is permissible, then either a very finely porous medium must be used (at the price of a higher medium resistance) or a **precoat** layer must be established before **filtration** begins.

Solids mass concentration.
[*See Solids concentration*]

Solids pressure. The pressure acting on the solid particles at any point in a **filter cake**.

Solids retaining disc (stack) centrifuge. A batch operated **disc stack centrifuge**, primarily intended for **liquid/liquid separation**, but able to **clarify** either liquid free of a small proportion of suspended solid, which collects in the **bowl** and has to be removed at the end of a cycle.
[*See also Solids ejecting disc (stack) centrifuge*]

Solids retention. [*See Dirt holding capacity*]

Solids throughput. The total amount of solids passing through an item of **separation equipment** in a given time. This is often expressed as specific throughput, by dividing the total figure by the **filtration area**.

Solids volume concentration.
[*See Solids concentration*]

Solids volume fraction. The proportion of solid matter in a porous mass, by volume, as a fraction. This fraction, plus the bed **porosity** as a fraction, equal unity.

Solka-Floc. A branded **cellulose** fibre **filter aid**.

Solute. Any material that can be dissolved in a **solvent** to give a **solution**. It can be inorganic, and is therefore probably **ionic**, or it can be organic, and so probably **non-ionic**.

Solution. A uniform mixture of one or more **solutes** in a liquid solvent. The solutes may be present as fully dissociated **ions**, or as undissociated molecules. For any given solvent/solute pair, there is a maximum concentration of solute that the solvent can hold at a given temperature (a maximum that normally increases with increasing temperature). The presence of a solute will affect the boiling point and the freezing point of the solvent, and may also affect its viscosity.

The solutes cannot be separated in a normal **filter**, but can with suitable **membranes**.

Solution pH. The relative acidity or alkalinity of a **solution** holding a solid in **suspension** has a significant bearing on the efficiency of **flocculation** of the suspension.
[*See also Electrical double layer*]

Solvation. The interaction between a **solvent** and a **solute**, the most common form being the hydration of solutes in water. Ions of a solute attract water molecules to form a hydrated ion, which is the actual form in which the ion exists in solution.

Solvent. Any liquid capable of dissolving another substance to form a **solution** of that substance in the solvent. In **filtration** practice by far the largest involvement is with water as the solvent, but many organic solutions are also handled.

Solvent sublation. A **flotation** process in which a layer of immiscible liquid sits on top of the flotation liquid, in order to trap the particles.

Sorting. The **separation** of a mixture of any kind into its constituent classes of component, by density, shape, colour, etc. (but not usually by size).
[*See also Segregation*]

Spacer. Spacers are used in a number of places to keep parts of **filters** separated in distance and so allow flow of fluid in those spaces. The specific use is in **membrane filtration** where the medium is wrapped around itself in a spiral configuration. This arrangement has a sheet of membrane medium, then a spacer sheet, then another sheet of membrane, and so on to the required thickness, which is then rolled up into a cylinder. The spacer here is a sheet of coarse mesh, to allow room for feed, **retentate** and **permeate** to flow between the sheets of membrane.
[*See also Spiral wound*]

Span. One of the relationships used to define a **particle size distribution**. It is defined by the ratio $(d_{90} - d_{10})/d_{50}$, where d is the particle size for 90, 50 and 10 % passing. The **pore size** span is an equivalent term for pore sizes.

Sparger, Sparging. A device intended to introduce a gas into a liquid in a finely divided state (and less commonly used for the introduction of a liquid into a gas in similar fashion). The sparger may consist of a pipe layout that covers the floor area of the liquid container, with small holes or nozzles at regular spacings along the pipes or around a pipe ring. Sparging of steam into a heating vat is one example, as is the sparging of air into a **dispersed air flotation** vessel.

Sparkler filter. A horizontal **pressure leaf filter**, with the leaves mounted on a frame within a horizontal cylindrical shell. At the end of the **filtration** cycle, the frame rotates through 90°, bringing the leaves into a vertical position, so that the accumulated **cake** falls off the leaves into the bottom of the shell. In some versions, the leaves may be vibrated to assist in cake release.

Specific cake resistance. A term used in **filter cake** formation theory for the resistance to flow through the cake per unit weight of cake solids.

Specific filtration resistance. The **flow resistance** of a **filter cake** of unit weight of dry solids per unit area of **filter medium**.

Specific surface (area). The actual area of the surface of a unit mass or volume of particles, which can be determined experimentally. The method of area determination must be stated – because gas adsorption techniques can give a larger area than those based on liquid flow. The volume-related specific area, S_v, is related to the **Sauter mean diameter**, d_{32}, of a mass of particles by:

$$S_v = 6/d_{32}$$

Specific volume. A measure of the **porosity** of a **filter cake**, defined as the total volume of the cake divided by the volume of the solids within it.

Spherical agglomeration. A relatively uncommon selective **aggregation** process for separating solids, which uses a liquid (e.g. an oil) as the aggregating agent.

Spherical latex particles. A favourite challenge material for **filter** testing with many workers, spherical latex particles can easily be generated in narrow size ranges.

Sphericity. For any real particle, this is the ratio of the surface area of a sphere that has the same volume as the particle, to the actual surface area of the particle. The reciprocal of **rugosity**.

Spigot. A term often used for the lower discharge nozzle of a **cyclone** or **hydrocyclone**, from which comes the coarser fraction of the fed solids.

Spin-Away dryer. A simple form of **centrifugal filter** used for **dewatering** polymer pellets, in which the slurry passes upwards through a stationary cylindrical perforated **screen**. Inside the screen is a rotating array of impeller blades, whose rotation imparts the necessary **centrifugal force** on the slurry.

Spin dry. A term used in **centrifugal filter** theory, for the drainage period after **filtration** is complete, and the accumulated **cake** is being **dewatered** by drainage under centrifugal force.

Spinneret. The working head of an **extrusion** process, in which a set of fine holes are machined (or a set of fine nozzles fitted), from which a molten material can be extruded under pressure as continuous **filaments**. An array of spinnerets may be mounted across a moving belt, so that the filaments can be collected, either unbroken, or broken into shorter lengths. The array would normally oscillate from side to side of the belt to give a random pattern to the laid material.
[*See also Melt blown, Spun bond*]

Spinning. In textile technology terms, spinning refers to the formation of a **yarn** from a bundle of **fibres** (or several **slivers**). A quite different meaning is also relevant to the present context, namely, the production of extruded **filaments** from a molten mass of polymer, by forcing the polymer through **spinnerets**.

Spin-on filter. A **utility filter** in which the **cartridge element** and its **housing** are an integral piece, which can be attached to the filter head by a screw thread. When the filter is fully loaded with dirt, it is unscrewed from the head, discarded and a new filter screwed back on. This design is much used in the automotive engine sector, where the filter head may be an integral part of the cylinder block.

Spiral. A device used for wet **gravity concentration**. It consists of a number of complete turns (usually 3 to 5) of a spiral channel, wrapped around a central supporting pillar. The channel, open at the top, is roughly semi-circular in cross-section. The **feed slurry** of mixed particles is fed to the top of the spiral and flows down the channel, separating the particles by size and density as it flows. The heavier particles concentrate in a band along the inner edge of the stream, and ports are placed at the bottom of the channel in each turn for the removal of these heavy particles. **Wash liquid** can be added from a separate channel at the inner edge, which flows outwards. Double spirals exist, which have two starts around one pillar.

Spiral classifier. A continuous **mechanical classifier** that uses a **screw conveyor**, operating in a semicircular trough, to scroll the separated solids out of the **pond** to their discharge point. They can be operated with steeper slopes (25–35%) than the dragging classifiers because there is less danger of solids slippage. There is also less turbulence in the pond so they are used for finer separations, and can give sharper cuts and drier **underflows**.
[*See also Drag classifier, Rake classifier*]

Spiral module. A replaceable **filter element** fabricated as a roll of **membrane** media, interspersed with **spacers**, with appropriate **end caps**, ready for insertion into a cylindrical **housing**, to make a membrane separation unit.
[*See also Spiral wound*]

Spiral wound. One of the four major structures used in creating a **membrane separation** device (the others being **flat sheet**, **tubular** and **hollow fibre**), with the particular aim of producing as large a **filtra-**

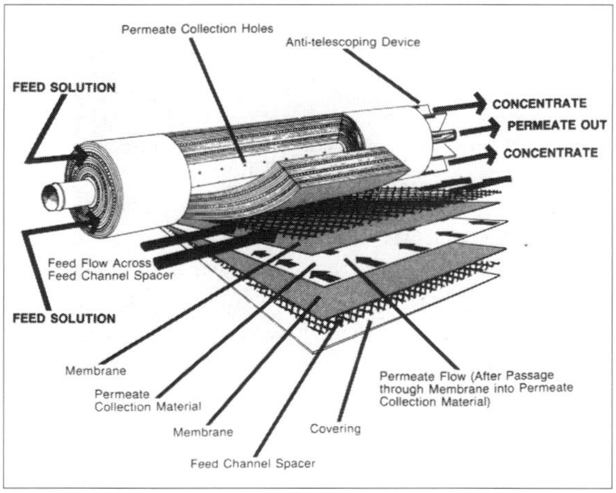

Spiral wound membrane medium

tion area in a given **housing** volume as possible. In common with the other types, the spiral wound system operates in **cross-flow** mode.

The replaceable membrane module consists of a roll of membrane medium, made up of a long sheet of membrane, with a **spacer** sheet on top of it, followed by another sheet of membrane, another spacer sheet, and so on, until the required thickness is reached. This multi-layer assembly is then rolled up tightly, and inserted into a cylindrical housing. The **feed slurry** and **retentate** flow axially along the roll, while the **permeate** follows the spiral path to the take-off tube at its centre.

The term "spiral wound" also refers to a specific version of the winding of a **yarn**, **wire** or **filament** on a **filter element core** in a spiral fashion, such that successive layers overlie previous layers at an angle, to form a common type of **filter cartridge element**.
[*See also Yarn wound*]

Split dam. A liquid discharge control device for a **decanter centrifuge** in the form of a **weir plate** for the **front hub**, made in two identical halves.

Spool wound. [*See Yarn wound*]

Spray discharge. A descriptive term for a condition of the fan-shaped discharge from the **underflow** nozzle of a **hydrocyclone**, which is thus operating fairly efficiently.

S-Press. A complex twin **belt press**, the Osprey S-Press, developed by A Klein, injects a strongly **flocculated** slurry into a serpentine path around a convoluted set of rollers, of different diameters, such that the cake has a long retention time and is submitted to quite intense **shearing**.

Spun bond, Spun bonded. The **extrusion** of molten polymeric materials as **filament** (melt spun), which filaments are then laid down, unbroken, on a moving belt, and further processed to ensure adequate **bonding** of the mass of filaments. The filaments may also be laid down on a rotating core to form a **cartridge** element.
[*See also Flash spun, Melt blown*]

Spun laced. [*See Hydroentanglement*]

Spun media. General term for materials made from extruded polymer **filaments**, laid down in a random fashion on a moving belt, such as **spun bonds**.

Square mesh, Square opening. A woven **mesh** (usually of **wire**) in which the individual **apertures** are square in shape (and therefore the **mesh count** is the same in both directions).

Squeeze belt. [*See Belt press*]

Stable suspension. A **suspension** that is resistant to being **agglomerated**, because of the electrostatic charges on the surface of the particles, causing the particles to repel one another.

Stability. A number of physical properties of a material that relate to its maintenance in use of its initial performance and design characteristics. In particular the stability of the material edges may be very important.

Stack. An array of a set of identical (or nearly identical) components – discs, rings, capsules – one above the other around a central former of some kind. One particular usage of the term is to describe the

Wire mesh with square apertures

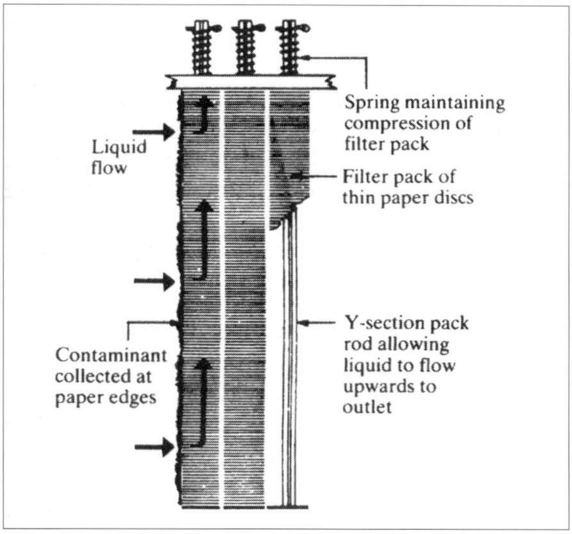

Cartridge of stacked paper discs

formation of a filter **cartridge** in this way, such that the spaces between the edges of the components form the **filter medium**. Another important usage is for the stack of conically-shaped discs that form the **separation zone** in a **disc stack centrifuge**.

Standard blocking. The basic mechanism of **depth filtration**, in which particles, smaller than the diameter of the pores of the **filter medium**, are assumed to build up on the walls of the pores, leading to a reduction in the effective **pore diameter**, and its eventual **plugging**.

Standard cake formation time (SCFT). A long established way of expressing the **filterability** of a **slurry**, as the time required to build a **filter cake** 1 centimetre thick under moderate pressure as driving force.

Standard filtration model. [*See t/V = f(t) method*]

Standard law condition.
[*See Hermans-Bredée filtration laws*]

Standard test dusts. [*See Air Cleaner Fine Dust*]

Staple (fibre). A term originally used of natural **fibres**, such as **cotton** or **wool**, which are short, as compared to the continuous **filament** of silk, or extruded synthetic **monofilaments**. The individual fibres may be 40 mm to 120 mm in length.

Yarns, and the **woven fabrics** made from them, produced from staple fibre are characterised by the great number of fibre ends that protrude from the yarn and give it a fuzzy surface. When such a material is used as a **filter medium**, the fuzziness may make it better at **separation**, but may also make it difficult to remove a **filter cake** from the medium's surface – a problem that is solved by proper treatment of that surface (such as by **singeing**). On the other hand, this same characteristic keeps a **precoat** layer firmly in place.

Staple fibres are also produced from monofilament sources by appropriate cutting of the filament into short lengths, but these will be smooth surfaced.

Starch. A natural product from a wide variety of vegetable matter (the production of which presents some challenging **separation** problems), which finds some use as a **coagulant** – especially in food industry operations.

Star feeder. A device for allowing the controlled discharge of a mass of particulate solids from a vessel under pressure. It comprises a number of vanes (usually six) mounted axially on a rotating shaft, inside a cylindrical housing, closed at each end. It is fixed, with its axis horizontal, below the vessel holding the solids, with a hole connecting them on the top of the cylinder, and a similar hole below the feeder connecting to the solids receiver or conveyor.

The rotating vanes are a close fit inside the cylinder, thus making a seal between the pressurised vessel and the outside. As the set of vanes rotates, the solids fill each compartment between two vanes in the feeder, and are then carried round to the lower hole and fall out of the feeder.

The star feeder works best with relatively dry (and certainly non-sticky) solids, and solids which are not very abrasive.

Start of line filter. A **pipeline filter**, usually a relatively coarse mesh **screen**, fitted at the start of a process or system line, to protect equipment in the line from **tramp material**.

Start-up. The process whereby a **filter** or **centrifuge** is put into use from its stationary state, for the first time or after any cause of **shut down**. Most items of **separation equipment** are able to start automatically, but it is a condition where out-of-balance forces are most likely to occur, especially with centrifuges, because of uneven loading of solids.

Stationary filter pan/tray. A cyclically operating **belt filter** in which a continuous belt of **filter medium** is passed over a long tray or set of trays. The belt is

stationary during the **filtration** cycle, and is then indexed across the tray to discharge the **filter cake**.
[*See also Adpec filter*]

Stationary leaf filter (Diastar). A vertical **pressure leaf filter**, with the leaves mounted radially (i.e. in a star shape) around a central feed pipe. This arrangement of leaves allows the inclusion of a large **filtration area** in a given shell volume. The **filter cake** is held on the leaves by gas pressure between **filtration** and **cake washing**.

Steam hood. A supplementary item of equipment for **vacuum filters**, in which the **steaming** of a **filter cake** can be undertaken to improve its **dewatering**.

Steaming. A supplementary process sometimes employed with **continuous vacuum filters**, whereby saturated or superheated steam is applied to the surface of the largely dewatered **filter cake** in a **steam hood**. The steam is sucked through the cake, heating the solids as it condenses. The heat reduces the **filtrate** viscosity and surface tension, making the **dewatering** easier. Some moisture is also lost from the warm cake after discharge.

Steam pressure filtration. A patented process of Bokela in which steam is used as the displacement medium for the entrained moisture in a **filter cake**. The steam acts not only to push the **mother liquor** out, but it also heats the cake, and the resultant condensate washes the cake as well – but only for cakes that are not affected by the higher temperatures.

Sterile filter, filtration. The removal of all harmful micro-organisms (**bacteria**, **viruses**, etc) from a fluid is an essential process in a number of **filtration** applications: respiration, living and working space ventilation, fermentation process venting, and so on. To be effective, a **filter medium** for this purpose must be able to remove particles completely down to at least 0.1 μm, and preferably to 0.01 μm (10 nm) – a filtration application that is on the borders between **microfiltration** and **ultrafiltration**. This is becoming a target that is more and more easily achieved by means of **nonwoven** media made from **nanofibres**, or by means of **membranes**.

Sterilisation. The process of rendering a fluid sterile, either by removing all of the harmful micro-organisms from it, or by destroying them. Destruction has been achieved by chemical treatment with chlorine or ozone, or by UV irradiation, or by heat (as in pasteurisation). Increasingly, however, sterilisation is being achieved by **separation** of the **microbes** by **filtration**, which has the great advantage of not having any other impact on the fluid (as might occur from heating it, or from the oxidising effect of chlorine). **Microfiltration** and **ultrafiltration** are now important processes in the provision of sterile air, or of fresh water.

Sterility. A fluid is said to be sterile if it is completely free of harmful organisms, either **pathogenic** or **pyrogenic**. This state is achieved by proper **sterilisation**, a process in which fine **filtration** is becoming an important means.

Stern layer. The layer of **ions** of an opposite charge (**counter ions**) that is firmly bonded to the surface of a charged particle. This layer is one, or at the most, two molecules thick.

Stern theory. Stern originated the **electrical double layer** theory, much used in the explanation of **coagulation** phenomena.
[*See also DLVO theory*]

Sticking factor. A term used in **cross-flow membrane** separations, defined as the amount of **fouling** material sticking at a membrane surface, divided by the amount of foulant brought to the surface by the suspension.

Sticky cake. A generally descriptive term for a **filter cake** that is liable to adhere to the **filter medium** and to other surfaces (such as those in the **casing** or discharge chute).

Stiffness. The ability of a material to prevent its deformation, usually tested in some way related to the resistance to bending. Standards are available for such tests for **paper** and **textiles**.

Stirred cell. An item of laboratory equipment used for measuring the performance of **cross-flow membranes**.

Stirred-cell filtration. A means of achieving **membrane filtration**, used only on the small scale (i.e. mostly in laboratory work), which employs **through flow**, rather than **cross-flow**. **Shear** is provided by the use of a stirrer in the slurry above a flat sheet membrane.

Stirred pressure nutsche filter. [*See Filter reactor*]

Stock. A term much used in the papermaking sector

to refer to the **feed slurry** of pulp fibres, ready for the papermaking machine.

Stokes. Sir George Stokes was born in Ireland in 1819, and died 84 years later after an astonishing career as physicist and mathematician, which saw him appointed Lucasian Professor of Mathematics at Cambridge (a post now held by Stephen Hawking) at the age of 30, elected to the Royal Society at 32 and its secretary at 35, which he remained for 30 years, until becoming President, a career path previously followed only by Sir Isaac Newton.

Not only was he an expert in fluid flow and dynamics, for which he is best known in the world of **filtration**, but he pioneered in such fields as UV light (originating the term fluorescence), the wave theory of light, the science of geodesy (publishing on gravity variations at the earth's surface), and the solar spectrum.

Stokes diameter. The diameter of a sphere showing the same behaviour as a particle **settling** in obedience with **Stokes' law**, i.e. the same density and the same velocity in the same fluid, under **laminar** flow conditions.

Stokes-Einstein equation. An equation used for the calculation of the **diffusion** coefficient for molecular species through a fluid:

$$D = RT/3\pi\mu x N_A$$

where R = universal gas constant
T = absolute temperature (°K)
μ = fluid viscosity
x = particle diameter
N_A = Avogadro's number

The calculated diffusion coefficient can then be used in the appropriate form of the **Peclet number** in order to calculate some capture parameters in considering **depth filtration** for **clarification** of a fluid.

Stokes' law. Stokes developed his law of viscous motion in 1851. The law states that, for a particle falling through a fluid of effectively infinite extent, under conditions of **laminar** flow, and under the influence of gravity, the velocity of fall will increase to a constant (terminal) value, dependent upon the **particle size**, the difference in densities of particle and fluid, and the dynamic viscosity of the fluid:

$$v = gd^2(\rho_p - \rho_f)/18\eta$$

where g = acceleration due to gravity
d = particle diameter
ρ = density of particle (p) and fluid (f)
η = viscosity of fluid

The infinite extent condition for the application of Stokes' law is a limiting and unreal one, but the calculation of **terminal velocity** this way is close enough provided that the settling is **unhindered** by any neighbouring particles, and that the flow is laminar.

Stokes number (Stokes-50 number). The basic Stokes number is a **dimensionless number** that characterises the inertia of a particle-fluid-target solid system, for use in particle **capture** calculations:

$$Stk = \rho_s x^2 u/18\mu d_i$$

where ρ_s = particle density
x = particle diameter
u = particle velocity
μ = fluid viscosity
d_i = a characteristic dimension of the target surface (e.g. a larger particle, a fibre, etc)

The Stokes-50 number, is a similar dimensionless number, used to describe particle movement in a **hydrocyclone**:

$$Stk_{50} = x_{50}{}^2(\rho_s - \rho)v_z/18\mu d_c$$

where ρ = liquid density
x_{50} = particle diameter at the hydrocyclone **cut point**
v_z = a characteristic liquid velocity inside the hydrocyclone
d_c = hydrocyclone diameter

Stokes settling velocity. The **terminal falling velocity** of a particle in a gravitational field, as predicted by **Stokes' Law**.

Stop distance. The horizontal distance travelled by a particle that has been projected horizontally into a viscous fluid, before it comes to rest.

Stop-start washing. If a **cake washing** process requires a long time, usually because the **diffusion** rates of the **solute** are low, then intermittent washing may be more efficient in use of wash water, with pauses between wash water flows, to allow diffusion to occur.

Strainer elements

Strain. In general, strain is the change in volume or shape of a material, resulting from applying a stress to that material. In **filter medium** terms this usually implies the extension of the medium under a load, divided by its original length.

Strainer. In general, this term applies to any **separator** used to remove coarse particles from a suspension that will then go on for subsequent processing. Thus it can refer to a **pre-filter**, or to a **filter** used to protect equipment downstream from it from the malign effect of **oversized** particles. Such a device will normally use a **screen** made of woven **wire mesh**, perhaps as a replaceable **element** or as a **self-cleaning filter**.

More specifically the word refers to a nearly horizontal, continuously rotating **drum**, made of a wire or plastic mesh, with a slurry fed to the inside of the drum, at its upper end. The strainer then acts as a gravity filter, **thickening** the slurry or even producing a **dewatered cake**, while the **filtrate** drains through the screen wall.
[*See also Microstrainer*]

Straining. A **filtration** process, very close to pure **surface filtration**, in which particles are trapped on the surface of the **filter medium** and do not build up to more than one or two particle layers.

Straining zone. The first, **feed** section of a **double belt press**, where the lower belt runs horizontally and strains most of the **filtrate** by gravity, to create a cake stiff enough to be drawn into the pressing zone of the **filter**.

Strata-Sand filter. A continuous **deep bed filter**, consisting of a vertical cylindrical tank, down through which the solid particles move into a hopper cone at the base, from the apex of which they are transported by an air-lift up through a tube at the axis of the tank, for cleaning and recycle to the top of the bed. The raw water is fed to the top of the bed, flows down through it, and the clean water leaves through a perforated screen cage near the bottom of the bed, fixed around the central pipe.

Stratification. The formation of any system of particles into horizontal layers, specifically referring to the layers formed in a **deep bed** as it settles back after being washed clear of trapped solids.

Streaming potential. The electric potential created in a liquid when it is forced to flow through a bed of fine charged **particles**.

Stream-line filter. A **cartridge filter** in which the **element** is made up of a stack of paper discs, without spacers, held together by a spring. The **filter medium** is then the edges of the paper discs, which form a smooth cylinder.

Streamline flow. A term synonymous with **laminar** or viscous flow.

Stream scanning. A group of indirect **particle size analysis techniques**, in which the flowing **suspension** is scanned by a beam from an instrument operating with such methods as light obstruction, light diffraction or light scattering. This normally determines some other parameter than diameter (such as actual or projected area, or volume) from which a mean diameter may be calculated.

Strength (of filter media). Generally refers to a group of mechanical properties of a material, but is specifically used for the property measured in a test to determine the pressure required to rupture the medium under controlled conditions.

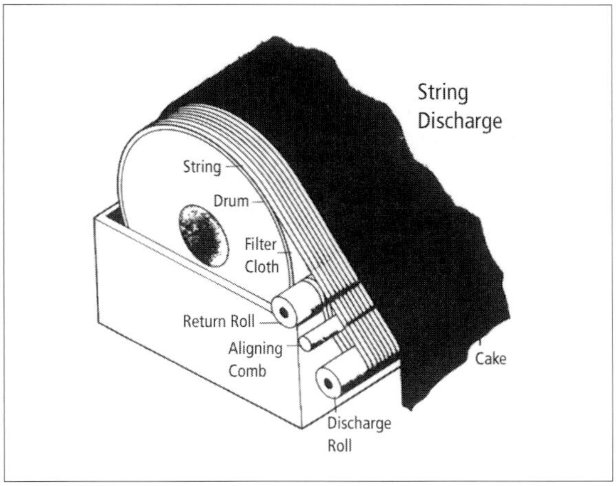

Drum filter with string discharge

Stretching. A method used to produce thin polymeric **membranes** out of an impervious sheet of the polymer, developed by W L Gore, for its **PTFE** membranes.

Stretch resistance. The resistance to stretching (or creep) shown by a material can be a very important factor in its suitability as a **filter medium**, especially as a **filter bag** or **belt**. Tests will therefore be undertaken to demonstrate tensile strengths in excess of what might be needed in **filtration** use.

Strindland filter. [*See Self-suctioning drum filter*]

String discharge filter. A means of effecting removal of the settled **cake** from the **filter medium** in a continuous **drum filter**. A set of strings runs parallel with the direction of movement of the medium around the drum, lying on the surface of the medium and spaced at about 1 cm apart.

The strings come into contact with the filter medium from a point just before the cells enter the feed slurry, and remain in contact with it through the **filtration**, **washing** and **dewatering** zones. Once the dewatering and any washing processes are complete, the medium passes over a roller to return to the filtration zone, but the strings carry on a further distance to another roller, in the process lifting the cake away from the medium. As the strings go round the second roller, the cake is broken away from the strings and is discharged from the filter. The strings return to the start of the cycle, through the teeth of an alignment comb.

String discharge is best suited to fibrous solids, which form a felted cake, and the strings interfere less with the filtration process than do chains.
[*See also Chain assisted cake removal*]

String wound. [*See Yarn wound*]

Subatmospheric operation. [*See Vacuum filtration*]

Submerged belt filter. A development of the **flat bed filter**, for the clarification of machine tool coolants, which has its feed roll of **filter medium** submerged at the bottom of the **slurry** tank. The medium unwinds from this roll, over another submerged roller, and then out of the slurry to a takeup roll. The edge of the medium seals against the tank wall to prevent contaminated liquid from bypassing the belt. This also enables the filter to operate at a partial vacuum. Clean **filtrate** is pumped out from under the medium, which is **indexed** forward when the liquid level builds up. The medium can be disposable **paper**, or cloth that is washed outside the filter, or within it.

Submerged disc filter. A continuous **rotary vacuum disc filter**, which produces a **thickened slurry** as the solid product. The discs are mounted vertically on a horizontal shaft, inside a rectangular casing full of slurry, so that the discs are fully submerged, requiring sealed bearings. The cake is discharged from the bottom of the discs by backflow of liquid, to collect in the hopper bottom of the vessel, from which it is removed as a thickened slurry. These filters can actually be designed to operate under vacuum or moderate pressure.

Submerged drum filter. One of several designs of simple, gravity driven, **rotating drum filters** used to thicken paper stock. In one version, most of the drum is submerged, although **filtration** occurs only near the bottom of the drum, because the liquid head is insufficient to overcome surface tension forces over the rest of it. Discharge of the **cake** (thickened to 4-8% fibre) can be by a **knife** or a **transfer roller**. A Rauma Repola version is similar but is fully enclosed and the cake is discharged by blowback with steam at 2 bar pressure. The drum can rotate continuously or intermittently.

Submerged membrane bioreactors. A **membrane bioreactor** that is completely submerged in the **slurry** to be treated. They usually operate at low **trans-membrane pressures**, and the slurry is agitated by gas bubbles to increase shear at the membrane surface.

Submergence, Submergence ratio. Most **rotary vacuum drum** or **disc filters** operate by submerging the **filter medium**, **cell** by cell, in a trough carrying the **feed slurry**. Whilst the medium is submerged, then **filtration** is possible.

The submergence ratio is the proportion of the total filtration area of the filter that is submerged at any one time. The most effective submergence will depend upon the proportion of the complete cycle that must be devoted to filtration, and how much is needed for **washing** and **dewatering**. It is normally possible to have about 40% submergence without requiring seals on the central shaft, but 75% is possible with appropriate mechanical structures.

Substrate. In general terms a substrate is the substance on which an enzyme or fermentation process acts. In filtration industry terms, however, it is the base layer of a **filter medium**, providing mechanical support to the active **filtration** layer (especially a **membrane**) above it.

Suction filter. [*See Vacuum filter*]

Sump. A term of many uses, mostly implying a holding tank of some kind. Most importantly, it refers to a collection zone below and after a **coalescer element**, into which the larger, coalesced droplets fall, prior to discharge.

Superficial velocity. A widely used term for the velocity of the fluid through the space occupied by a **filter medium,** in the absence of that medium. It is given by the fluid flowrate divided by the whole cross sectional area of the **filter**.

Supernatant. Refers to the clear liquid that forms above the **settling** particles of a **suspension** undergoing a **sedimentation** process. As more appears from the settled feed slurry, eventually the level of the supernatant rises to the point where it can overflow (as product if from a **clarifier**).

Surface active agent. A specific chemical substance (occasionally called a tenside), or mixture of substances, added to a liquid in order to change its interfacial properties, in particular its surface tension. These changes may be particularly beneficial in achieving **separation** of **emulsions**.

Surfactants are complex organic molecules, containing a hydrophobic group, such as a long hydrocarbon chain, or an acrylate group, with a hydrophilic (ionic – typically sulphonic) tail. The molecules orient themselves in the liquid surface.

The use of surfactants has made a whole range of industrial (and domestic) activities much easier – at the cost of foam in effluent streams, and a problem of slow biodegradation of surfactant chemicals entering natural waters.

Surface area. The surface area of a mass of particles or fibres can be determined by **adsorption** measurements, and may then be used as a means of calculating the average dimension of a particle or fibre.

Surface diameter. One definition of the size of a real particle, given by the diameter of the sphere whose surface area is in the same ratio to its volume, as is that of the particle.

Surface filter, Surface filtration. A major type of **filtration** mechanism, in which the **separation** is decided entirely by the perforations of a **filter medium**'s surface, and no separation occurs as a result of the thickness of the medium. It would be better called surface **straining**, because the separation is a matter only of the size (and, to a lesser extent, shape) of the particles in the suspension or mixture to be separated.

The closest real processes to surface filtration are **cross-flow filtration** (in which any potential for **cake** formation is resisted by the scouring action of the liquid flow), and **screening** (or **sieving**). The reality of surface filtration is that no filter medium can be negligibly thin while retaining adequate strength, and that what actually happens is that the particles in the suspension form bridges across the apertures in the screen, and a **filter cake** then forms above the surface.
[*See also Depth filtration*]

Surface potential. A characteristic of any **particle** in a **suspension** is that it is extremely likely to be electrically charged (and mainly negatively so), and this charge will be seen as an electric potential at its surface. The reaction to this potential is that a layer of

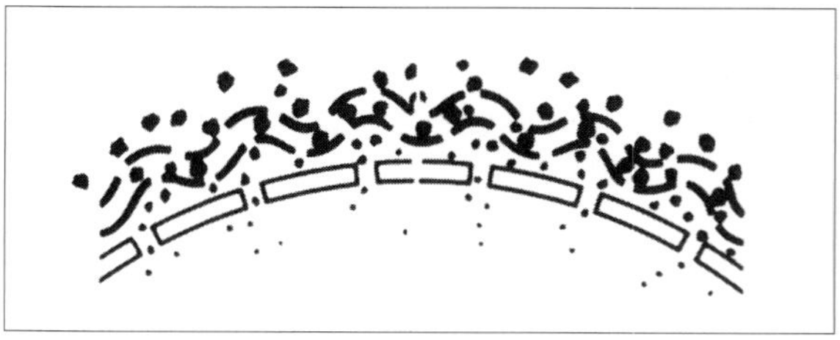

Surface filtration

counterions will form in the liquid surrounding the particle. For **agglomeration** of particles in a suspension to be effective, the surface potential must be neutralised to the extent that the non-electric **Van der Waal**'s forces can result in the attraction of one particle for another.
[*See also Electrical double layer, Zeta potential*]

Surface shape coefficient. A parameter used to describe non-spherical particles, given by the ratio of the surface area of the particle to the square of its **effective diameter**.

Surfaces (inclined). [*See Lamella separator*]

Surface tension. The property of a liquid that causes the exposed surface to contract to the smallest possible area, the result of an imbalance at the surface in inter-molecular attraction. The surface tension of the residual liquid in a **filter cake** has a significant effect upon the ease of final **dewatering**.
[*See also Interfacial tension*]

Surface-volume shape coefficient. A parameter used to characterise a non-spherical particle, given by the ratio between the **surface shape coefficient** and the **volume shape coefficient**.

Surfactant. [*See Surface active agent*]

Suspension. A mixture of distinct, insoluble **particles** with a **fluid**, whose motion is sufficient to keep the particles apart and supported by the fluid. If the fluid stops moving, then the particles will settle under the influence of gravity. The fluid can be called the **continuous phase**, while the particles represent the dispersed or **discontinuous phase**. In a dilute suspension, the particles are separated by a distance that may be large by comparison with their diameter. In a concentrated suspension, the particles will be much closer together, and if the suspension becomes so concentrated that the particles touch one another, this state is then called a **sludge**.
[*See also Sedimentation*]

Suspension density. The average or bulk density of solids and fluid combined in a **suspension**:

$$\rho_{sl} = c_v \rho_s + (1 - c_v)\rho_l$$

where ρ = density of suspension (sl), solid particles (s) and fluid (l)
c_v = **solids volume concentration**

Suspension thickening. A term used specifically in **membrane** process descriptions, for the increase in solids concentration between **feed** and **retentate**.

Svarovsky model. A model for the design and scale-up of **hydrocyclones**, employing **dimensionless numbers**.

Swarf. The small particles removed from a piece of metal being machined to shape, which are carried away from the work zone by a specially formulated liquid, the machine tool coolant or **cutting fluid**. This liquid is too valuable to waste, but the swarf would interfere with its re-use, so the cutting fluids must be **filtered**.

Swarm sedimentation. The settling behaviour of a mass of suspended solids, under the influence of gravity or **centrifugal force**, when particle size, shape and density all lose their effect on the rate of **sedimentation**. A sharp division can be seen between the top of the sedimentation zone, and the clear liquid above it, and the main parameter influencing the settling rate becomes the **suspension concentration**. Some **segregation** is observable over the whole range of swarm sedimentation conditions, especially under centrifugal sedimentation, but mainly the settling occurs without **classification**. At the highest concentrations, the mass does settle homogeneously – a condition termed **zone sedimentation**.

Sweep coagulation. The collection of fine particles in a **suspension** of low solids content, by producing insoluble fluffy precipitates of metal hydroxides in the suspension, which then enmesh or sweep up the fine particles.

Sweetland filter. A version of a **pressure leaf filter**, patented by Sweetland in 1905, with **leaves** hanging vertically in a horizontal tank. In early versions, the lower part of the tank hinged open for solids discharge and cleaning. In later versions, access is through the upper part of the tank.

Swelling. The swelling of a natural fibre fabric can change the **filtration** properties of a **filter medium** made from it. The wet behaviour must therefore be determined in preference to dry performance for media to be used for liquid filtration. One significant effect is to close the inter-**yarn** spaces in a **woven material**, so increasing the proportion of flow between the fibres of the yarn.

Swivel beaker. A container used in a **beaker centrifuge**, which contains a small sample of a **suspension**. The beaker hangs vertically when the centrifuge is stationary, but as the centrifuge starts to rotate the beaker swings into a horizontal orientation for the settlement to take place.

Symmetric membrane. A **membrane** whose pore size structure is the same across the whole width of the material. Such a membrane is likely to be used for **microfiltration**.
[*See also Asymmetric membrane*]

Synthetic (fibre, filament). Synthetic materials are widely used as **filter media**, the term mainly being used for polymeric **fibres** and **filaments**, as compared with those from natural sources. The main feature of synthetic materials is the ability to build valuable performance properties into the fibres or filaments, which cannot easily be matched by natural materials.

T

Table filter. A continuously operating horizontal **vacuum filter**, in the form of a circular array of **filter elements**, each a segment of the circle, thus making a table. Each segment has **filter medium** secured to its top, and there are vertical walls around the periphery of the table, and around the central axis. The slurry to be filtered is introduced from above, by means of a set of weir boxes, across the whole diameter of the table, which rotates about its central axis, so that the segments can be progressively filtered, **washed** and **dewatered**. The **filtrate** is drawn by vacuum through the filter medium and discharged from below the table.

The accumulated **cake** is usually removed by means of a rotating screw, which lifts the solids over the peripheral wall, but leaves a heel behind on the medium surface. Some designs have a flexible wall that flexes away from the table in the discharge zone so that the solids may be more easily removed. The table filter is used for solids that are easily filtered.
[*See also Pan filter*]

Table flotation. An **agglomeration skin flotation** process that uses a **shaking table** to perform the **separation**.

Tail. That part of a **wash curve** for a **filter cake** corresponding to high **wash ratios**. A curve with a long tail implies that large volumes of wash liquid are required to remove the **solute**.

Tailings. A term used in the **minerals processing** sector to refer to the concentrated waste (**gangue**) left after the removal of the valuable ore component, by means of a **classification** process, although it is also used for the **underflow** in froth flotation processing, where, confusingly, it may refer to either product or waste.

Tailings pond. The term used in the minerals industry for the lagoon in which **tailings** are collected (and to some extent **dewatered**).

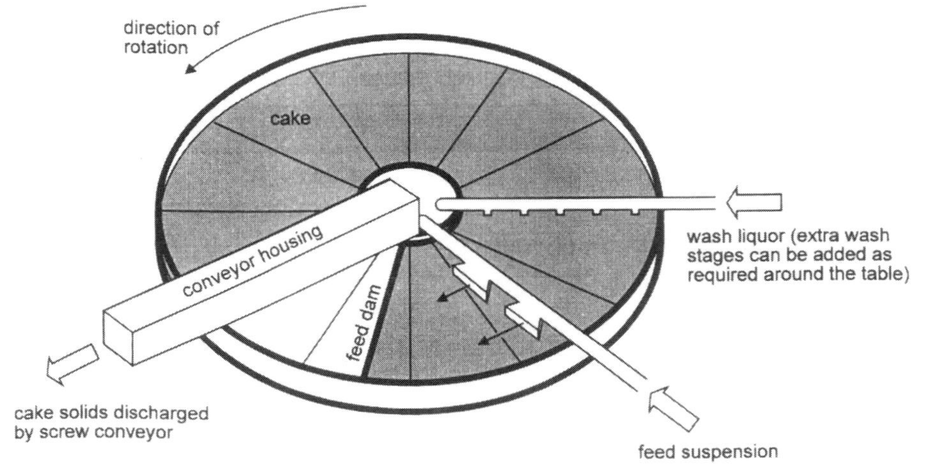

Rotary table filter

Talmage and Fitch method. A method used in the designing of **gravity thickeners**.

Tangential flow. [*See Cross-flow*]

Tangential velocity. In a normal **hydrocyclone** the feed **slurry** is introduced at the top of the unit, at a tangent to the periphery. The tangential velocity of a suspended particle is an important design parameter because it is the way in which a particle following the liquid flow will experience the **centrifugal force** necessary to carry the particle to the hydrocyclone wall.

Tank. The outer containment vessel for a multi-**element filter**, which holds the slurry to be filtered, and is also able to hold the fluid pressure needed to drive the filtration. (Also called a shell.)
[*See also Slurry heel*]

Tank filter. A term sometimes used for a **batch filter**, usually operating under pressure, which has multiple filtering **elements** suspended in a single vessel.
[*See also Candle filter, Leaf filter*]

Tap density. The bulk (or apparent) density, i.e. the height occupied by a given weight of a powder, after a vertical cylinder containing it has been tapped in a given manner.

Taper coagulation. A process of **coagulation** of fine particles, which takes place in a series of stages with decreasing **shear** used from one stage to the next.

Tapered elutriator. A **hydraulic classifier** is one that operates in the **hindered settling** mode, in which a fluidised bed is formed in the separating zone, leading to better classification. The bed is produced by placing a constriction at the bottom of the classifier. In the tapered (or conical) **elutriator** the constriction is achieved by tapering the vessel, in the form of an inverted cone, with the elutriating liquid entering through the narrow neck, in countercurrent with the leaving heavy or larger fraction.

TDS. [*See Total dissolved solids*]

Tear factor, Tear resistance. The resistance of a material (usually a paper or paper-like material) to tearing can be determined by standard test methods, well established in the paper and textile industries. The tear factor of a paper is the tear resistance divided by its **grammage**.

Tedman separator. An inertial separator resembling a **hydrocyclone**, in which the rotation of the fluid is promoted by a spiral groove cut in the wall. An additional water flow is used as well as the slurry feed. It operates at low velocity, which minimises erosion.

Tefaire (bag filter). A **filter bag** material for dust removal from gases at moderate temperatures made from a **batt** of 25% glass fibre and 75% PTFE fibre, **needled** onto a **scrim** and heat set.

TEM. [*See Transmission electron microscope*]

Temperature adjustment. The operating temperature of a **filtration** process has some effects on filtration performance, generally the higher the temperature the easier the filtration, mainly because of a decrease in liquid viscosity.

Temperature resistance. The resistance to higher temperatures of a **filter medium** is one of its more important characteristics, especially as operating conditions tend to get more severe with passing time, rather than less. This has led to an increasing use of fluorinated polymers (PTFE, PVDF) and ceramics as media materials.

Tenacity. A term sometimes used for the breaking load of a fibre: the stress applied to the fibre at its breaking point.

Tenside. [*See Surfactant*]

Tensile strength. The resistance of a material to breaking when subjected to pulling forces. Measured as the load on the material at its yield point, divided by its original area.

Tensiometer. An instrument used to measure the **surface tension** of a liquid, or to measure the **interfacial tension** between two immiscible liquids.

Tensioned wire. A method of **cake discharge**, similar to **chain** or **string discharge**.

Tension roller. A roller carrying the **filter medium** in continuous filters such as **belt filters** or **presses**, or **tower presses**, with a mechanical adjustment capability allowing it to tighten the recirculating medium.

Tenten (sand) filter. A type of moving bed **granular medium filter**, in which the solids flow down inside a perforated cage, out of which the clean water flows, into a conical hopper bottom, from which an air lift carries it back to the top of the bed after cleaning.

Terminal pressure. The final pressure in a system just before it is shut down, or at which it becomes necessary, say, to change a **filter element**.

Terminal settling velocity, Terminal velocity. Any body falling in a fluid under the influence of gravity will initially accelerate under that influence. As its velocity increases, the viscous forces opposing its movement will also increase, until they equal the gravitational force, when net acceleration reduces to zero. The body will then continue to fall at this maximum velocity, known as its terminal velocity, which is calculable from **Stokes' law**.

Tertiary treatment. The third main stage in municipal sewage treatment intended to remove all materials from the water not removed by **primary** or **secondary treatment**, increasingly necessary as water discharge consents become progressively more demanding. The removal, of particles finer than 1 μm and of residual soluble components, is achieved by **microfiltration** or **ultrafiltration** (and sometimes even **reverse osmosis**), biological processing (for nitrate and nitrite removal), chemical precipitation (for phosphate removal), or **adsorption** (colour removal). New processes are developed as demands on final purity get higher.

Terylene. A trade name for synthetic **polyester** fibre, which is used extensively in the manufacture of **filter media**.

Terzaghi model. A theoretical model for predicting the rate of **compression** of **filter cakes**. Also known as the Terzaghi-Voigt model, it is based on rheological considerations of cake consolidation.
[*See also Voigt element*]

Test contaminants, Test dusts. Any material used in challenge tests to determine the efficiency of a **filter** or of a sample of **filter medium**.
[*See also, for example: Arizona Test Dust, Ballotini*]

Test sieve. A **screen** made to precise **aperture** dimensions and mounted in a frame so that it can be used as one of a set of sieves, in the determination of a **particle size analysis**.

Tex system. A standard method for the measurement of **fibre** or **filament** fineness. The tex itself is expressed as the weight in grams of 1000 metres of the material (and so is dependent upon the material density). The tex is mostly used for yarns, while the **decitex** is used for fibres.
[*See also Denier*]

Textile. Any natural or synthetic **fibre** or **filament**, or **yarn**, suitable for making up into fabric or cloth, including the made up materials as well. Covers woven, knitted and nonwoven fabrics, as well as threads, cords, ropes, braids, lace, embroidery, and nets. Paper is not considered to be a textile, although some non-woven materials are made from fibres in the same way as paper. Nor is wire mesh so considered.

Thermal conditioning. A term that can be applied to thermal treatments for **filter media** (such as **calendering**), it is more often used of holding a compressible **sludge** at a temperature well above the ambient figure for a period to make it more easily filtered.

Thermal convection. Flow movements in a fluid that are caused by the density differences within the fluid, created by a temperature gradient. They are a usual reason why very small particles (of the order of 1 μm or less) will not settle out of a suspension under the influence of gravity alone.

Thermal drying. If a **sludge** is required effectively free of retained moisture, then it is unlikely that this state will be achievable by mechanical means alone, and the final removal will need to be done thermally, in some kind of dryer. Thermal energy is much more expensive than mechanical energy, so the maximum amount of **dewatering** is sensible before the sludge is thermally dried.

Thermal fixation. A term sometimes used to describe the treatment of **woven materials** made from synthetic **fibres** or **filaments**, usually with hot air, in order to condition the fabric to give it dimensional stability.

Thermally bonded. The **adhesion** of **fibres**, powders, etc, by heating under pressure, so that softening occurs and the material fuses together at the points of contact. This is the same process as **sintering**, but that term is generally reserved for inorganic materials, whilst thermal bonding applies to thermoplastic polymers.

Thermal precipitation. A device used in the analysis of **aerosols**. An aerosol sample is drawn through a narrow channel, along the centre of which is a wire that can be heated. Particles are repelled by the hot wire, and are collected upon microscope slides positioned on either side of the wire. At an air flow of 7 ml/min, all particles of 5 μm or less are precipitated.

Thermal stability. The thermal stability of a **filter medium**, i.e. its ability to retain its physical characteristics at higher temperatures, depends very largely on the chemical nature of its basic material. The purer that material is, then the easier it is to forecast the stability.

Thickener. [*See Gravity thickener*]

Thickening. The increase of the solids concentration of a liquid **suspension** by the removal of part of the liquid by means of a **separation** device. The thickening can be achieved by **sedimentation** (gravity or centrifugal) or by a **cross-flow filter**. The thickened slurry may be an adequate product, or its formation may be the first stage of a **dewatering** process, with one or more stages to follow.

Thickening factor. In any **separator** operated so as to achieve the **thickening** of a feed slurry, the thickening factor is the ratio between the solids concentration in the **underflow** and the feed concentration.

Thickening filter. Filters operating in **through-flow mode** do not produce a flow of thickened slurry, but a **cake**. If thickening by means of **filtration** is required, then a **cross-flow** type of filter must be used, and the term is used generally of any filter so operated.

Thimble, Thimble filter. A **filter**, most often made of stiffened paper, shaped like a dressmaker's thimble, and mainly used for analytical purposes.

Thin channel module. A less common format for a **membrane** material intended as a **filter medium**, which is a tubular **element** with a rectangular (or trapezoidal) cross section.

Thin layer filtration. A term, normally synonymous with **cross-flow filtration**, to denote a **filtration** process that is accelerated by the minimisation of the cake depth, and so it can have a more general meaning: that of any kind of sieving separation, i.e. a **surface filtration** process.
[*See also Delayed cake formation, Dynamic filter*]

Thin layer separator. [*See Lamella separator*]

Thixotrope, Thixotropy. A thixotrope is a **non-Newtonian** liquid, i.e. one that does not obey the Newtonian law of the independence of viscosity upon shear rate. In this case, viscosity decreases with increasing shear, so that thixotropy is equivalent to **shear-thinning**. Such a material does not recover its consistency immediately after removing the shear stress.

Thread count. In any **woven** material, the thread count is the number of **warp** or **weft** threads per centimetre along or across the material, respectively.

Three-column centrifuge. A batch-operated **centrifuge**, with a perforated or solid **basket** rotating about a vertical axis. The basket is rigidly mounted in its housing, but the latter is mounted on three vertical spring-dampened legs, to deal with any machine vibrations.

Three phase boundary. A point or line in space where three distinct phases come into contact, either solid/liquid/gas, or solid/liquid/liquid, or liquid/liquid/gas.

Three-phase decanter. A **decanter centrifuge** able to separate suspended solids from a liquid mixture in the normal way, but at the same time able to separate two immiscible liquids, one from the other, as a result of their difference in density, and to discharge them independently, usually by means of **peeling discs**. This machine has applications in waste treatment (for mixtures of oil and water that contain some solids) or in vegetable and animal oil and fat processing.

Three point suspension. A **basket centrifuge**, rotating about a vertical axis, which is suspended from a supporting frame at three points, to minimise instabilities.

Through-flow. The operating mode of a **filter** in which all of the **filtrate** passes through the **filter medium**, as distinct from the **cross-flow** mode.
[*See also Dead-end filtration*]

Through put. A term used to describe the processing capacity of any item of equipment. In **separation** terms it may refer to the total **slurry** quantity processed, or the amount of solids or the quantity of **filtrate**, and it can be stated in volume or mass terms.

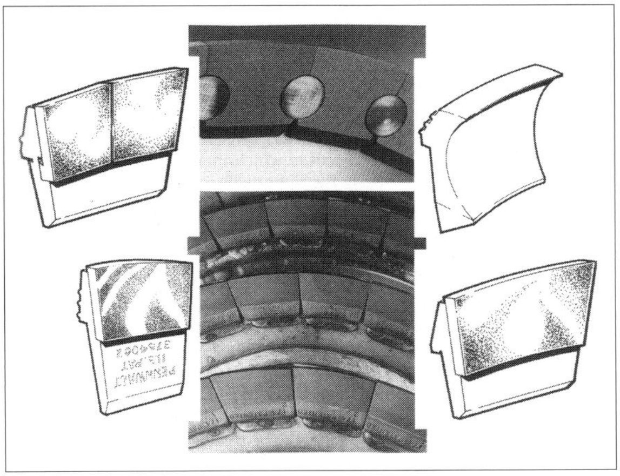

Decanter flight tile designs

When used in **filtration**, it can also be expressed as a specific throughput, per unit area of the filtration surface.

Through washing. The **cake washing** process in some **filter presses**, in which wash liquid is introduced behind one cloth, flows through the full thickness of cake, and flows out through the other cloth.

Tile. In a **decanter centrifuge**, the edge of the **flight** of the **conveyor scroll** that moves the separated solids along the **bowl** wall is thereby subjected to considerable erosion. A set of replaceable tiles is fixed to the surface of the flight to resist the erosion.

Tiller-Shirato theory. A model developed to assist in the understanding of the behaviour of the formation of compressible **filter cakes**, which takes a similar form to **Darcy's law.**
[*See also J factor*]

Tilted-plate clarifier, separator.
[*See Lamella separator*]

Tilting pan filter. A long established form of **vacuum filter** in which a set of suction boxes or pans is arranged in a horizontal circle around a central control valve and **filtrate** offtake. Each pan is mounted on a hollow radial arm, connected to the centre. The set of pans rotates in an **indexing** fashion, so that each pan progressively receives slurry, filters it, dewaters the cake, washes it, dewaters it finally, and then tilts about its own radial arm to empty the cake into the collection zone below the filter. The filter thus operates effectively continuously. (Sometimes called a rotary tilting pan filter, to distinguish it from a tippable nutsche filter.)

Time cycle. A large number of **filters** operate in the semi-continuous mode, where the various phases of the **filtration** process: feed, preliminary **drainage**, filtration, **dewatering**, **washing**, etc, proceed in different locations in the filter. The design of the filter must allow all of these phases to occur, and the overall time cycle is a significant determinant of the economics of the operation of the filter. A major factor is **cake thickness** – a thin cake is quickly formed and easily removed, whereas a thick cake takes longer to form, but proportionally less time is taken up with the unproductive cake release phase.

Tipping pan filter. [*See Tilting pan filter*]

Titus filter. A type of **nutsche filter**, equipped with an internal agitator, and able to invert for **cake release**.

T-line strainer. [*See T-type strainer*]

Top fed filter. A term that can be used of any **filter** in which the **feed slurry** is fed from above onto a horizontal filtering surface, but the term is usually reserved for vacuum filters (e.g. **table** and **belt filters**).
[*See also Top feed drum*]

Top feed drum. A **drum (filter**) in which the feed slurry is fed onto the drum at the top of its rotation, rather than have the drum dip into a trough of slurry.

Torque. If a force acts in the circumference of a circle, then the product of that force and its distance from the centre of the circle is the moment of the force, or its torque. Torque can apply to a driving force, causing rotation, or to the force resisting rotation, and is of particular relevance to the design of driving systems for **centrifuges**.

Torque overload arm. In any drive mechanism, there may be a need to slip the drive if the driven resistance gets too high. This is certainly true with a **decanter centrifuge**, where the **scroll** is driven by a gear box operating off the main drive motor. If the resistance to the rotation of the scroll becomes too high, then the torque overload arm will be released. This is an arm, mounted on the gearbox pinion shaft, which is kept at rest by a torque overload mechanism. The mechanism releases the arm when an overload occurs.

Tortuosity. The degree of deviation of a continuous pore through a porous mass, from the straight-through path, is called its tortuosity. It is the ratio between the distance that liquid has to flow through the pores of a **medium** or **cake** and the thickness of the medium or cake.

Total dissolved solids (TDS). A term used particularly in the analysis of water, especially as a determinant in the re-use of treated waters. It gives the total amount of solids dissolved in the water, and can be expressed as mg/l, or parts per million, or as milliequivalents per litre of cations.

The term is also used occasionally to refer to the total dry solids content of a **sludge**.

Total efficiency. In any separation device, this is the mass of solids in the **underflow**, or retained on the **filter**, divided by the mass in the feed.
[*See also Total separation degree*]

Total flotation. A **flotation** process designed to **clarify** a dilute suspension, in which all the particles attach themselves to bubbles.

Total separation degree. A term used to describe the total amount of **separation** of solids from a feed suspension achieved in an item of separation equipment:

$$\Theta = m_s/m_0$$

where m_s = mass of solids separated from the suspension
m_0 = mass of solids in feed suspension

Tower press. A **filter press** arranged as a vertical stack of plates, each lying horizontally on the one below. The **filter medium** is a continuous belt moving from one plate to the next, and being parted from its accumulated **filter cake** in between plates. The tower press exists in **chamber press** and **diaphragm** chamber forms, and all of the normal **washing** and **dewatering** steps can be carried out in each cycle. Large filtration areas, up to 150 m^2, are possible, and the advantage of the horizontal orientation is that distribution of the suspension over the plates is more even, and so therefore is the **cake thickness**. This, in turn, leads to more efficient **washing** of the cake.

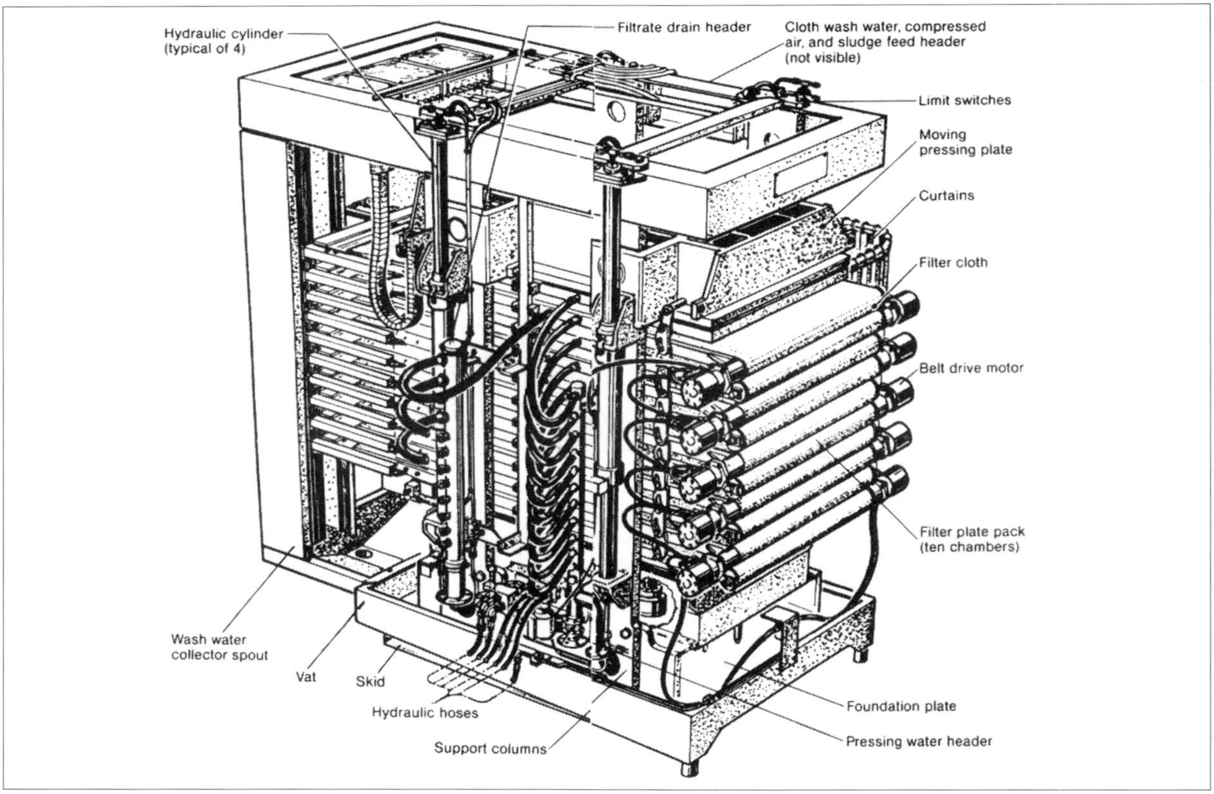

Automatic tower press

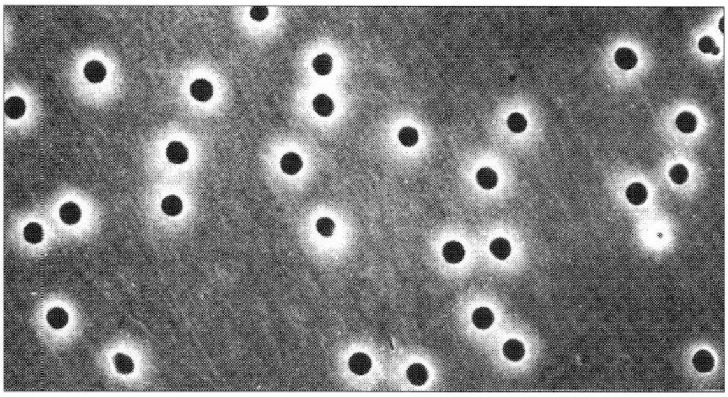

Track etched membrane

That is the most common use of the term, but it is also used for a continuous **belt press** whose tapering section is vertical. The two belts run on parallel sets of rollers, with the distance between the filtering surfaces decreasing down the length of the filter, with the **drainage** zone at the top, followed by the **compression** zone. The Manor tower press belts are sealed at the sides to prevent extrusion of cake. The feed slurry is not usually preconditioned but needs to be fairly firm.

Tracer. A soluble material deliberately added to the feed slurry to simulate an impurity or solute in a **washing** experiment.

Track-etching, Track-etched membrane. A process for the creation of **membranes**, involving irradiation of a polymer film by high energy particles, to create the initial pores, mainly cylindrical in shape, followed by chemical etching to enlarge the pores to the required size.

Tracking belts. In a twin **belt filter**, it is frequently vital that the belts lie accurately in relation to one another. This requires a device at the feed end of the filter to ensure that both belts return to the feed position accurately in register.

Traction thickener. A **gravity thickener** in a cylindrical tank, whose **rake** at the bottom of the tank is in the form of a single arm with one end fixed to a central column and with wheels at the other end which are driven around the circumference of the tank.

Tramp material. Large objects that have found their way into the feed stream of a **separator** accidentally and need to be removed (such as by **straining**) to protect the separator.

Tramp oil. A term referring to the free oil that is separate from an emulsion being used as a machine tool coolant, which may come from oil leakage from machinery, or from breakdown of the emulsifying agents in the coolant. Tramp oil may impair the efficiency of a **cutting oil filter**, and is often removed from the cutting fluid flow by a rotating steel drum, or a belt skimmer.

Transfer roller. [*See Roller discharge*]

Transfer unit. A measure of performance sometimes used of a **clarification** operation, in a similar manner to the transfer unit of the phase change separation processes such as distillation. The number of transfer units equals the natural logarithm of (1 – **total efficiency**).

Transmembrane. Descriptive of the direction across a **membrane** perpendicular to the membrane surface, and moving from the feed/**retentate** side to the **permeate** side.

Transmembrane pressure. The average pressure differential across a **semi-permeable membrane medium** at any point in its operation. This is a measure of the driving force for the process in question, which can be expected to increase as the membrane surface becomes **fouled**. [NB it is not the same as the pressure upstream of the membrane, because the pressure in the **permeate** may also be above the ambient value, nor is it the pressure drop along the length of the filter on the **retentate** side.]

Transmission. A term sometimes used, as an alternative for **penetration**, i.e. for the proportion of a contaminant that passes through a **filter** or a piece of **filter medium**.

Transmission electron microscope. A microscope that uses, instead of a beam of light, a beam of electrons. The electrons that pass through the sample being studied are focused to give an image on a fluorescent screen. Its resolution is about 1000 times better than an optical microscope.

Transporter belt. A term sometimes used for the **rubber belt** of a **rubber belt filter**.

Transtubular. The direction along the axis of any tubular **filter housing**, but refers in particular to a **membrane** system such as a single tube or a **spiral wound module**, moving from the entry point of the feed slurry to the exit of the **retentate**.
[*See also Transmembrane*]

Travelling band. A continuous **clarifier** in which a coarse **screen** in the form of an endless belt moves horizontally between two rollers, with solid collected during its upper traverse, to be washed off when inverted in the lower zone.

Travelling belt filter. [*See Rubber belt filter*]

Travelling pan filter. An obsolete design of **filter** consisting of a series of pans moving in a belt configuration.

Travelling screen, Travelling screen strainer. A coarse **strainer** used at the inlet to raw water systems, such as the intake to a power plant's cooling system. It consists of an endless belt moving vertically through the raw water zone, the belt being made of a series of panels, each having a **filter element** within it plus a collecting bucket – the latter being washed clean of accumulated solids as it reaches the top of the belt's movement. (Sometimes called a vertical band filter or strainer.)

Traversing knife. A device for discharging accumulated solids from a **basket centrifuge**, consisting of a short knife that moves up and down parallel to the basket's axis, and through the **cake**.

Tray. A **filter plate** fitted with a rim for holding **wash liquid**. Also refers to the vacuum boxes fitted below the belt of some **vacuum belt filters**.

Tray belt filter. A continuously operating **vacuum belt filter** in which the vacuum is applied through sets of trays mounted under the **filter medium**, which moves continuously over the tops of the trays.

A traditional form of filter, originated by Dorr Oliver, it offers good possibilities for **dewatering** and **washing** of the **filter cake**.

Tray filter. A pressure **plate filter**, using **trays** instead of **plates**.

Tray thickener. A **gravity thickener**, with a very shallow cone bottom, repeated in several stages or trays, one above the other. Each tray has a radial **rake**, driven by a common central shaft, which drags collected solids to a central **underflow** point, which then becomes the feed point for the tray below it.

Tricanter. A trade name for a **three-phase decanter** as made by Flottweg.

Trickle filter, Trickling filter. [*See Biological filter*]

Triple lead. An uncommon arrangement of a screw **conveyor**, which has three separate and equally spaced **flights** mounted on the **hub**. Is used occasionally in a **decanter centrifuge**.

Trommel, Trommel screen. A **screening** machine, in which the screen surface is made in the form of a cylinder, which is then mounted with its axis inclined to the horizontal, and rotated slowly. The material to be screened, which may be a slurry or a dry mixture, is fed inside the screen at its upper end, and the smaller particles pass through the screen leaving the larger ones to pass out at the bottom. The screen is made of coarse mesh, or perforated plate, or wedge-shaped bars, and may have constant aperture size along its length, or the apertures may be graded, with the smallest at the upper end and the largest at the lower end.

Trommel filter. A less common use of the term trommel refers to a particular type of **rotary drum vacuum filter**, which uses low-grade paper as its **filter medium**. The paper is fed off a large roll onto the drum just ahead of the feed point, or where the drum dips into the slurry, and the paper is taken off the drum when the cake is dewatered, and easily discharged from the paper, which rolls up into a used roll, to be discarded.

Tromp curve. [*See Fractional grade efficiency*]

Trough. A term used for the container for the feed slurry on a bottom fed **vacuum filter**.

True density. A term occasionally used for "**density**", to compare with "**bulk density**". The true density of a porous mass is the ratio of its mass to the volume of solids in it, with all voids and pores excluded.

T (type) strainer. An in-line **strainer** in which the inlet and outlet connections are in the same line, with the strainer **element housing** at right angles to them.
[*See also Y (type) strainer*]

Tube. One of the standard formats of a **filter**, in which the **filter medium** is made as a long cylinder, usually operating with **filtrate** flow from the outside of the tube to its interior. This is one of the formats for **membrane** media, but is also found in **cartridges** made of other media.
[*See also Tubular element*]

Tube press. A variable volume **pressure filter**, operating batchwise, under very high pressure, developed by the then English China Clay Company, for the filtering of china clay suspensions. A vertical cylindrical filter **element** is enclosed by a cylindrical elastomeric sleeve, outside which is a cylindrical pressure vessel. The slurry to be filtered is introduced between filter element and the surrounding sleeve, at relatively low, although increasing, delivery pressures. Once the maximum feed pressure has been reached, depending on the nature of the feed pump, then the feed system is closed and the tube press is isolated. Hydraulic oil is now introduced between the elastomeric sleeve and the pressure vessel outside it, so as to squeeze the filter cake by means of the flexible sleeve.

Squeezing pressures up to 10 MPa or more are used, to **dewater** the **filter cake**, often to a degree of dryness where further thermal drying is unnecessary. The sleeve is then sucked off the cake, and the inner tube is lifted free of its housing, to where the cake can be blown off the filter medium.

Tube settler. A **lamella separator** whose separating means, instead of parallel flat sheets, is made up of inclined tube modules.

Tube sheet. A tube sheet was originally the flat plate at each end of a bundle of heat exchanger tubes into which each tube was sealed. The term is now also used for the flat sheet, mounted horizontally in a **fabric filter**, from which all of the **bags** or **candles** hang.

Tube tests. Simple tests for the rate of **sedimentation** of a test slurry. A sample is placed in a stoppered tube, which is then shaken before being mounted vertical for a period of time while the settlement of the solids is observed – perhaps the thickness of the settled layer against settling time. A **long tube test** uses a tube of length roughly comparable with the height of the settling tank for batch sedimentation, while a **short tube test** is used for testing of continuous sedimentation.

Tubular (bowl) centrifuge. A **sedimenting centrifuge** in the form of a cylindrical tube that is relatively long compared to its diameter (while most other centrifuges have a fairly small length/diameter ratio). It runs at high rotational speeds for the **separation** of immiscible liquids, mainly in the processing of vegetable oils. It has no solids discharge system, so any solids in the feed collect at the bowl wall, and must be removed before they become too large in volume. Such removal requires the centrifuge to be stopped, demounted and washed free of accumulated solids, before being replaced to continue operation.

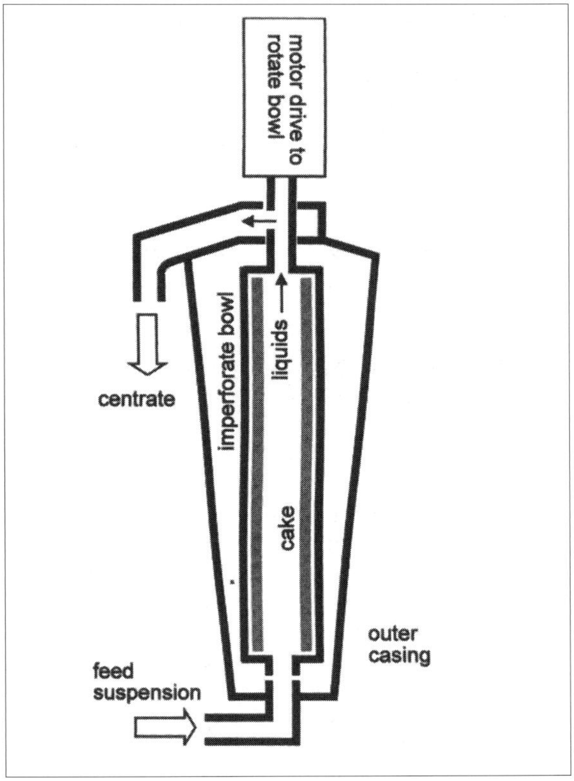

Tubular bowl centrifuge

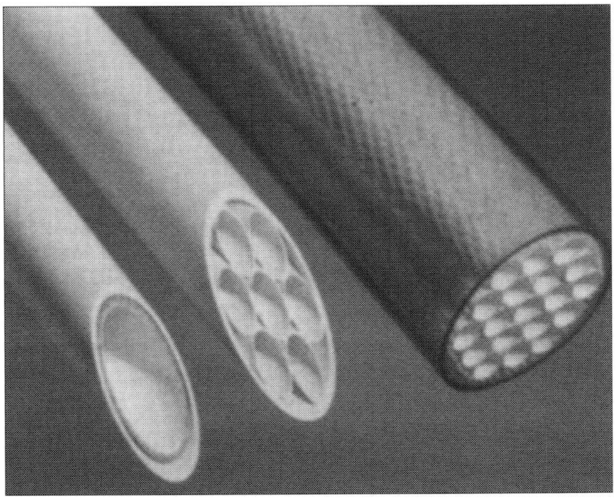
Tubular membrane formats

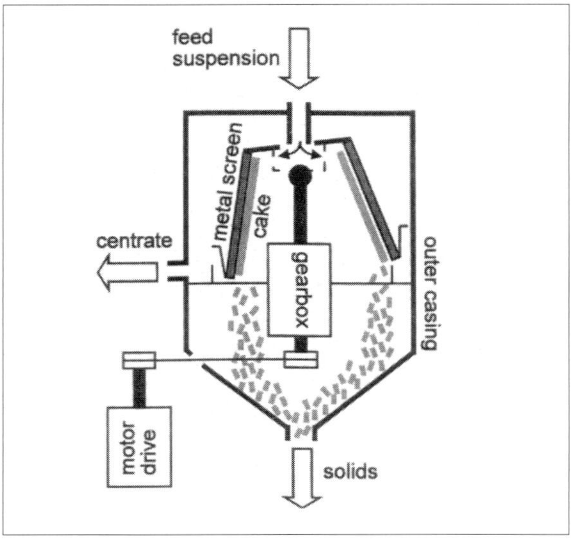

Tumbler cone-screen centrifuge

Tubular element. Any filtering **element** in the form of a long rigid tube, but more specifically used of a wide diameter (in the region of 10 mm), thin walled tube used for **cross-flow filtration**.

Tubular pinch. A term descriptive of the behaviour in a **particle capture** mechanism where the moving particle covers several fluid streamlines, and is thus subject to wandering across the fluid flow prior to capture.

Tumbler centrifuge. A version of the **cone screen centrifuge**, in which the solids are propelled along the length of the screen to the discharge point by a tumbling motion of the screen. The centrifuge is in the form of a cone, with its larger diameter further from the feed end. The slope of the wall of the cone is such that the dewatered solids will not flow across the screen unless the gyratory, tumbling motion is applied. The required motion is achieved by allowing the axis of rotation of the centrifuge to gyrate about 5° about a central position.

Turbidimeter. An instrument used to measure the surface area of a powder from the scattering of a light beam passed through a liquid **suspension** of the powder.
[*See also Nephelometer*]

Turbidity. Cloudiness in a liquid, caused by a **suspension** of very fine **particles**, or of **colloidal** particles. Measured by a number of methods, and usually represented by an index of some kind.
[*See also Formazin turbity index, Jackson turbity unit*]

Turbodrain. A proprietary **gravity belt thickener**, made by Bellmer, for the thickening of thin, **flocculated** slurries, ahead of, for example, a **belt press**. The slurry flows across a continuous porous belt, and plough-share shaped baffles move in the thickening slurry, to open new channels for the **filtrate** to drain through.

Turbo hydrocyclone. A **hydrocyclone** in which the movement of the liquid is created by an internal rotor, rather than by means of a tangential inlet. Typified by the Voith classifier in which the feed **slurry** is introduced through a top axial inlet. The overflow leaves through a **vortex finder** pointing upwards through the apex of the cone. The underflow discharges through an annulus around the vortex finder. This design is used for **clarification** but turbo hydrocyclones can be designed for **thickening** or **classification**.

Turbulence. The non-laminar flow behaviour seen in fluids that are flowing through a tube or open channel at high enough velocities to take the **Reynolds number** above a critical value, and into the turbulent region.

Turbulence promoter. A means of creating turbulence, for example, in a **cross-flow membrane** filter as a means of reducing **fouling**.

Turbulent boundary. A term used in describing the fluid flow near to a **membrane** surface. It is the boundary dividing the fully turbulent fluid flow from the progressively more streamlined flow nearer to the membrane surface.

Turbulent flow. Fluid flow at sufficiently high velocity (and hence high **Reynolds numbers**) such that eddies generated in the fluid are not dissipated, and cause movement of the fluid at angles to the main flow direction – and so cause mixing of the fluid. The flow characteristics are governed mainly by the inertia of the fluid.

Turbulent flow is desired in some **separation** processes – to ensure efficient destruction of concentration polarisation at a **membrane** surface, for example; but it is to be minimised in others – such as to avoid the resuspension of settled solids into the supernatant liquid in a **clarifier**.

Turpinson separator. A **separator** consisting of an array of **hydrocyclones**, mounted in a circle inside an inverted conical shell, with the cyclones at the periphery of the widest part of the cone, each being mounted parallel to the wall of the cone (and hence at an angle to vertical). All the units are fed from a common feed source, and all discharge to the main **underflow** at the narrow part of the cone.

t/V = f(V) method. An experimental procedure used to determine both the resistance to flow of a formed **filter cake**, and also the resistance of the **filter medium** on which it is formed. The test is performed in a laboratory scale **nutsche filter**, through which a well-mixed **suspension** is filtered under constant pressure difference across cake and medium. A log is kept of the cumulative volume of **filtrate**, V, passing through the filter, against the elapsed **filtration time**, t. A graph is then plotted of the ratio t/V versus V, which, for an **incompressible** cake, should be a straight line, of positive slope, showing an intercept at the t/V axis for zero V – this intercept gives the resistance of the empty filter medium:

$$t/V = a + bV.$$

The medium resistance is calculated from the intercept, a:

$$a = R_m \eta / A\Delta p$$

where R_m = filter medium resistance
η = liquid (dynamic) viscosity
A = filtration area
Δp = constant applied pressure drop

while the specific cake resistance, r_c, can be found from the slope, b:

$$b = \kappa \eta r_c / 2A^2 \Delta p$$

where κ = **kappa factor**, a measure of slurry concentration.

Cake resistances have values between $10^{11}m^{-2}$ and $10^{16}m^{-2}$, the lower value applying to easily filtered materials such as coarse crystals or minerals, and the higher to fine slowly filtering materials, such as pigments or bacteria.

The basic equation relating t/V to V is obtained by integrating the general cake formation equation under the conditions of constant pressure filtration.

Twill weave

Twill weave. One of the main types of **weave** pattern for woven fabrics, in which the **weft** thread passes over two (or sometimes more) **warp** threads, and then under the same number, in repetition across the width of the material. On the return pass, the weft repeats this pattern, but displaced one warp thread from the first pass, and so on. The result is to form a strong diagonal pattern on the material, whose angle is determined by the number of warp threads passed at a time by the weft.
[*See also Plain weave, Satin weave*]

Twin belt filter. [*See Belt filter*]

Two-regime model. A model used to explain the behaviour of a **filter cake** in a **centrifugal filter**, in the **filtrate drainage** state, distinguishing between the full flow drainage while the cake is flooded, followed by a slower film drainage process as the cake empties.

Two-phase flow. In general engineering terms, two-phase flow refers to the flow of gas (or vapour)-liquid mixtures through a pipe or processing zone. In the more particular case of **filtration** processes, it refers to the flow of a gas-liquid mixture through the pores of a **filter cake** or **filter medium**. The two phases will then move effectively independently of one another, because the capillary pressure holding the liquid phase in the pores is much stronger than the resistance to the gas flow.

U

ULPA. [*See: Ultra-low penetration air (filter)*]

Ultimate elongation. The maximum increase in length of a material sample just before it ruptures in a rupture test.

Ultimate Oxygen Demand (UOD). The amount of oxygen necessary fully to oxidise all of the organic material in a liquid suspension. Derived by carrying out a **BOD** test over a very long time.

Ultimate pulp density test. A laboratory test undertaken to determine the depth of the **compression** zone required in a **gravity thickener**.

Ultra centrifuge. A very high speed **centrifuge**, usually used for laboratory tests, and operating at **G-forces** of 60,000 or more.

Ultrafiltration. A **membrane filtration process** that covers the range from the lowest end of **microfiltration** to the upper end of **reverse osmosis** (or, increasingly, **nanofiltration**). Ultrafiltration operates in **cross-flow** mode, and is primarily aimed at the retention of large, dissolved or colloidal, molecules, rather than of solid particles. This boundary has, however, become rather blurred, as ultrafiltration membranes are proving to be economically operable at lower pressures and higher **cutoff** points, so as to compete with the lower reaches of the microfiltration spectrum. In concert with its earlier specialisation in molecular separations, UF membranes are often specified in terms of **molecular weight cutoff**, rather than diameters, although the UF range is also thought of as covering 1–100 nm.

Ultraflotation. A **flotation** separation process for **colloids**, which operates in two stages: the colloids are first adsorbed onto larger auxiliary solid particles, which are then removed by flotation.

Ultra-low penetration air (filter) (ULPA). A grade of **air cleaning filter**, giving the finest grades of **separation** of **dust** particles. On the **CEN** classification scale in standard number EN779, the ULPA range covers the three finest classes, U15 to U17.

Ultrasonics, Ultrasonic filter. Ultrasonics is the technology of using ultrasound, with a frequency in excess of 16kHz, for a number of useful purposes. In the case of a **suspension** of solid particles, ultrasound can assist in the **agglomeration** process. In a **filter**, ultrasound can assist in the separation process, and it is suggested that a combination of ultrasound and electric potential can be very effective.

Umbrella discharge. The underflow from a hydrocyclone occurring at low solids concentration and characterised by an umbrella shape.

Underdrainage. Once a **filtrate** has passed through a **filter medium**, it must find its way to the channels by means of which it will leave the **filter**. This means that the medium cannot be supported on an impervious flat surface, but on some kind of open grid or slatted surface, strong enough the support the medium, but open enough to provide the minimum resistance to drainage of the filtrate.

Underflow. The product from a **sedimentation** operation that contains a higher concentration of solids than the **feed**. It usually comes from the base of the **sedimenter**, hence its name.

Undersize. The portion of the particles in a feed slurry that passes through the apertures of a specified **sieve**. [*See also: Fines*]

Uniform film model. A model of **filter cake washing** behaviour that assumes that the initial **drainage** of mother liquor from the pores of the cake leaves all of the pores empty, except for a uniform layer of mother liquor on the pore surfaces.

Uniformity coefficient. A measure of the spread or width of any **particle size distribution** in a granular mass, although it is a term used to specify the sizing of sand particles as used in **deep bed filters**, especially in the treatment of fresh and waste waters. It is defined as the ratio of two sieve aperture sizes, the first allowing 60% of the particles to pass, in a test sieving, and the second that allowing 10% to pass. Typically, sand particles for this application have a size in the range 0.4 to 0.6 mm, at which level the uniformity coefficient should not be greater than 1.7. (Also known as the Hazen uniformity coefficient.)

Universal dam. [*See: Eccentric dam*]

Universal filter. The Gaudfrin Universal **filter** has intermittently rotating vertical **leaves** arranged radially. The leaves are first submerged in feed slurry and gas pressure is used to form a **cake**. As the carousel rotates, the leaf at the end of the feed slurry compartment is lifted out and lowered into a **wash liquid** compartment, which also uses gas pressure to drive the liquid through the cake. The leaf is then lifted out of this compartment to a cake discharge compartment where it can be cleaned. This design has lower utilisation of volume than other pressure leaf filters so is expensive. However, operation is semi-continuous.

Unloading. Refers to the release of separated solids from a **filter** into the downstream flow (**filtrate**), as a result of a change in liquid flow rate, or of mechanical shock or vibration, or because of an excessive pressure build-up, or failure of the **filter medium**.

Upflow/upward flow (sand) filter. A **deep bed filter** for water **filtration**, in which the flow of raw water is upwards through the **filter medium**. The medium is usually stratified, with the coarsest particles at the bottom, so efficient **depth filtration** is achieved. In modern water treatment practice, an upflow filter is followed by a downflow filter to remove the finest particles.

Upper casing. The upper half of a **centrifuge** casing, especially for a **decanter centrifuge**, usually hinged to allow access to the **bowl**, and with a sealed joint between it and the fixed, lower casing.

Utility filter. In any manufacturing process, there are items of equipment that are an essential part of the production process (such as the crystalliser in the manufacture of salt or sugar), and there are those that form part of the services necessary to keep the main process going, but not an essential part of it (such as the boiler house, or the hydraulic oil system). **Filters** that form part of these ancillary systems are termed utility filters, as distinct from the **process filters** in the main equipment line.

V

Vacuum belt filter

Vacuband filter. [*See Miller filter*]

Vacuum belt filter. Any one of a wide range of **belt filters** with vacuum as the **filtration** driving force. A **filter medium**, in the form of an endless belt, is supported on a strong rubber belt below it, to which a **vacuum tray** fastens while the vacuum is connected. These filters are available in a wide variety of sizes, and are very well suited to the filtration of fine, slow filtering solids, where thorough washing is required.

Vacuum box. A chamber in a **vacuum filter** for collecting **filtrate** below a rubber belt or rigid belt filter.
[*See also Vacuum tray*]

Vacuum disc filter. A **disc filter** in which the disc is dipped into the slurry to be filtered. The **filtrate** is sucked through the **filter medium** on either side of the disc by a vacuum drawn on its inside. The accumulated **cake** is scraped off the surface as the disc rotates out of the slurry trough (or the disc may be moved to different stations for washing and cake discharge).

Vacuum drum filter. [*See Drum filter, Rotary vacuum filter*]

Vacuum filter. Any item of **filtration** equipment that operates with a vacuum below (downstream of) the **filter medium** as the driving force to cause the flow of **filtrate** through the medium. Full vacuums are not possible, since the system is limited by the vapour pressure of the liquid being filtered (in the case of aqueous systems, about 0.2 bar absolute). Vacuum filters are almost always used for the filtration of relatively thick slurries, with a view to the recovery of the solids, rather than for **clarification** purposes. They may operate batchwise, or continuously – in the latter case there needs to be a system, often quite complicated, for the removal of the **cake** from the filtration zone to the solids discharge.

For a long period of time, the continuous vacuum filter was the major tool for solid separations in mineral processing, waste treatment and other large volume applications, but they are being increasingly replaced by **pressure filters**, which, though mechanically more complicated still, are capable of much greater throughputs, because of the potentially higher driving forces.

Vacuum flotation. A **dissolved air flotation** process in which the air is dissolved at atmospheric pressure and the flotation cell operated under vacuum.

Vacuum leaf test. A simple test used for measuring the **filterability** of a **slurry**, which involves the immersion of a small **leaf filter element** in a sample of the slurry to be tested, connecting the element to a vacuum source, and observing the manner of **cake formation** on the leaf after regular intervals of time.

Vacuum tray. A chamber in a **vacuum filter** for collecting **filtrate** below a reciprocating tray or intermittent belt filter.
[*See also Vacuum box*]

Vallez filter. A **pressure leaf filter** patented by Vallez in 1916, in which circular filter **leaves** are mounted one above the other on a central hollow shaft, all in a pressure vessel **housing**. Once **cake formation** is complete, the shaft rotates, turning the leaves under spray nozzles that wash the cake off the leaves into the base of the pressure vessel. This was the first design of pressure leaf filter in which the elements could be cleaned without the need to open the filter.

Valve nozzle disc stack centrifuge. A **disc stack centrifuge** in which the collected solids are discharged at the periphery of the **bowl** through **nozzles**, each of which is fitted with a valve that is opened intermittently, either under manual control, or automatically, triggered by the depth of the solids layer in the bowl.

Van der Waals forces. Weak bonding forces between **particles** that are not the result of unbalanced electric charges on the particle surface, but are produced by interactions between the oscillating dipoles within the solid structure. They only come into play once the stronger electric charge forces have been neutralised and electrostatic repulsion has been eliminated. (Also known as London forces.)

Vane. A longitudinal plate, running in the axial direction, fixed to the **conveyor hub** of a **decanter centrifuge**, to support the conveyor **flights**. Used to achieve full **axial flow**, in the hubless conveyor, and for **Sigma enhancement**.

Vanner. An endless **belt** that slopes away to each side, used for gravity **concentration**.

Vapour. Although specifically referring to any gas at a temperature below its critical value, it is mostly used of a gas that is condensable under ambient conditions – so forming a liquid that might interfere with a subsequent process.

Vapour permeation. A **membrane separation** process in which a vapour is separated either from a non-condensible gas or from a mixture of vapours. The membrane is non-porous and movement of the vapour is by diffusion of its molecules through the material of the membrane.
[*See also Gas permeation, Pervaportaion*]

Variable chamber filter. [*See VC Filter*]

Variable chamber press. A **pressure filter** in which the cake space can be reduced mechanically to **compress** the **cake** more strongly than the feed pressure alone can achieve. This is usually done by means of flexible diaphragms that can be squeezed onto the cake, either mechanically or hydraulically.
[*See also Diaphragm press, Tube press, VC Filter*]

Variable pressure filtration.
[See *Constant rate filtration*]

Variable rate and pressure filtration. The behaviour of a **pressure filtration** system that is fed by a pump other than a positive displacement one. The delivery pressure slowly increases as the **cake thickness** increases, and the **filtrate** rate decreases at the same time.

Variable volume filter. A term sometimes used for a batch **compression filter**.

Varnish. A term used in hydraulic systems for the breakdown products that form a brown deposit on metallic surfaces. They are generated by oxidation of the hydraulic fluid, or by its thermal breakdown, and they deposit on the **media** of **filters** in the system, thereby causing a blockage.

VC (Variable Chamber) Filter. A horizontally mounted, batch operated **pressure tube filter** with a flexible impervious **diaphragm** surrounding the **filtration** zone. Once the batch is filtered, the diaphragm is hydraulically squeezed on its outside so that it squeezes the **filter cake** in turn, and thus **dewaters** the cake quite effectively. Developed and used in the dyestuffs industry.

V Disc press. A **compression** machine consisting of two large diameter rotating discs, mounted at a slight

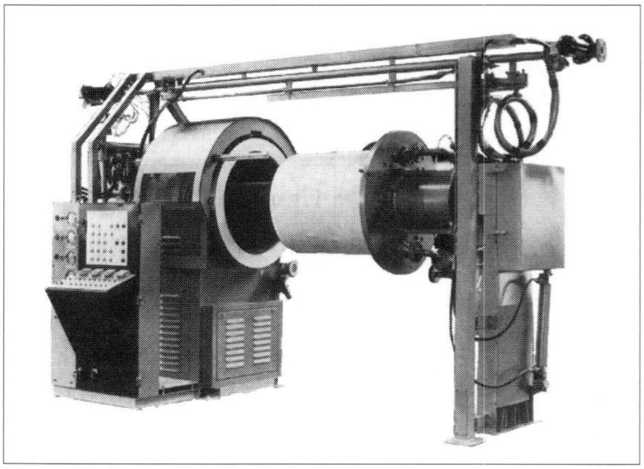

Variable chamber filter

angle to one another, so that the gap between them reduces from the top of the discs to the bottom. The material to be compressed is fed in at the open end at the top of the gap and can be compressed by a factor of four as the discs rotate. The expressed liquid leaves at the bottom, and the compressed solids are scraped out of the gap towards the top of the completed rotation.

Veco screens. A range of **screens**, made by Stork, with a very precise aperture, both in size and shape, made by electroforming of thin sheets of metal. They are an increasingly important form of **filter media** in **centrifugal filters**.

Velocity head. The kinetic pressure in a fluid created by virtue of its movement. If the fluid is incompressible, then the velocity head is the difference between the total and static pressures. The velocity head:

$$p_v = \tfrac{1}{2}\rho v^2$$

where ρ = fluid density
v = fluid velocity

Ventilation filter. Any **filter** used in a room ventilation scheme, almost always operating by **depth filtration**. They are classified by the **CEN/EUROVENT** system.

Venturi scrubber. A **wet scrubber** for dirty gases, in which the gas passes at high velocity through a constriction in the pipe, and is there contacted with a liquid spray. The high relative velocity between dust particles and spray droplets results in a high impaction efficiency, and the particles are thus trapped in the droplet, thereby increasing in size, and so become much more easily separated in the following wet **collector**. A high pressure drop across the venturi is necessary for the removal of sub-micrometre particles.

Vertical leaf filter. A **tank filter**, operating under vacuum or pressure, in which the **leaves** are vertical, **cake** forms on both sides and is discharged by a variety of mechanisms, either wet or dry. Versions exist with the **shell** either vertical or horizontal.

Vertical pendulum centrifuge. A **centrifugal filter** in the form of a perforated **basket**, with its axis of rotation vertical, and suspended on its drive shaft, with drive and gears above the basket. Any tendency to vibrate is diminished by the suspension. The **centrifuge** operates batchwise, with solids discharge through the open bottom of the basket, usually effected by means of a **peeler knife**. Its main application is in the dewatering of sugar crystals.

Vertical pressure filter. [*See Tower press*]

Vertical travelling screen. [*See Travelling screen*]

Vibrating centrifuge. A continuous **centrifugal filter** in the form of a horizontally mounted **screen**, the latter being in the form of a cone cantilevered at its smaller diameter, with dewatered solids being discharged over the wider end of the cone. The angle of the cone is such that the component of the **centrifugal force** parallel to the surface of the cone is insufficient to move the solids along the screen (thus allowing a good residence time for dewatering). The screen in addition receives a torsional vibration (reciprocation in the direction of rotation), which causes the solids to move over the screen towards the discharge end, at a speed controlled by the frequency and amplitude of the imposed vibration. Versions also exist in which the vibration is parallel to the axis of rotation.

This machine is well suited to easily dewatered suspensions, and it is able to operate at high flow rates, although it does need a fairly thick feed concentration to avoid **wash out**.
[*See also Cone screen centrifuge*]

Vibrating screen. A rectangular **screen** or **sieve**, mounted effectively flat, but with its long side at a slight angle to the horizontal. The wet or dry feed mixture enters at the upper end and moves down along the screen because the screen is mechanically vibrated, with the reciprocating action working parallel

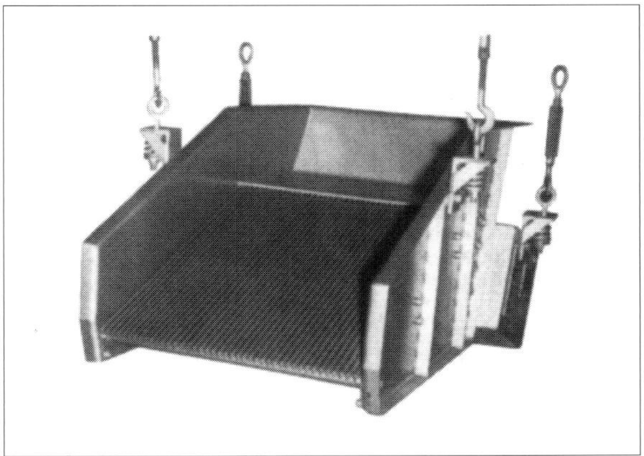

Vibrating screen

to the long side. This form of **separator** is much used in the mineral processing industry, especially for sand and aggregate grading, or for the **dewatering** of such materials.
[*See also Gyratory screen*]

Vibration absorption plate. A massive plate of steel or concrete used to give weight to a piece of equipment and hence reduce the amplitude of any vibrations generated by the equipment. Such a plate is often used in a **centrifuge** assembly (such as a **pusher**), and is then mounted on **vibration dampers**.

Vibration damper. Strong springs mounted between the base of a piece of equipment prone to vibration and the foundations of that equipment, to protect them from the effects of the vibration. Especially are these necessary for most **centrifuge** mountings, because uneven **cake** formation can cause quite severe out-of-balance forces and hence machine vibration.

Vibration isolator. [*See Anti-vibration mount*]

Virus. An infectious agent that can live and multiply only in the cells of animals, plants or bacteria. It is extremely small (ranging in size from 20 to 400 nm), and has a very simple structure. Its small size makes it very difficult to remove, say from water suspension, and the makers of **microfiltration** and **ultra-filtration** equipment continue to face a significant challenge to provide separation processes that will remove virus particles.

Viscosity. The measure of the ability of a liquid to flow freely, reported either as **dynamic viscosity** (or just, viscosity) or **kinematic viscosity**.

Viscosity ratio. A term sometimes used for the ratio of the **viscosity** of the **mother liquor** in a **filtration** process to that of the **wash liquid**.

Viscous drag. A force acting on a particle moving through a fluid, due to skin friction – the friction between the solid and the fluid.

Viscous filter. An **air conditioning filter** which has a coarse **filter medium** (**wire mesh** or coarse fibre) coated in a viscous liquid, such as an oil, which greatly assists in the trapping of dust particles in the air. Those media made of metal can then be washed clean with steam or detergents, so as to be re-usable.

Viscous flow. The flow regime in which the flow characteristics are governed mainly by the **viscosity** of the fluid. Also known as laminar or streamline flow.

VOC. A commonly used abbreviation for Volatile Organic Compound(s), used in the measurement of working and atmospheric air quality, and in the control of emissions to the atmosphere. It is of growing concern in industrial operations, and one means for its control (once elimination and/or minimisation techniques have been fully employed) is the **combination filter**, providing gas phase separations as well as gas/solid – using **activated carbon** media, for example.

Void. Any empty space within a solid mass, which may be connected to other voids, or may be completely surrounded by solid material.

Voidage. The empty space in a mass of porous solid. It is made up of pores of all shapes and sizes, some of which provide complete flow channels through the mass, and others of which are closed pockets of space.
[*See also Porosity, Specific volume*]

Voidage fraction. The ratio of the volume of the voids in a porous mass to the total volume of the mass (voids plus solids):

$$e_f = V_v/(V_v + V_s)$$

[*See also Porosity, Void ratio*]

Voidage function. A term developed by **Richardson and Zaki** in a study of **hindered settling** of non-**flocculated** systems. They showed that the settling rate in the suspension is related to the **terminal velocity** multiplied by the **porosity** raised to a power dependent upon the **particle Reynolds number**. This power factor they termed the voidage function.

Void ratio. A measure of the porosity of a cake or any mass of porous solid, given by the volume of voids divided by volume of solids:

$$e = V_v/V_s$$

[*See also Porosity, Voidage fraction*]

Void volume. A measure of the actual volume of voids in a porous mass, given by the total volume of the mass (i.e. area times thickness) multiplied by its **porosity**.
[*See also Voidage fraction*]

Voigt element. In the study of **expression** of moisture from a **filter cake** by **compression**, the simple equations derive from **Darcy's law**. These are inadequate to describe the later stages of consolidation, for which the concept of secondary consolidation, caused by creep effects in the cake, was introduced. The Voigt element is a component of this quite complex model.
[*See also Kelvin Voigt component*]

Volatility. A measure of the likelihood that a liquid will vaporise at given conditions of temperature and pressure. The more volatile the liquid (i.e. the higher its vapour pressure), the more easily it is vaporised.

Volume concentration.
[*See Suspension concentration*]

Volume diameter. A means of representing the size of an irregular particle. It is the diameter of the sphere having exactly the same volume as the particle in question.

Volume distribution. The relative frequency distribution within a mass of particles, as expressed by particle volume.

Volume median diameter. The particle diameter in a mixture of particles below which there is 50% by volume of the particles in the mixture.

Volume reduction. A term used in **cross-flow membrane filtration** for the ratio between the flow of feed and the flow of **retentate**.

Volume shape coefficient. The ratio between the volume of a particle and the cube of its measured diameter.

Volume split. A term used by Trawinski for the ratio between **overflow** (or **underflow**) rate and **feed-rate**. More often just referred to as the overflow/feed ratio.

Volumetric flow rate. The fluid flow rate expressed in terms of a volume flowing through a pipe, or space, in unit time.

Vor Siv. An inverted conical **wedge wire screen** with the flow following a spiral path from the broadest part of the screen to its narrowest. The feed slurry is projected tangential to the inner periphery of the top of the cone, which gives it the spiral motion. Fines and liquid pass through the screen, while the coarse solids discharge through the apex.

Vortex. A whirling flow of fluid in a cylindrical container, the flow being induced by the velocity of the fluid as it enters the container. This is the basis of the separation process achieved in a **cyclone** or **hydrocyclone**.

Vortex cleaner. A simple **cyclone separator**, also known as the straight-through cyclone.

Vortex finder. The outlet pipe from a **cyclone** or **hydrocyclone**, mounted at the top of the unit, adjacent to the entry pipe. It picks up the **secondary vortex** from inside the body of the cyclone, and is the means of exit from the cyclone of the main flow of fluid, now freed of most of its suspended solids. A residual quantity of fine solids will pass out with this flow, whose quantity is dictated by the diameter of the vortex finder.

V Tank. A long rectangular tank with a shallow V-shaped bottom, used for **classification**. There may be a series of discharge points for the **underflow**. The **overflow** leaves from the opposite end of the tank to the feed. They can handle larger flows than a **settling cone** but the classification is not so sharp.

W

Wadell factor. A **shape factor** used to describe the irregularity in shape of a **particle** compared with a sphere. A perfect sphere has the lowest ratio of surface area to volume of any shape of particle, with the same volume. The Wadell factor is the ratio of the surface area of a sphere of the same volume as the particle, to the actual surface area of the particle.
[*See also Sphericity*]

Wall friction test. A means of measuring the ability of a filter **cake** to flow across a surface. The static friction can be measured by placing the cake on a pivotted plate of the same kind of material as the wall to be used, and then tilting the plate until the cake begins to slide. With care the plate can then be swung back until the cake ceases to slide. This gives the dynamic wall friction.

Wall shear rate. A component of the film theory for modelling **membrane separation** processes. It is used in an equation for calculating flux rate through the membrane.

Walton test filter. A laboratory test **filter** used for designing **pressure leaf** filters.

Warp. The name given to the direction on a weaving loom matching the lengthening of the woven material, and also to the threads (**yarn** or **filament**) that run the length of the loom. (For materials that are not woven, but which come from a machine in a similar way, the term "**machine direction**" may be used.)
[*See also Weave, Weft*]

Wash belt, Wash blanket. An open weave or absorbent belt, laid in contact with a **filter cake**, usually in a **vacuum filter**, over the whole of the washing zone, and perhaps some of the cake dewatering zone as well. Its purpose is to give a better distribution of the wash liquor than is achieved with spray nozzles alone. The belt is normally continuous, running over a set of rollers, in the case of a **drum filter** leaving the cake surface before the cake discharge point, and returning just after the drum has left the slurry trough.

Sometimes a relatively short static belt is used, which is then referred to as a drag belt.

Wash box. A wash box is a trough into which the **filter medium** from a **rotary drum belt discharge filter** is cleaned by high pressure water jets, prior to its being taken back to the drum.

However, the term also refers to a trough extending across the width of a **drum filter**, with a weir on one side, over which the wash liquor flows, to give a uniform wash flow over the whole width.

Wash column. A vertical tower column used for **washing** or **leaching** granular solids, in which the solids and wash liquid flow in countercurrent flow (liquid up and solids down).

Wash curve. A graphical plot of a washing parameter against **wash ratio**. The cumulative wash curve is the plot of the fraction of **solute** left in the **cake** against wash ratio, while the instantaneous wash curve is the plot of the concentration of solute in the **wash filtrate**.

Wash degree. A term used to describe the amount of solute in a **mother liquor** that is to be washed out of a **filter cake**. It has values of 1 before any washing occurs, and 0 when washing has achieved total removal of the unwanted material.

Wash efficiency. The fraction of **mother liquor** removed after the passage of 1 void volume of **wash liquid** during a **displacement washing** of a **filter cake**.

Wash filtrate. The **filtrate** collected from the **filter** during the **cake washing** stage.

Washing. In **filtration** or **centrifugation** terms, washing is a process that follows filtration and **cake formation**, when it is necessary to displace the residual moisture in the cake with another liquid, usually pure water. Washing is most often necessary when the **suspension** to be separated has come from a crystallisation or precipitation process, and the resultant **mother liquor** contains impurities that need removal from the separated solids. The washing can be achieved by spraying the cake with wash liquor in place on the **filter medium** (**displacement washing**), or by **reslurry washing** after removal of the cake to a separate vessel.

The word is also used of the cleaning of a continuous filter medium, after the removal of an accumulated filter cake, to remove residual solids, before the medium returns to the filtration zone, as may be employed in a drum or belt filter.

A third meaning relates to the improvement of a **classification** process by using an injected liquid to separate small particles from coarser particles.

Washing liquid, Wash liquid. The relatively clean liquid used to wash a **filter cake**, once **filtration** and **drainage** of the **mother liquor** are complete. This liquid is most often water.
[*See also Cake washing, Washing*]

Washing pipe, Washing spray. Devices that are used to apply washing liquids to the surface of a **filter cake** in a **filter** or **centrifuge**. A washing pipe will usually have a single spray nozzle at its end, but may stretch across the cake with several nozzles.

Washing ratio, Wash ratio. The ratio of the volume of **wash liquid** required to achieve adequate washing, to the volume of the pores in the **filter cake**. Ideally the ratio should be close to unity, but in practice may be much larger. Good filter design should keep the ratio as low as possible, in order to reduce the quantity of washing effluent, unless the latter can be recycled.

Wash out. A term used of unwanted behaviour in a range of continuous **centrifugal filters**, where the feed is too low in **concentration**, and so flows quickly across the separating **screen** of the centrifuge and leaves the centrifuge without being dewatered. Wash out is avoided by pre-**thickening** of the feed suspension to a minimum value of the concentration, dependent upon the type of centrifuge.

Wash out is also possible in a **decanter centrifuge**, when the unseparated feed leaves from the cake discharge ports.
[*See also Basket centrifuge, Thickening*]

Wash plate. The plate in a **filter press** through which the **wash liquid** enters during **through washing** of the accumulated **cakes**.

Wash time. The time taken for the **washing** of a **filter cake** to the required level of freedom from the **mother liquor**. This time must be allowed for in the **filtration** cycle, and may be critical in **centrifugal filters**, where there is little time available

Water. In volume terms, water is by far the most common liquid found in **separation** processes. This is so whether the water is the basis for the **mother liquor** carrying valuable solids to be recovered, or whether the water itself is the objective of the treatment process. The treatment of raw water is vital to human life, and much of the required purification is done by **filtration** or **sedimentation**.

Water filter. Any **filter** whose purpose is the cleaning of water. The largest volume of such treatment is done in **deep bed filters**, although many other types of **clarifying filter** are used.

Water flux. The flux of clean water through a clean **membrane**. Also known as pure water flux.
[*See also Water value*]

Water injected hydrocyclone. A normal hydrocyclone with water injection into the flow in order to improve the sharpness of **classification**, although the cut size usually increases because the injected water disturbs the flow patterns. Injection can be into the cone or the cone can be divided by a short cylindrical section into which the extra water is injected. Some performance improvement can be achieved by retrofitting water injection into an existing hydrocyclone, so long as this does not disturb the flow patterns too much.

Water recovery factor. A term used in **cross-flow filtration**, defined as the fraction of the feed stream recovered in the **permeate**.

Water value. A term sometimes used to indicate the **permeability** of a fresh **filter medium**, as tested with a pure water flow, and is usually enumerated as flow per unit area. It can be used to compare the

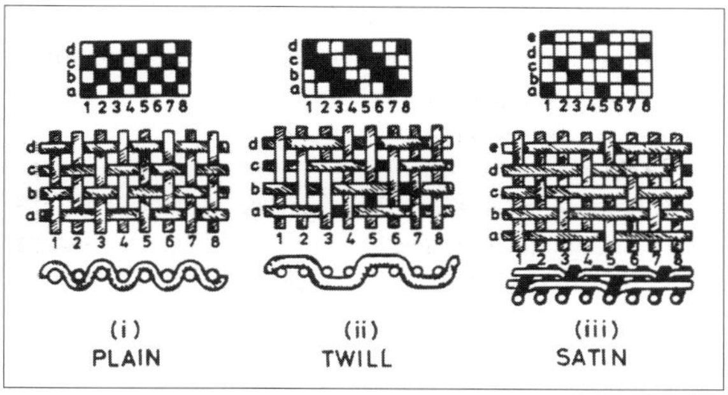

Types of weave

permeabilities of a range of filter media, but has nothing to say about the behaviour of the medium when filtering a **suspension**.
[*See also Water flux*]

Wear insert. Any fabricated material that is inserted into another to combat erosion. Especially used in the **decanter centrifuge** for the protective material in the **feed zone**, around the cake discharge holes and other in areas prone to wear.

Wear plate, A part of the complicated rotary valve at the heart of a **rotary drum** or **disc filter**, which is used to connect each **filter cell** in turn to the main **filtrate** offtake pipe. It is a thin replaceable disc of suitable material, sitting between the rotating components of the valve, and its stationary body.

Weave. The pattern of **warp** and **weft** threads in a **woven** fabric, the three main weaves being **plain**, **twill** and **satin**. The weave of a **filter medium** is an important characteristic in determining its **filtration** efficiency, its tendency to blind, and even its operating life.

Web. A term used in the fabrication of **nonwoven** materials, for a thin layer of **fibres**, laid down at random, and usually in the dry state. If the web is carded, then there will be a preferential orientation for the fibres. Several such layers, piled one on top of the other (normally with the orientations of successive layers at right angles), make up a **felt**.

Wedge wire. A relatively inflexible rod-like wire, made with four flat sides, two of which are parallel, one wider than the other, these two sides being considerably shorter than the other two sides. These can then be assembled into a **screen**, by fixing them to backing rods, with the wider faces adjacent to one another, to give a very precise screening surface, whose apertures expand away from the surface of the screen, so minimising blockage of the apertures. A screen of wedge wire bars is the basis for the **filter medium** of most **centrifugal filters**, and for a number of **straining filters**.

Wedge wire filter. A coarse **clarifying filter**, featuring a cylindrical filter **element** made from wedge wire bars fixed together to give long narrow apertures, and with filtration flow through the screen from outside in, so that any separated solids collect on the outside of the element. The screen can easily be cleaned by use of a scraper or brush, running up and down the element, frequently as part of a **self-cleaning filter**.

Wedge wire sieve. [*See Sieve bend*]

Wedge zone. The part of a **double belt press** where the **filter cake** is compressed (and hence its thickness diminishes, giving a shallow wedge shape to the cake as it moves through the filter). The wedge zone follows the **straining zone**, in which the cake is dewatered by gravity, before the pressure is applied by the upper belt. The zone is created by having several pairs of rollers placed closer and closer together.

Weft. In a **woven** material, the weft threads (**yarn** or **filament**) are those that run across the loom, interweaving with the **warp** threads, and usually at right angles to them. The set of warp threads are laid onto the loom, and the weft is then interlaced with one or more warp threads to give the pattern of the required **weave**.
[*See also Cross machine direction*]

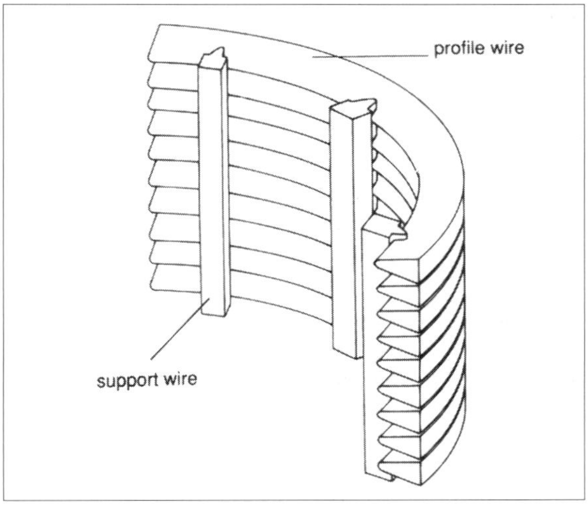

Wedge wire screen

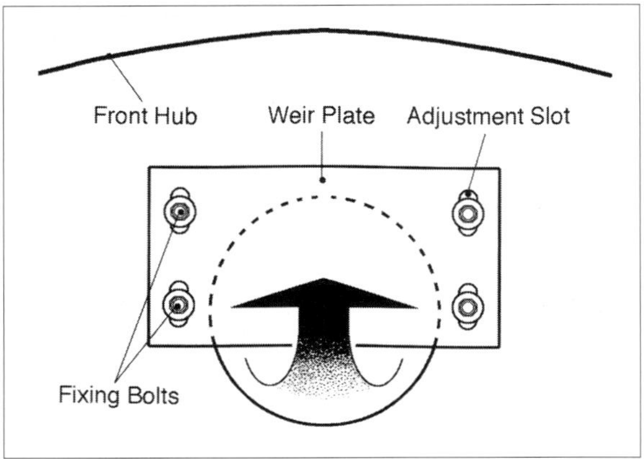

Simple centrate weir

Weir, Weir disc, Weir plate. A weir is a control device in the flow path of any liquid, whereby the liquid flows over a raised lip across the flow. In **separation** terms, the weir is the annular disc mounted at the liquid discharge end (**front hub**) of a **decanter centrifuge**, which controls the overflow of clarified liquid, and, by its height (above the centrifuge **bowl wall**) determines the depth of the **pond** in the decanter. It may have a fixed height, or be adjustable in height, or be serrated in a way that allows faster discharge the deeper the pond, and it is usually replaceable.

A weir can also be the perimeter wall of a **gravity sedimentation** system, over which the clarified liquid flows into the discharge channel.

Weir box. A device for the controlled distribution of feed slurry or wash liquid over part or all of the **filtration** area, in which the liquid flows over a weir.

Well screen. A coarse **strainer** fitted over the bottom end of a pipe sunk into the earth for the extraction of oil, gas or water. It is intended to prevent the escape into the well pipe of any large particles of the surrounding rock.

Wet beach. That part of the **beach** in a **decanter centrifuge** that is below the **pond** level.

Wet bursting strength. The pressure at which a standard disc of pre-wetted **paper** medium ruptures when subject to compressed air or hydraulic pressure.

Wet collector, Wet scrubber. A device for washing a gas clean of suspended dust, usually in the form of a counter-current washer, with air flowing upwards and a flow of water falling down through it. It can be in the form of a spray tower, with the liquid in the form of atomised droplets, or the liquid can flow over various types of solids packing.
[*See also Venturi scrubber*]

Wet discharge. The method of cake discharge that uses a liquid to wash the cake off the **filter medium** (by **backflushing** or **sluicing**) and so produces a slurry product. This would then lead naturally into **reslurry washing**.

Wet-laid. A **nonwoven** material that is made by preparing a well-mixed slurry of the appropriate (relatively short) fibres, usually suspended in water, and then spreading this over a wire mesh belt (or a belt of any other coarsely porous material), so that the suspending liquid drains rapidly through the belt to leave a damp **web** of fibres, which is subsequently passed through a set of heated rolls to dry it and/or prepare its surfaces. The most common form is **paper**, made from cellulose fibres (or, increasingly, glass fibres for **filter media**), but plastic and metal fibre materials can also be made in this way.
[*See also Dry laid*]

Wet strength. The tensile strength of a fabric or paper when it is completely wet. Additives may be used to strengthen the wet material.
[*See also Wet bursting strength*]

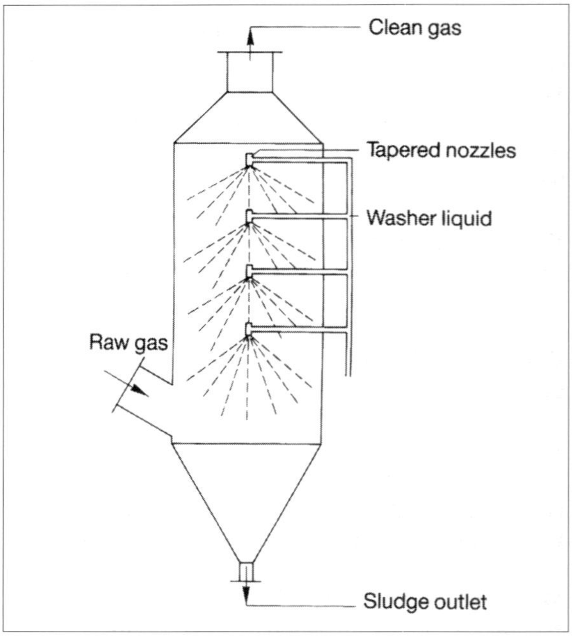

Wet gas scrubber

Wettability. The characteristic of a material that describes its ability to be wetted by a liquid (usually water, although the principles apply to any other liquid). This means that a drop of liquid placed on the surface of the material will spread across that surface if it is easily wetted, or remain as a droplet, without spreading, if it is not. This behaviour is particularly relevant to the ease, or otherwise, of passage of a liquid through a **filter medium** or **filter cake**, especially with respect to flow through **membranes** – the more easily wetted the substance, the easier the liquid will flow through a porous mass of that solid.
[*See also Contact angle, Hydrophilic/Hydrophobic, Surface tension*]

Wetting. The spreading of a liquid over the surface of a material with which it comes into contact. If the liquid does this spontaneously, it is said to be a completely wetting (or spreading) liquid, and its contact angle will be zero. Glass, water and air form such a system, while the glass/mercury/air system is the opposite. Wetting has a strong influence on **capillary** pressures in pores, and hence on the **dewatering** of filter **cakes**.
[*See also Wettability*]

Wetting agent. [*See Surfactant*]

Wetting angle. [*See Contact angle, Wetting*]

Whirlsizer. An open topped **hydrocyclone**, with water injection, used for **classification**. Feed slurry enters tangentially under gravity or through a pump. **Wash water** is injected tangentially at the base of the cylindrical section. Fines discharge over a weir at the top while the coarse material discharges axially at the bottom.

Wide angle cone screen centrifuge. A term sometimes used for the **slip discharge cone screen centrifuge**.

Wilhelm and Naide method. A design method for **gravity thickeners**.

Wilkinson's equation. A design equation for a liquid in laminar flow, defining the liquid flow rate. It is the equivalent for **non-Newtonian liquids** to the **Hagen-Poiseuille equation** for **Newtonian fluids**.

Willus and Fitch method. A laboratory test used for estimating the torque on a **thickener rake**.

Wimmer centrifuge. A **sedimenting centrifuge** in the form of a solid **bowl**, rotating about a horizontal axis. Accumulated solids are discharged cyclically, by drawing back the bowl to allow the solids to flow into the casing. The pipe-cone version (RKZ) has a mainly cylindrical bowl, while the earlier double cone version (DKZ) is in the shape of two facing cones.

Windage. The flow of air induced in the **casing** of a **centrifuge** by the rotation of the centrifuge **bowl**. This air flow may cause some of the **centrate**, or even the discharging solid, to be deflected from its proper path, and so the casing is provided with **baffles** to reduce the air flow and minimise such movement.

Windage and friction. The term used for that component of the power input to a **centrifuge**, especially a **decanter centrifuge**, arising from **windage** and the various frictional forces in the system.

Window. A hole in the **conveyor flight** of a **decanter centrifuge**, which allows the clarified liquor to flow axially (otherwise, it would flow helically, between the flights).

Wing-shear strength. A characteristic of thick **sludges** intended for disposal to landfill. It is determined by immersing a wing-shaped probe into a mass of sludge, and measuring the torque necessary to rotate the probe. The wing-shear strength depends upon the nature of the solids and their moisture content.

Wire. A continuous **filament** of metal, made by **extrusion**, available in a wide variety of diameters (and also of cross-sectional shapes), from a few nanometers up to a few millimetres. Wire is normally considered to be flexible, whereas metal **rod**, distinguishable only by its larger diameter, is not easily bent.

Wire mesh. Material made by the weaving of **wire**, usually with **warp** and **weft** as a single **filament**, into a mesh, most often characterised by a high degree of precision in the **apertures** between the wires. Wire mesh is the basis of a wide range of **filters**, especially the **sieves** used for particle size characterisation. It may be made into continuous **belts** by the careful welding of the ends of a strip of mesh. It is frequently used, as a **filter medium**, as a set of meshes, one immediately above the other, one

Woven wire mesh

or more to provide the necessary separation, the other(s) to provide mechanical strength to the whole medium. Wire meshes can give a very wide range of cut points, down to 5 μm. (Fine wire meshes are often referred to as wire cloth.)

The individual wires may be **crimped** to stabilise the mesh at the points where one wire crosses the other, thus creating a very stable and accurate separation medium. Stability can also be conferred by the **sintering** of the mesh, either at the contact points of a single layer, or as a multi-layer mesh.

In making a wire mesh, a balance must be struck between using a wire fine enough to give the required degree of separation, yet thick enough to give the mesh adequate strength. **Dutch weaves** are often used to achieve this optimum behaviour.
[*See also Bolting cloth*]

Wire wound. In addition to its use as a **filament** in woven meshes, **wire** can be wound round a core, usually in spiral fashion, to provide a filter **element**.

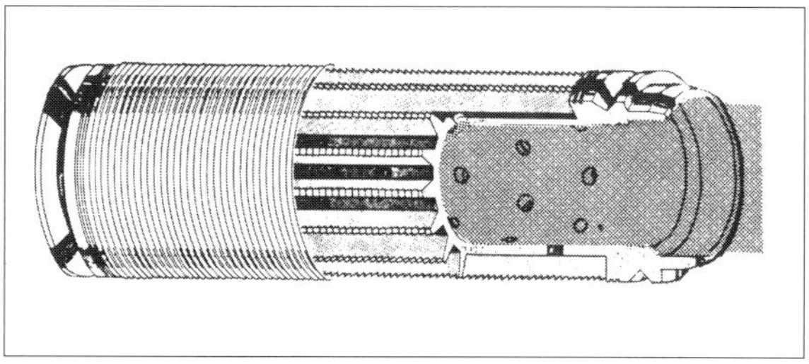

Wire wound cartridge element

The wire often sits in grooves cut in the core, so as to give a regular aperture between successive turns of the wire. The wire may be circular in cross section, or of a wedge shape (to give a flat outer surface).
[*See also Yarn wound*]

Wood flour. Finely ground wood that is used as a **precoat** or **filter aid** material. It is a cheap material, but not so useful as other filter aids.

Woof. A term used occasionally in place of **weft** in a weaving process.

Wool. Natural fibres from animal coats, mainly those of sheep. Fine wool fibre ranges from 4 to 7.5 cm in length, while coarse fibre can be up to 35 cm.

Work-place filtration. An increasingly important application for **filtration**, intended both to provide workers with healthy breathing air, and to prevent contamination of that air by leakage from within a process environment.

Worm extruder. [*See Screw press*]

Worm screen centrifuge. A **cone screen centrifugal filter**, in the form of a **basket centrifuge** with a horizontal axis of rotation, cantilevered from its drive at one end. The basket is in the shape of a cone, expanding in diameter from the driven end, where the slurry is fed. The angle of the cone is too small to allow the dewatered solids to slide across the screen to be discharged, and solid movement is effected by means of a worm screw inside the basket, rotating at a speed differential to the basket. In common with **pusher centrifuges** and other versions of self transporting centrifuges, the feed needs to be adequately thick enough to prevent **wash out**. Best suited to very easily filtered and drained solids, because washing is not easy

Wound cartridge. [*See Yarn wound*]

Wound module, Wrapped module.
[*See Spiral module, Yarn wound*]

Woven. Any material made on a loom from **warp** and **weft** threads, normally crossing at right angles, and characterized by a particular **weave**.

Woven wire, Woven wire cloth. [*See Wire mesh*]

Fabric woven from staple fibre

Wyckoff & Botset equation. A relationship describing the behaviour of fluids flowing through a wet **filter cake**. Both liquid and gas phase **permeabilities**, in a gas pressure dewatering process, are dependent upon the degree of **saturation** of the cake. The gas phase permeability, as a function of the saturation, is related to the single phase permeability by a relative gas permeability, which ranges from 0 to 100%:

$$p_g(S) = p_c p_{rg}(S)$$

where p = permeability of gas (g) and single phase (c)
S = degree of saturation
p_{rg} = relative gas permeability.

Similarly for the liquid phase:

$$p_l(S) = p_c p_{rl}(S)$$

where $p_{rl}(S)$ is described by the **Brutsaert equation**.

X

Xi index. A measure of the **permeability** of **precoats**, similar to a **beta index**.

Y

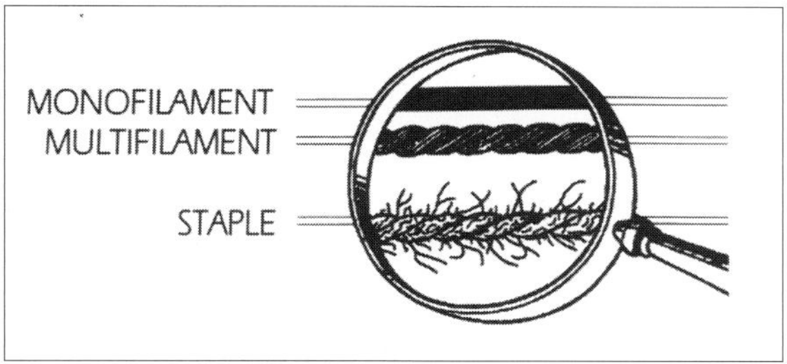

Yarn types

Yarn. An assembly of **fibres** or **filaments**, natural or synthetic in origin, twisted or laid together so as to form a continuous strand, which can be used in the weaving or knitting of a textile fabric.

Also an old term for the measure of the weight of a cotton cloth, where 1 yarn = 840 yd/lb.

Yarn solids loading. In a **filter medium** made from a fibrous or multifilament yarn, the slurry is free to flow through the spaces between the yarns, and also between the fibres of the yarn itself. The inter-fibre spaces are usually much smaller than those between the yarns, and so they plug much more quickly, if there is a range of particle sizes in the feed. Solids held within the yarns are much more difficult to remove by such techniques as back flushing.

Yarn wound. A **yarn** wound **filter cartridge** is formed by the laying down of one or more continuous yarns on a central **core**, spirally along the core, in several layers. There are two types of yarn winding: spool wound, in which the successive turns of the yarn lie close together, and spiral wound, with widely spaced turns, creating a honeycomb-like structure.

Yea. An old fashioned measure of the weight of a linen cloth, where 1 yea = 300 yards of fibre per pound.

Yield. A term sometimes used for the recovery of solids in a **separation** process.

Yield point. An important physical characteristic of continuous material to describe its behaviour when being stretched. The yield point under applied strain is that tension force that causes plastic deformation to begin (i.e. with measurable velocity).

Yield strength. The stress applied to a material at which a defined (e.g. 0.2%) permanent deformation occurs.

Yoke. [*See End cap*]

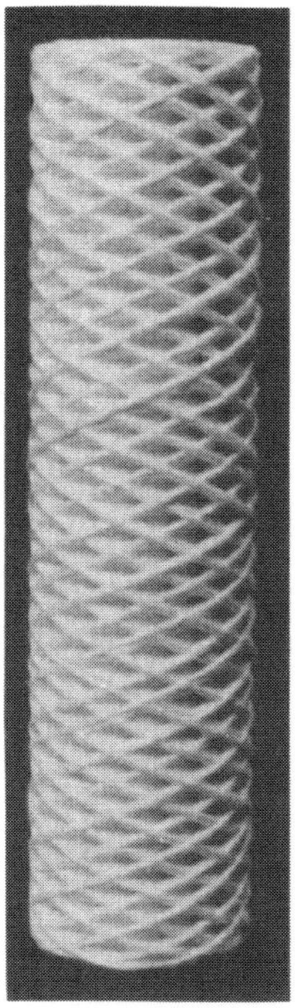

Yarn wound cartridge

Yoshioka construction. A graphical construction from empirical data used in the design of **batch settling** equipment. A series of tests are made of the settling behaviour, and the results are plotted as a curve relating **flux** to **underflow** discharge concentration. A straight line drawn as a tangent to this curve to give the limiting flux, from which the equipment area can be found.

Young-Laplace law. [*See Laplace equation*]

Y (type) strainer. A type of in-line **strainer** in which the **housing** for the strainer element forms an angle of about 45° with the direction of the pipeline to which the strainer is attached.

Z

Zaki equation. [*See Richardson & Zaki equation*]

Zeitsch number. A **dimensionless number** developed by Zeitsch as part of an analysis of **capillary drainage** in **centrifugal filters**.

Zeolite. A complex silicate material, whose crystal lattice contains holes of a size capable of holding a water molecule. The water may be driven off by mild heating, and, on exposure to moist air, water is then re-adsorbed, so that the air is dried. Zeolites, both natural and synthetic, are used as desiccants and **molecular sieves,** and for **ion exchange**.

Zeolite membrane filter. A relatively novel form of **membrane**, zeolites have the advantage of being able to be grown *in situ* on a strong support. It can be made with pore diameters of less than 1 nm, and is highly **hydrophilic**.

Zero charge. [*See Point of zero charge*]

Zero vertical velocity locus.
[*See Locus zero vertical velocity*]

Zeta-Plus medium. A range of **wet-laid filter media**, made by Cuno, which have strong electrical effects built into the material.

Zeta potential. A characteristic of the zone of a liquid immediately adjacent to the surface of a **particle**, which describes some of the effects of electric charge at the particle surface. The surface of a particle suspended in a liquid is normally electrically charged, usually negatively so. This attracts a layer of ions from the dissociated molecules in the liquid, of opposite electric charge. This layer largely, but never entirely, balances the charge on the surface of the particle, and a second, more diffuse layer of the ions forms outside it, to complete the neutralisation. The zeta potential is the electric potential gradient across this double layer into the neutral liquid. The phenomenon is much used in the theories of fluid flow through porous masses and is the basis for Cuno's **Zeta-Plus filter media**. It also is a measure of the suitability of **electrophoretic** methods of separation and of resistance to **coagulation**.
[*See also Electrical double layer*]

Zinc powder. A fine powder, formed by the devolatilisation of zinc vapour, or the atomisation of molten zinc, which makes a good **challenge test** material for **filter media**.

Zonal centrifuge. A multi-chamber centrifuge in which solids are collected at a point of neutral density by using heavy liquids to create a density gradient. The separating liquids and feed enter at low speed before the centrifuge is accelerated to achieve the separation. Recovery of the solids is usually done by introducing a dense liquid to displace the components. This can be done with the bowl stationary or rotating at low speed. The zonal centrifuge is mainly used as a piece of test equipment.

Zone sedimentation, settling. Describes the **sedimentation** behaviour of natural or flocculated **suspensions** once they change from **dilute** suspension behaviour (single particle, **Stokes' law** settling) into **hindered sedimentation** (or **swarm sedimentation**). The particles' settling behaviour is now strongly influenced by their neighbouring particles, and the settling rates are given by **Richardson & Zaki's equation**. Zone settling results in the formation of a sharp front (or, more correctly, rear).